A Specialist Periodical Report

Aliphatic Chemistry
Volume 4

A Review of the Literature Published during 1974

Senior Reporter
A. McKillop, School of Chemical Sciences,
University of East Anglia

Reporters
E. W. Colvin, University of Glasgow
D. W. Dunwell, Lilly Research Centre Ltd., Windlesham, Surrey
G. Pattenden, University of Nottingham
J. C. Saunders, Lilly Research Centre Ltd., Windlesham, Surrey
B. P. Swann, Lilly Research Centre Ltd., Windlesham, Surrey

© Copyright 1976

The Chemical Society
Burlington House, London, W1V 0BN

ISBN: 0 85186 572 0
ISSN: 0305-618X
Library of Congress Catalog Card No. 74-82046

Printed in Northern Ireland at The Universities Press (Belfast) Ltd.

Foreword

This report on Aliphatic Chemistry is similar in scope and format to the previous three volumes in the series. A team from the Lilly Research Centre (D. W. Dunwell, J. C. Saunders and B. P. Swann) have reviewed developments in the chemistry of acetylenes, alkanes, allenes and olefins (Chapter 1), while E. W. Colvin makes his fourth consecutive contribution with a major report on the chemistry of functional groups other than those covered by the Lilly team (Chapter 2). Chapters 3 and 4, which deal with the chemistry of naturally occurring polyolefinic and polyacetylenic compounds and the prostaglandins respectively, have been written once again by G. Pattenden. As in the previous volumes, coverage of the literature has been on a January-to-December basis.

July 1975 A. McKillop

Contents

Chapter 1 Acetylenes, Alkanes, Allenes, and Olefins   1
By D. W. Dunwell, J. C. Saunders, and B. P. Swann

1 Acetylenes	1
Synthesis of Acetylenes	1
Use of Acetylenes in Synthesis	5
Cycloaddition Reactions of Acetylenes	10
Cycloadditions with Dimethyl Acetylenedicarboxylate	10
Other Cycloaddition and Orbital-symmetry-allowed Reactions	14
Other Additions to the Acetylenic Bond	17
Electrophilic Additions	17
Nucleophilic Additions	19
Metal-catalysed Reactions of Acetylenes	21
Other Reactions of Acetylenes	22
Physical Properties of Acetylenes	27
Acetylene–Allene Interconversions	28
2 Alkanes	30
3 Allenes	32
Synthesis of Allenes	32
Reactions of Allenes	35
4 Olefins	39
Synthesis of Olefins	39
Cycloaddition Reactions of Olefins	47
Diels–Alder Cycloadditions	47
1,3-Dipolar Additions	51
Other Olefin Cycloadditions	51
The Ene Reaction	56
Sigmatropic Rearrangements	59
Electrophilic Additions	59
Hydroboration	64
Oxymercuration and Oxythallation	64
Metal–Olefin Reactions	66
Oxidation of Olefins	71
Other Reactions of Olefins	73
Theoretical and Spectral Properties of Olefins	79

Chapter 2 Functional Groups other than Alkanes, Acetylenes, Allenes and Olefins 81

By E. W. Colvin

1	**Carboxylic Acids**	**81**
	Preparation	81
	Protection	85
	Properties	86
	Derivatives and Reactions	86
	Peracids	88
2	**Carboxylic Acid Anhydrides**	**89**
3	**Lactones**	**89**
	Preparation	89
	α-Methylene-γ- and -δ-lactones	93
	Syntheses of Naturally Occurring Lactones	95
	Properties	96
4	**Carboxylic Acid Esters**	**97**
	Preparation	97
	By Esterification	97
	Of β-Hydroxy- and $\alpha\beta$-Unsaturated Esters	99
	General	100
	Alkylation	103
	Cleavage	106
	General Properties and Reactions	108
5	**α-Amino-acids**	**108**
	Preparation	108
	Protection and Deprotection	109
	Properties and Reactions	110
6	**Carboxylic Acid Amides**	**111**
	Amide (Peptide) Bond Formation	111
	General Preparation	112
	Hydrolysis and Protonation	115
	Properties and Reactions	117
7	**Nitriles and Isocyanides**	**119**
	Preparation	119
	α-Anions	120
	Properties and Reactions	122

8 Aldehydes and Ketones — 123

- Preparation — 123
  - By Oxidation — 123
  - By Addition, Insertion, and Rearrangement — 127
  - By Regeneration — 132
  - General — 134
- Carbonyl Umpolung and Related Techniques — 140
  - Sulphur-based Examples — 140
  - Other Cases — 142
- Alkylation — 146
- Introduction of $\alpha\beta$-Unsaturation — 150
- Conjugate Addition (Alkylation) and Reduction — 152
- Reduction and Reductive Coupling — 156
- Condensations — 159
  - Directed Aldol Condensations — 159
  - Organometallics — 160
  - Wittig Types — 163
  - General — 163
- Halogen Derivatives and Oxyallyl Zwitterions — 165
- General Properties and Reactions — 167

9 Alcohols — 172

- Preparation — 172
- Properties — 176
- Reactions — 177

10 Amines — 178

- Preparation — 178
- Properties and Reactions — 182

11 Alkyl Halides — 184

- Preparation — 184
- Halogenoalkyl and Halonium Ions — 187
- Halogenoalkyl Radicals — 189
- Reduction to Hydrocarbons and Oxidative Coupling — 190
- General Properties and Reactions — 190

12 Ethers — 190

- Preparation and Reactions — 190

13 Sulphur — 192

- Sulphoxides — 192
- Sulphones — 194
- General — 194

14	**Miscellaneous**	**197**
	Diazoalkanes	197
	Organometallics and Metal Complexes	199
	Phase Transfer Catalysts and Crown Ethers	200
	Chiroptical Properties	201
	Computer-designed Synthesis	202
	Bases	202
	General	202
15	**Reviews**	**203**

Chapter 3 Naturally Occurring Polyolefinic and Polyacetylenic Compounds 205
By G. Pattenden

1	**Introduction**	**205**
2	**Naturally Occurring Polyacetylenes**	**205**
	Introduction	205
	New and Known Polyacetylenes from Nature	205
	Synthesis of Natural Polyacetylenes	208
3	**Naturally Occurring Allenes**	**209**
4	**Natural Acetylenes and Olefins from Marine Sources**	**210**
	Introduction	210
	Polyolefins from *Dictyopteris*	210
	Bromo-olefins from *Verongia* Sponges	211
	Enyne Cyclic Bromo-ethers from *Laurencia* (family *Rhodomelaceae*)	212
	Halogen Compounds from Sea Hare, *Aplysia*	212
	Miscellaneous Compounds from Marine Sources	213
5	**Polyolefinic Microbial Metabolites**	**214**
	Polyolefinic Macrolides	214
	Primycin	214
	Amphotericin B	214
	Leucomycins	214
	Lankacidins	215
	Antibiotic XL-704	215
	Vermiculine	215
	Other Polyolefinic Macrolides	215
	Polyolefinic Macrocycles (Non-macrolide)	216
	The Cytochalasans	216

	Ansamycin Antibiotics	217
	Maytansines (Maytansinoides)	217
	Rifamycins	217
	Halomycins	219
	Geldanamycin	219
	Pyran-Pyranoid Compounds	220
	Aurovertin B	220
	Aureothin	220
	Pestalotin (LL P880α)	220
	Phacidin	221
	Metabolite LL-Z 1220	222
	Other pyran-pyranoid compounds	222
	N-Heterocyclic Polyolefinic Compounds	222
	Antibiotic X-5108 (Goldinodox) and Mocimycin (Delvomycin)	222
	Tenellin and Bassiamin	222
	Butenolide Metabolites	223
	Penicillic Acid	223
	Multicolic and Multicolosic Acids	223
	Patulin	224
	Miscellaneous Polyolefinic Microbial Metabolites	225
6	**Cyclopentenone Polyolefinic Compounds**	**227**
	Jasmone	227
	Terrein	231
	Cryptosporiopsin	231
	Rethrolones	231
	Hop Constituents	231
7	**Naturally Occurring Polyolefinic Degraded and/or Modified Isoprenoid Compounds**	**234**
	Edulans	234
	Constituents of Tobacco	234
	Other Compounds	236
8	**Polyolefinic Insect Pheromones**	**236**
	Pheromonal Secretion of Queen Butterfly	236
	Propylure	236
	Pheromonal Secretions of Codling Moth and Red Bollworm Moth	236
	Pheromone of the Lesser Peachtree Borer	237
	Pheromone of the Angoumois Grain Moth	237
	Other Pheromones	238

9	Miscellaneous Natural Polyolefins	238
	Flexirubins Pigments	238
	Polyene Amides	238
	Kawa Lactones and Related Compounds	240
	Other Compounds	241

Chapter 4 Chemistry of the Prostaglandins 243
By G. Pattenden

1	Introduction	243
2	Nomenclature	243
3	Synthesis of Primary Prostaglandins	243
	Corey's Bicyclo[2,2,1]heptane Route	244
	Refinements	244
	New Routes to Key Intermediates	244
	Routes by the Merck Group	249
	Routes involving Cyclopentane Ring Synthesis	249
	Routes incorporating Conjugate Addition of Vinylcopper Reagents to a Functionalized Cyclopentenone	250
4	Synthesis of A-Prostaglandins	253
5	Synthesis of C-Prostaglandins	254
6	Synthesis of Modified Prostaglandins	255
	11-Desoxyprostaglandins	255
	Oxaprostaglandins	257
	9-Oxaprostaglandins	257
	10-Oxaprostaglandins	257
	11-Oxaprostaglandins	257
	Thiaprostaglandins	258
	9-Thiaprostaglandins	259
	7-Thiaprostaglandins	261
	Prostaglandin D_2	263
	Methylprostaglandins	263
	Other Modified Prostaglandins	263
7	Epiprostaglandins	263
8	Prostaglandins in Coral	263
9	Metabolism of Prostaglandins	264
10	Biosynthesis of Prostaglandins	265
11	General	266

Author Index 267

1
Acetylenes, Alkanes, Allenes, and Olefins

BY D. W. DUNWELL, J. C. SAUNDERS, AND B. P. SWANN

1 Acetylenes

Synthesis of Acetylenes.—Full experimental details have been given for the preparation of alkynes and dialkynes by the reaction of mono- and di-halogenoalkanes with lithium acetylide–ethylenediamine complex. Dimethyl sulphoxide (DMSO) is the preferred solvent for this reaction.[1] Alkylcopper compounds RCu react with copper acetylides to give disubstituted acetylenes, the products of unsymmetrical coupling.[2]

Lithium ethynyl borates, which can be readily prepared from trialkylboranes and lithium acetylide–ethylenediamine complex, are decomposed by iodine to produce the terminal alkylacetylene in high yield. This represents an extension of the previously reported synthesis of internal acetylenes by an analogous procedure (Scheme 1).[3] A similar type of transformation can also be carried out using methanesulphinyl chloride in place of iodine.[4]

$$R_3B + LiC\equiv CH \longrightarrow Li^+[R_3BC\equiv CH]^-$$

$$Li^+[R_3BC\equiv CH]^- + I_2 \longrightarrow RC\equiv CH + R_2BI + LiI$$

Scheme 1

Only the strong base combination of sodamide and sodium t-butoxide was found to be capable of producing the required *syn*-elimination of hydrogen bromide from 1-bromocycloalkenes to give the corresponding cycloalkynes.[5]

An ingenious new synthesis of acetylenes starts with bis(tri-n-butylstannyl)-ethylene. Treatment with one equivalent of butyl-lithium converts this compound into the corresponding monolithium derivative, which can then be alkylated and the tri-n-butylstannyl group cleaved with lead tetra-acetate in acetonitrile to give the acetylene (Scheme 2).[6] An advantage in the use of

---
[1] W. N. Smith and O. F. Beumel, jun., *Synthesis*, 1974, 441.
[2] R. Levene, J. Y. Becker, and J. Klein, *J. Organometallic Chem.*, 1974, **67**, 467.
[3] M. M. Midland, J. A. Sinclair, and H. C. Brown, *J. Org. Chem.*, 1974, **39**, 731.
[4] N. Naruse, K. Utimoto, and H. Nozaki, *Tetrahedron*, 1974, **30**, 2159.
[5] P. Caubere and G. Coudert, *Bull. Soc. chim. France*, 1973, 3067.
[6] E. J. Corey and R. H. Wollenberg, *J. Amer. Chem. Soc.*, 1974, **96**, 5581.

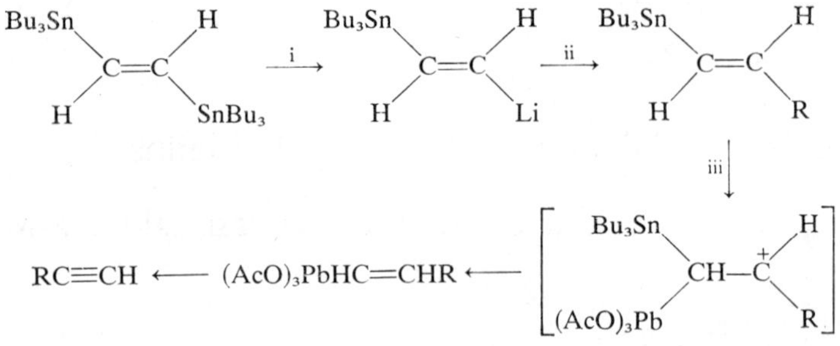

Scheme 2

Reagents: i, BuLi; ii, RX; iii, Pb(OAc)$_4$

the lithium intermediate is that addition of this compound to a copper acetylide gives an R$_2$CuLi species, the tri-n-butylstannylethenyl group of which can be transferred to an αβ-unsaturated ketone. Oxidation of the resulting vinylstannane derivative with lead tetra-acetate gives the terminal acetylene, and the overall process constitutes a method for the preparation of ethynyl ketones [*e.g.* (1)] which are not accessible by conjugate addition using dialkynylcopper-lithium species.

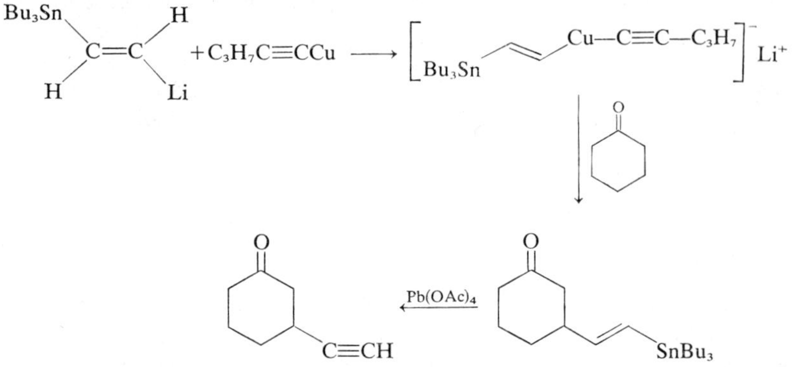

Two convenient syntheses of t-butylacetylene from pinacolone have been reported, which involve the use of either phosphorus pentachloride–potassium t-butoxide in DMSO[7] or trifluoromethanesulphonic anhydride in carbon tetrachloride containing pyridine.[8]

Potassium fluoride and tetraethylammonium fluoride are known to be good agents for the promotion of dehydrohalogenation reactions. Good yields of acetylenes are obtained from vinyl halides when the *anti* configuration

---

[7] P. J. Kocienski, *J. Org. Chem.*, 1974, **39**, 3285.
[8] R. J. Hargrove and P. J. Stang, *J. Org. Chem.*, 1974, **39**, 581.

is present. The *syn*-compounds generally give good yields of the corresponding allenes. The reactions are catalysed by crown ethers.[9]

A simple preparative procedure for acetylenic carboxylic acids by alkylation of the lithium salts of acetylenes by ω-bromoalkanoic acids in hexamethylphosphortriamide has been described.[10]

$$\begin{bmatrix} C_4H_9C{\equiv}CCu & CO_2Et \\ & \diagup\!\!\!\diagdown \\ & CH_2 \end{bmatrix}^- Li^+ + HC{\equiv}CCH_2Br \longrightarrow HC{\equiv}CCH_2\underset{CO_2Et}{C}{=}CH_2$$

(2)

The novel vinylcuprate reagent (2) has been found to be a useful three-carbon fragment. A survey of its reactivity with organic halides shows that the reagent is very specific for propargyl and allyl halides. Benzyl bromide is not attacked.[11]

The reaction of 1-bromoalk-1-ynes with two equivalents of butyl-lithium gives 3-butylalk-1-ynes (3) after prolonged treatment. The intermediate is a dilithioalkyne, which can also be prepared directly from terminal alkynes, and which undergoes alkylation at the 3-position.[12]

$$RCH_2C{\equiv}CBr + BuLi \longrightarrow [R\bar{C}HC{\equiv}C^-]2Li^+ \longrightarrow \underset{Bu}{RCHC{\equiv}CH}$$
$$+$$
$$BuBr$$

(3)

αβ-Unsaturated ketones react with chlorovinyltriphenylphosphorane to give the corresponding vinyl chloride. Treatment of these compounds with butyl-lithium gives an unsaturated carbene, which rearranges to give the corresponding alkenylacetylene.[13] The reaction of dimethyl phosphonodiazomethane with aldehydes has been extended to include sugars and sugar aldehydes, which are converted into acetylenes containing one more carbon atom.[14]

Trimethylsilylacetylenes are converted into αβ-acetylenic amides on treatment with dialkylcarbamoyl chlorides and aluminium chloride.[15] Use of Castro–Stephens conditions for the coupling of *m*-iodophenylacetylene-copper(I) salt in pyridine gives a 4.6% yield of the cyclic hexamer.[16]

---

[9] F. Naso and L. Ronzini, *J.C.S. Perkin I*, 1974, 340.
[10] N. Gilman and B. Holland, *Synthetic Comm.*, 1974, **4**, 199.
[11] J. P. Marino and D. M. Floyd, *J. Amer. Chem. Soc.*, 1974, **96**, 7139.
[12] A. J. Quillinan, E. A. Khan, and F. Scheinmann, *J.C.S. Chem. Comm.*, 1974, 1030.
[13] H. Fienemann and G. Köbrich, *Chem. Ber.*, 1974, **107**, 2797.
[14] A. Gonzalez, J.-B. Zumwald, F. Perret, and J. M. J. Tronchet, *Helv. Chim. Acta*, 1974, **57**, 1505.
[15] P. Bourgeois, G. Merault, and R. Calas, *J. Organometallic Chem.*, 1973, **59**, C4.
[16] H. A. Staab and K. Neunhoeffer, *Synthesis*, 1974, 424.

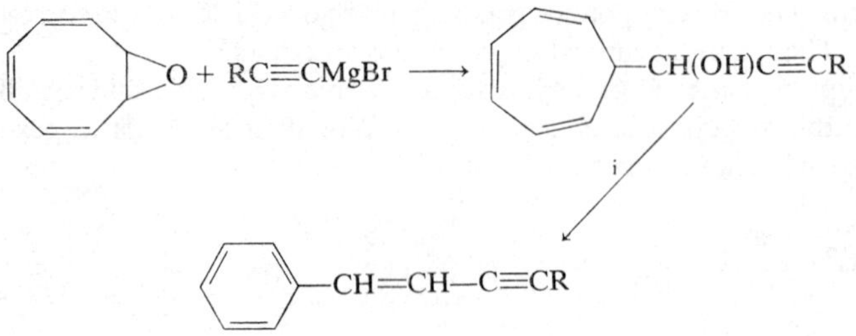

Scheme 3

Reagent: i, SOCl$_2$

Cyclo-octatetraene oxide can be converted into enynes by the sequence shown in Scheme 3.[17]

3,3,6,6-Tetramethyl-1-thia-4-cycloheptyne (4), the first isolable cycloheptyne, has been synthesized by lead tetra-acetate oxidation of the corresponding dihydrazone. The large deformation of the C—C≡C—C angles causes the ^{13}C≡C resonances to shift to lower field, and the reactivity of (4) in addition processes is enhanced. Thus (4) will react with phenyl azide,

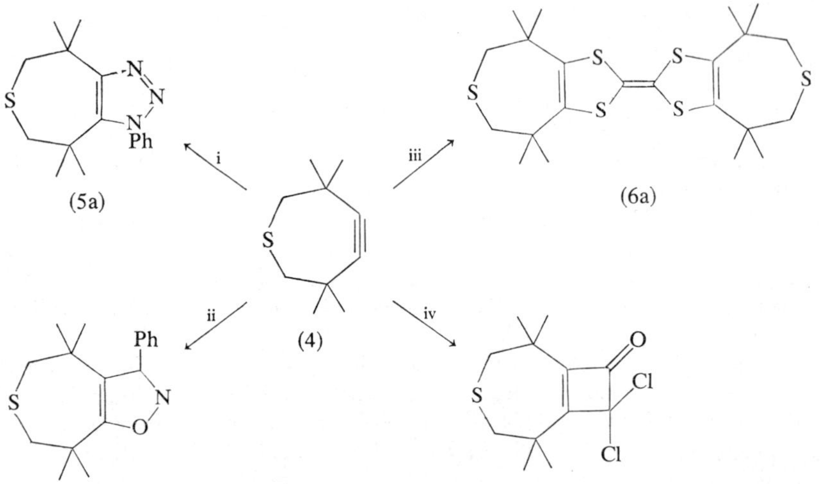

Reagents: i, PhN$_3$; ii, PhCH=N(O$^-$)Me; iii, CS$_2$; iv, CCl$_2$CO

Scheme 4

[17] J. Hambrecht, H. Straub, and E. Müller, *Chem. Ber.*, 1974, **107**, 2985.

nitrones, carbon disulphide, and dichloroketen to give the products (5a), (5b), (6a), and (6b), respectively (Scheme 4).[18]

The first synthesis of cyclononyn-5-one involves the fragmentation of the tosylhydrazone (7) with base.[19]

(7)

Use of Acetylenes in Synthesis.—Interest in the use of organoborane derivatives of acetylenes as synthetic reagents has continued unabated. The alkynyltrialkylborates (8), which are prepared by the addition of alkynyl-lithium salts to trialkylboranes, are particularly useful. Protonolysis by acetic acid gives olefins in good yield,[20] and use of the thexylborane derivatives enables good stereospecificity to be maintained. The reaction of (8) with acetyl chloride followed by heating gives the 2-oxa-3-borolens (9), which can be oxidized by the Jones reagent to give $\alpha\beta$-unsaturated ketones. This method has proved to be particularly useful for the preparation of hindered ketones.[21] Similarly, the reaction of chloroalkyl alkyl ethers with (8) gives a mixture of isomers (10) and (11), depending on the solvent and the structure of the borane. Addition of water to the mixture results in selective hydrolysis of (10) to (12). Hydrolysis of the residue with acetic acid converts (11) into (13), and the overall transformation constitutes a stereospecific synthesis of allyl alkyl ethers.[22]

The reaction of (8) with oxirans gives the six-membered cyclic compounds (14). These intermediates can be cleaved with hydrogen peroxide to give γ-hydroxy-ketones (15); protonolysis by acetic acid gives trisubstituted alkenes of the homoallylic alcohol (16) stereospecifically; and the tetrasubstituted derivatives (17) can be obtained by use of a sodium hydroxide–iodine work-up procedure. These transformations are summarized in Scheme 5.[23]

In a similar series of reactions, dialkylboranes can be converted into alcohols[24] and into amines[25] (Scheme 6).

Thexylborane reacts with two equivalents of 1-iodo-alk-1-ynes to give (18). Addition of two mole equivalents of sodium methoxide to (18; R = Bu^n)

---

[18] A. Krebs and H. Kimling, *Annalen*, 1974, 2074.
[19] G. L. Lange and T.-W. Hall, *J. Org. Chem.*, 1974, **39**, 3819.
[20] N. Miyaura, T. Yoshinari, M. Itoh, and A. Suzuki, *Tetrahedron Letters*, 1974, 2961.
[21] M. Naruse, T. Tomita, K. Utimoto, and H. Nozaki, *Tetrahedron*, 1974, **30**, 835.
[22] P. Binger and R. Köster, *Synthesis*, 1974, 350.
[23] M. Naruse, K. Utimoto, and H. Nozaki, *Tetrahedron*, 1974, **30**, 3037.
[24] G. Zweifel and R. P. Fisher, *Synthesis*, 1974, 339.
[25] A. Pelter, A. Arase, and M. G. Hutchings, *J.C.S. Chem. Comm.*, 1974, 346.

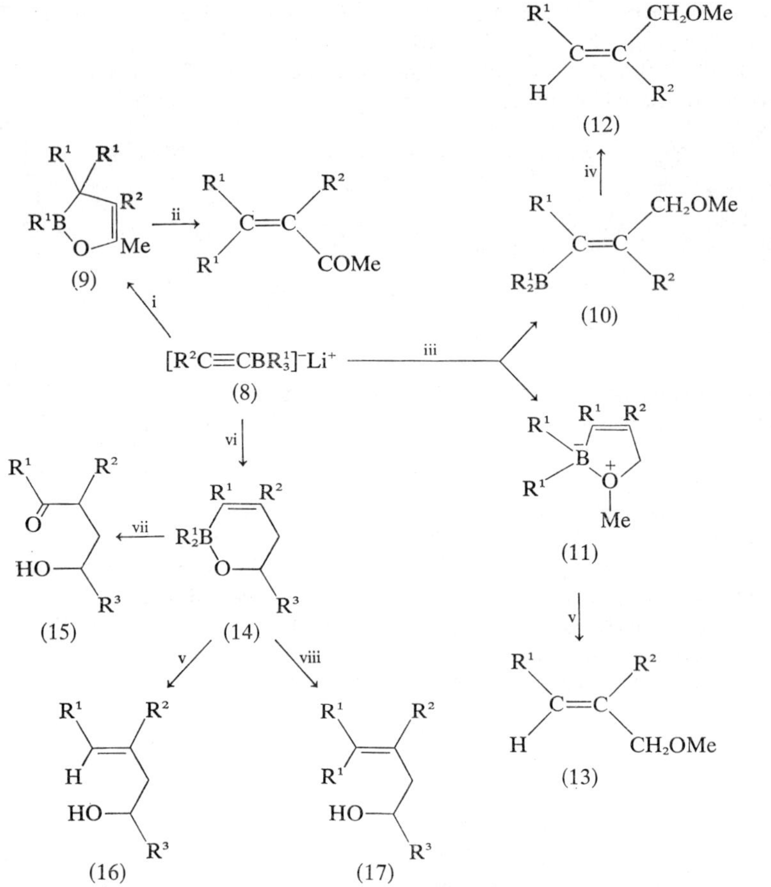

Reagents: i, MeCOCl; ii, CrO$_3$; iii, MeOCH$_2$Cl; iv, H$_2$O; v, AcOH; vi, O⟨R^3 ; vii, H$_2$O$_2$; viii, NaOH–I$_2$

Scheme 5

gives *trans*-1,4-di-n-butylbuta-1,2,3-triene.[26] Hydroboration of the product with di-isoamylborane and protonolysis gives pure *cis,trans*-dodeca-5,7-diene by addition across the central double bond.

Perhaps the most interesting use of borate salts of acetylenes which has been described has been the functionalization of the 4-position of pyridine. Many derivatives can thus be obtained which are difficult to prepare by other routes (Scheme 7).[27]

[26] T. Yoshida, R. M. Williams, and E. Negishi, *J. Amer. Chem. Soc.*, 1974, **96**, 3689.
[27] A. Pelter and K. J. Gould, *J.C.S. Chem. Comm.*, 1974, 347.

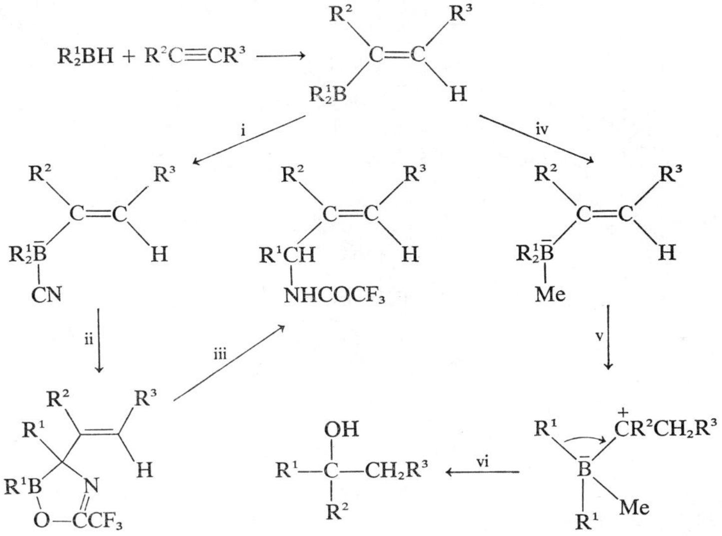

Reagents: i, KCN; ii, (CF₃CO)₂O; iii, NaOH; iv, MeLi; v, H⁺; vi, H₂O₂

Scheme 6

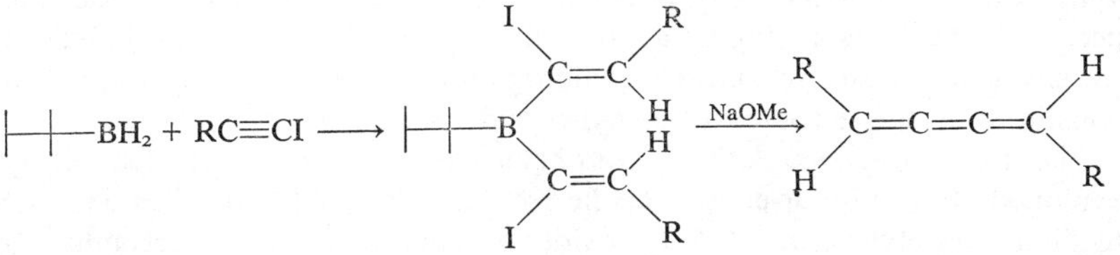

Acetylenic acetals can be converted stereoselectively into α-keto-ethers or cis-allylic ethers by treatment with R₂BH compounds followed by oxidation or protonolysis.[28]

Addition of alkyl-lithium reagents to allylic alcohols and diphenylacetylenes has been studied previously. Reaction with propargyl alcohols gives good yields of the 2-butyl allyl alcohol. When a methoxy-group is present, elimination occurs and an allenic alcohol is formed, e.g. (19).[29]

Lithio-1-trimethylsilylpropyne reacts with cuprous iodide in ether at −78 °C to give the corresponding organocopper compound, which reacts

---

[28] G. Zweifel, A. Horng, and J. E. Plamondon, *J. Amer. Chem. Soc.*, 1974, **96**, 316.
[29] L.-I. Olsson and A. Claesson, *Tetrahedron Letters*, 1974, 2161.

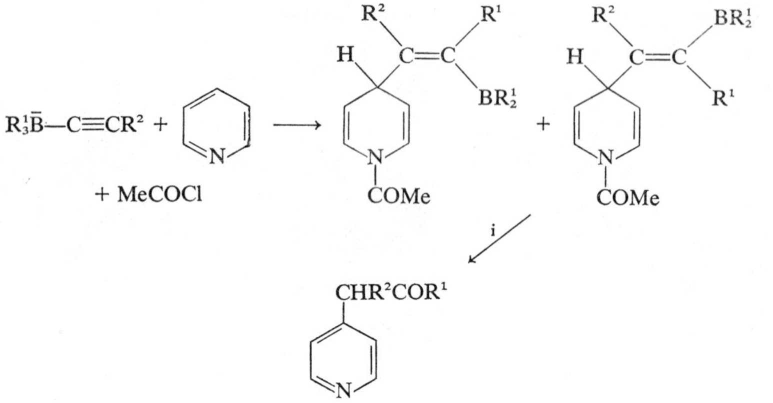

Scheme 7

$R_3^1\bar{B}$—C≡CR² + [pyridine] + MeCOCl ⟶ [R², H, C=C, R¹, BR₂¹ substituted dihydropyridine with COMe] + [R², H, C=C, BR₂¹, R¹ substituted dihydropyridine with COMe]

↓ i

CHR²COR¹ (on pyridine)

Reagent: i, H₂O₂

$$C_3H_7\underset{Me}{\overset{OMe}{C}}\!\!-\!C\!\equiv\!C\!-\!CH_2OH \xrightarrow[Et_2O]{BuLi} C_3H_7\underset{Me}{C}\!=\!C\!=\!\underset{C_4H_9}{C}\!-\!CH_2OH$$

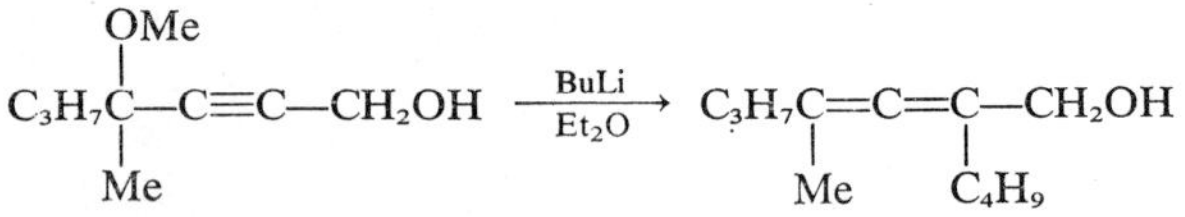

(19)

with ethyl *trans*-penta-2,4-dienoate to give a 4:1 mixture of allenic and acetylenic products arising by a 1,6 addition across the system. Removal of the silyl group with silver nitrate–sodium cyanide has shown that this process constitutes a simple route to 1,5-enynes and 1,4,5-trienes.[30]

The use of acetylene derivatives in heterocyclic synthesis has been widely exploited. For sulphur-containing heterocycles the addition of thioacetate to diphenylacetylene in DMSO provides a convenient route to tetraphenylthiophen.[31] Thiourea adds to penta-1,4-diyn-3-ones to give 2,6-dialkyl-4H-thiapyran-4-ones.[32] Disulphur dichloride adds to ethyl phenylpropiolate to give a product which, on oxidation with hydrogen peroxide, undergoes cyclization to give the benzothiophen 1,1-dioxide (20).[33]

Irradiation of 4-oxo-2,6-diphenyl-4H-thiapyran 1,1-dioxide in the presence of phenylalkynes gives moderate yields of thiepin 1,1-dioxides (21).[34]

Diels–Alder addition of ynamines to pyrimidines followed by a retro elimination of a nitrile moiety constitutes a method for the preparation of substituted pyridines, and rules have been established for prediction of the orientation of

[30] B. Ganem, *Tetrahedron Letters*, 1974, 4467.
[31] S. V. Amosova, B. A. Trofimov, N. N. Skatova, O. A. Tarasova, and A. G. Trofimova, *Doklady Akad. Nauk S.S.S.R.*, 1974, **215**, 95 (*Chem. Abs.*, 1974, **80**, 145 638a).
[32] K. G. Migliorese and S. I. Miller, *J. Org. Chem.*, 1974, **39**, 843.
[33] W. Ried and W. Ochs, *Chem. Ber.*, 1974, **107**, 1334.
[34] N. Ishibe, K. Hashimoto, and M. Sunami, *J. Org. Chem.*, 1974, **39**, 103.

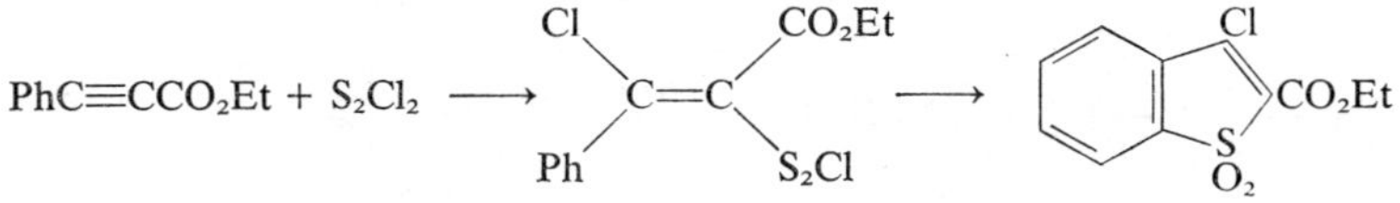

Acetylenes, Alkanes, Allenes, and Olefins

[Scheme showing reactions with structures (20) and (21)]

PhC≡CCO$_2$Et + S$_2$Cl$_2$ ⟶ [vinyl intermediate with Cl, Ph, CO$_2$Et, S$_2$Cl] ⟶ [benzothiophene-S,S-dioxide with Cl, CO$_2$Et]

(20)

[Thiopyranone-S,S-dioxide] + PhC≡CR $\xrightarrow{h\nu}$ [cycloheptatriene-S,S-dioxide product]

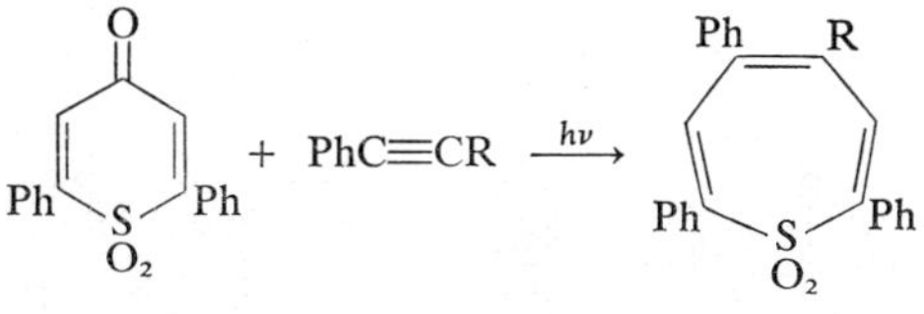

(21)

substituents in the products.[35] Nucleophilic addition of hydrazines to penta-1,4-diyn-3-ones gives pyrazoles.[36] A similar type of reaction between o-phenylenediamine and diarylprop-1-yn-3-ones gives benzodiazepines.[37] Methyl propiolate reacts with enamines of β-keto-esters to give *trans*-dienamino-esters (22). Deuteriation experiments show that a 1,5-hydrogen shift occurs in this step. The products can be further transformed into 2-pyridones[38] (Scheme 8).

Propiolic acid dianion has been used as a nucleophilic acyl synthon in the total synthesis of (±)-pestalotin.[39]

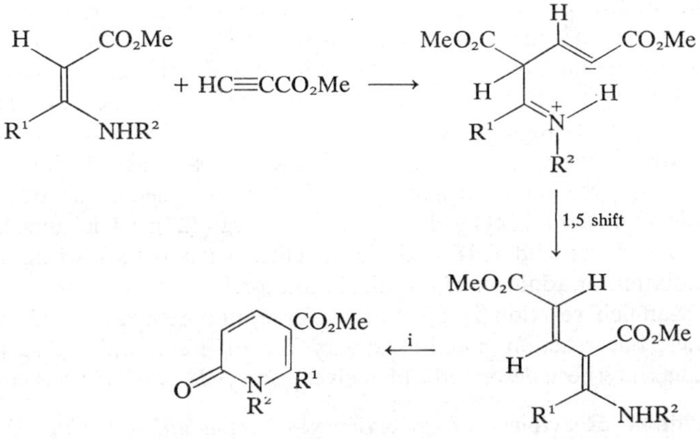

[Scheme 8: enamine + HC≡CCO$_2$Me → zwitterionic intermediate → (via 1,5 shift) dienamino-ester (22) → (reagent i) 2-pyridone]

(22)

Reagents: i, Δ, DMF

Scheme 8

[35] H. Neunhoeffer and G. Werner, *Annalen*, 1974, 1190.
[36] V. N. Yandovski and T. K. Klindukhova, *Zhur. org. Khim.*, 1974, **10**, 136 (*Chem. Abs.*, 1974, **80**, 108 432h).
[37] S. P. Korshimov, V. M. Kazantseva, L. A. Vopilina, V. S. Pisareva, and N. V. Utekhina, *Khim. geterotsikl. Soedinenii*, 1973, 1421 (*Chem. Abs.*, 1974, **80**, 27 225r).
[38] N. Anghelide, C. Draghici, and D. Railenau, *Tetrahedron*, 1974, **30**, 623.
[39] R. M. Carlson and A. R. Oyler, *Tetrahedron Letters*, 1974, 2615.

$$\underset{\text{Ph}_2\overset{|}{\text{C}}-\text{C}\equiv\text{C}-\text{C}\equiv\text{C}-\text{CH}_2\text{CH}_2\text{OH}}{\text{OH}} \xrightarrow{\text{H}^+} \text{Ph}_2\text{C}=\text{CH}-\text{[2,3-dihydro-}\gamma\text{-pyrone]}$$

(23)

$$\underset{(24)}{\text{HOCH}_2\text{CH}=\overset{\text{Me}}{\underset{|}{\text{C}}}-\text{C}\equiv\text{C}-\overset{\text{R}}{\underset{|}{\text{C}}}\text{HOMe}} \xrightarrow[\text{DMSO}]{\text{KOBu}^t} \underset{(25)}{\text{[Me-furan]}-\text{CH}=\text{CHR}}$$

[THF] + NC—C≡C—CN ⟶ [THF-vinyl] C=C(CN)(NC)(H)

(26)

Ynamines are acylated by diketen and the intermediates thus formed can be cyclized to derivatives of 2-amino-4-pyrone.[40] Irradiation of non-cisoid 1,2-dioxo-compounds with 1-alkylthioprop-1-ynes in benzene gives *cis*- and *trans*-2-alkylthiobut-2-ene-1,4-diones *via* an oxeten intermediate. The *cis*-compound can be ring-closed to a furan derivative with stannic chloride.[41] Acid-catalysed ring closure of 1,1-diphenylhepta-2,4-diyne-1,7-diol gives the 2,3-dihydro-γ-pyrone (23) in good yield.[42] Base-catalysed cyclization of the acetylenic allyl alcohol (24) yields the alkenyl furan (25).[43] Dicyanoacetylene forms 1:1 adducts with THF and diethyl ether, such as (26), without either u.v. irradiation or addition of a radical initiator.[44]

The Mannich reaction is often used for aminomethylation of terminal acetylenes, but reaction conditions vary for different amines. A general method has now been described which gives good yields with various amines.[45]

Cycloaddition Reactions of Acetylenes.—*Cycloadditions with Dimethyl Acetylenedicarboxylate.* Table 1[46-58] summarizes some of the cycloaddition reactions undergone by dimethyl acetylenedicarboxylate (DMAD).

---

[40] J. Ficini and J. P. Genêt, *Bull. Soc. chim. France*, 1974, 2086.
[41] A. Mosterd, H. J. Matser, and H. J. T. Bos, *Tetrahedron Letters*, 1974, 4179.
[42] A. S. Medvedeva, M. M. Demina, and M. G. Voronkov, *Izvest. Akad. Nauk S.S.S.R* 1974, 366 (*Chem. Abs.*, 1974, **81**, 49 517).
[43] A. J. de Jong, L. Brandsma, and J. F. Arens, *Rec. Trav. chim.*, 1974, **93**, 15.
[44] R. J. Bushby, *J.C.S. Perkin I*, 1974, 274.
[45] R. Mornet and L. Gouin, *Bull. Soc. chim. France*, 1974, 206.

Acetylenes, Alkanes, Allenes, and Olefins

Table 1 *Products of cycloaddition reactions involving dimethyl acetylenedicarboxylate*

Starting Material	Product (E = CO_2Me)	Ref.
2-diazocyclohexanone + Other rings	bicyclic pyrazolo-azepinone with E, E substituents	46
dibenzo[c,f]cinnolinium N-ylide (−NCH₂R)	tetracyclic diazepine product with H, E, E, R, H substituents	47
Me₃CNC	cage structure with four E groups and two C=N-CMe₃ substituents	48
Ph-C(O⁻)=C-NPh / PhN=CPh (mesoionic)	Ph, NPh, NPh, Ph-substituted bicyclic intermediate with E, E → Ph-substituted pyrrole (Ph, E, E, Ph, N-Ph)	49

[46] M. Martin and M. Regitz, *Annalen*, 1974, 1702.
[47] M. J. Rance, C. W. Rees, P. Spagnolo, and R. C. Storr, *J.C.S. Chem. Comm.*, 1974, 658.
[48] H. J. Dillinger, G. Fengler, D. Schumann, and E. Winterfeld, *Tetrahedron*, 1974, **30**, 2553.
[49] G. Singh and P. S. Pande, *Tetrahedron Letters*, 1974, 2169.

Table 1 (Contd.)

Starting Material	Product (E = CO$_2$Me)	Ref.
thiazolium enolate (R^2, R^1, MeN)	bicyclic adduct → pyrrole with R^1, R^2, E, E, NMe; R^1 = Ph, R^2 = H	50
thiazolium enolate (R^3, R^{2}N, R^1)	bicyclic adduct → thiophene (R^1, R^3, E, E) + pyridinone (R^3, E, E, R^1, R^2)	51
N-R dihydropyridine	bicyclic adduct → azocine (E, E, NR)	52
(MeO)$_2$C:	MeO, MeO, OMe dihydrofuran with C≡CCO$_2$Me, E, E	53
N-CO$_2$Me pyrroline bridged	N-CO$_2$Me azabicyclic diene, E, E	54

[50] K. T. Potts, J. Baum, E. Houghton, D. N. Roy, and U. P. Singh, *J. Org. Chem.*, 1974, **39**, 3619.
[51] K. T. Potts, E. Houghton, and U. P. Singh, *J. Org. Chem.*, 1974, **39**, 3627; K. T. Potts, J. Baum, and E. Houghton, *ibid.*, p. 3631.
[52] R. M. Acheson, G. Paglietti, and P. A. Tasker, *J.C.S. Perkin I*, 1974, 2496.
[53] R. W. Hoffman, W. Lilenblum, and B. Dittrich, *Chem. Ber.*, 1974, **107**, 3395.
[54] S. R. Tanny and F. W. Fowler, *J. Org. Chem.*, 1974, **39**, 2715.

Acetylenes, Alkanes, Allenes, and Olefins

Table 1 (Contd.)

Starting Material	Product (E = CO₂Me)	Ref.
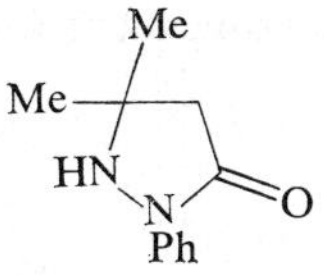	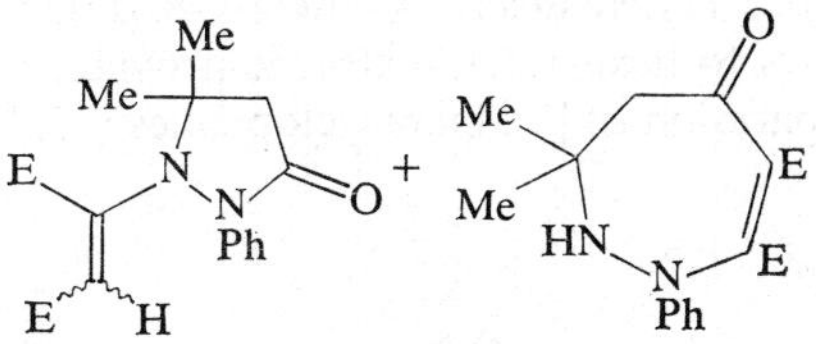	55
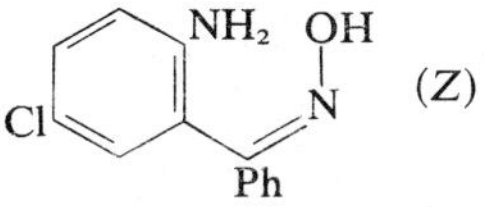	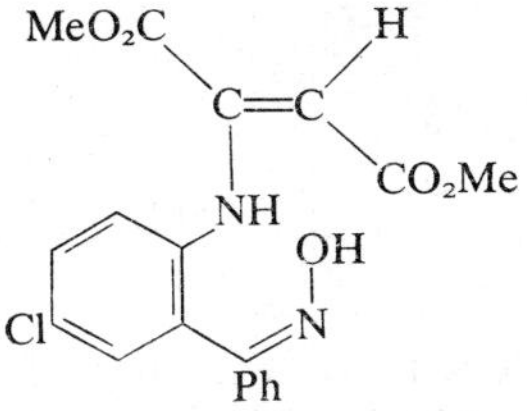	56
		57
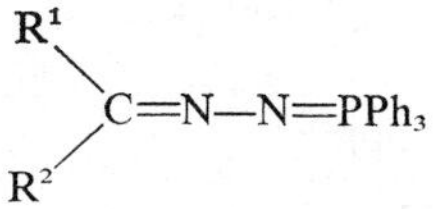	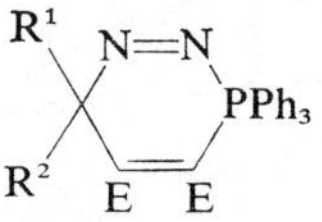	58

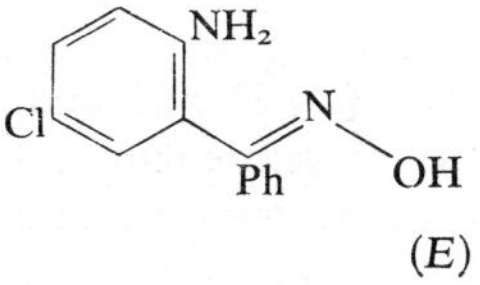

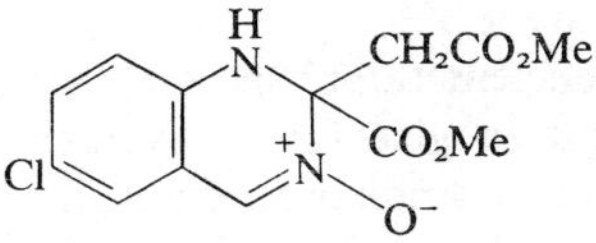

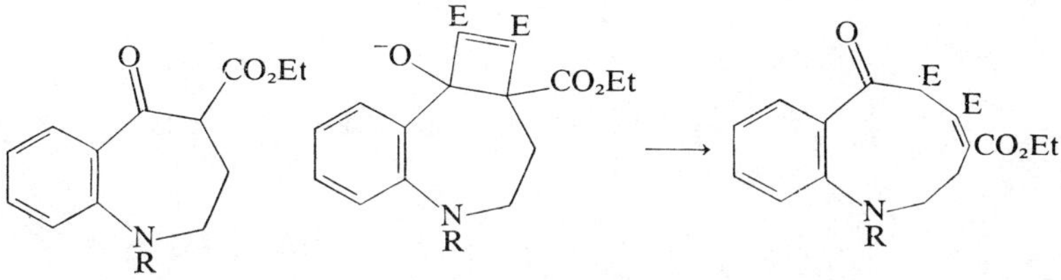

R = –SO₂C₆H₄Me

---

[55] S. N. Ege, E. Y. Tsui, R. L. Spencer, B. E. Potter, B. K. Eagleson, and H. Z. Friedman, *J.C.S. Chem. Comm.*, 1974, 216.
[56] J. B. Hester, jun., *J. Org. Chem.*, 1974, **39**, 2137.
[57] M. Lennon, A. McLean, I. McWatt, and G. R. Proctor, *J.C.S. Perkin I*, 1974, 1828.
[58] I. Shahak and Y. Sasson, *Israel J. Chem.*, 1973, **11**, 729.

Other Cycloaddition and Orbital-symmetry-allowed Reactions. Phenyl trifluoromethanesulphonyl acetylene undergoes extremely facile Diels–Alder reactions with dienes, and has generally been found to be more reactive than dimethyl acetylenedicarboxylate (DMAD).[59] Cycloaddition of symmetrical acetylenes to hexa-1,2,4,5-tetraene provides a new and convenient route for the preparation of [2,2]paracyclophanes (27).[60]

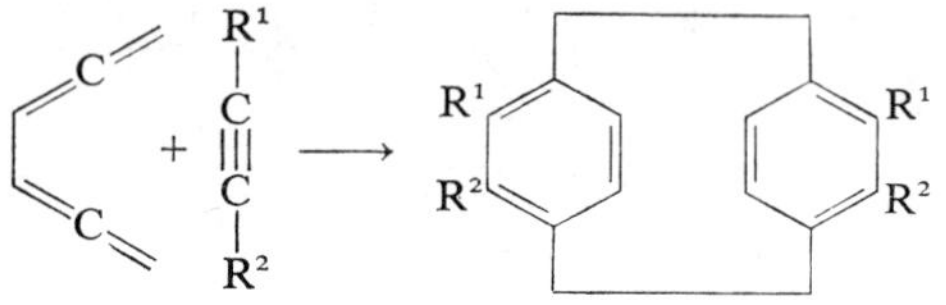

$R^1 = R^2 = CO_2Me$, CN, or CF_3

(27)

Dicyanoacetylene reacts with steroidal systems such as ergosteryl acetate to give products of both Diels–Alder addition and an 'ene' reaction.[61]

2,5-Dimethoxycarbonyl-3,4-diphenylcyclopentadienone reacts with acetylenes to give arenes.[62]

Electron-deficient thiophens such as (28) undergo a [2+2] cycloaddition with ynamines to give (29).[63] A high yield of the bicyclo-enamine (30), which has three asymmetric centres, has been achieved by ynamine addition to 5-methyl-2-cyclohexenone.[64] The addition of 1-diethylaminopropyne to the

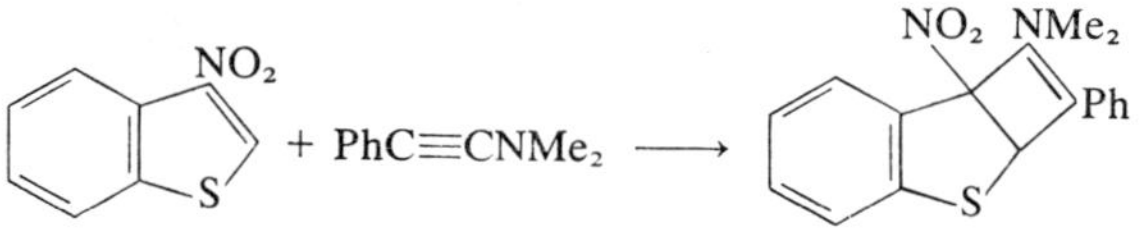

(28) + PhC≡CNMe₂ ⟶ (29)

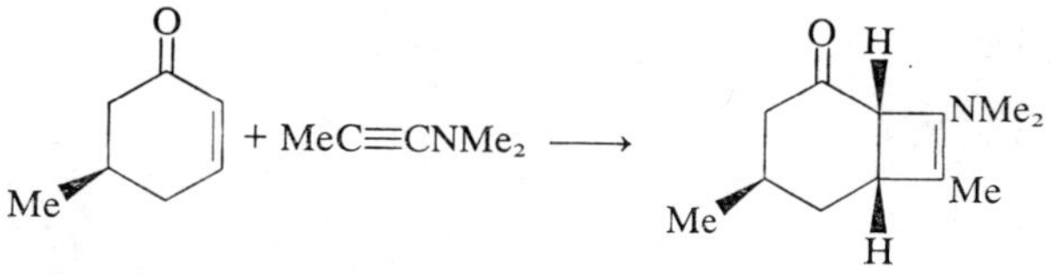

+ MeC≡CNMe₂ ⟶ (30)

---

[59] R. S. Glass and D. L. Smith, *J. Org. Chem.*, 1974, **39**, 3712.
[60] H. Hopf and F. T. Lenich, *Chem. Ber.*, 1974, **107**, 1891.
[61] A. Abramovitch and P. W. Le Quesne, *J. Org. Chem.*, 1974, **39**, 2197.
[62] D. M. White, *J. Org. Chem.*, 1974, **39**, 1951.
[63] D. N. Reinhoudt and C. G. Kouwenhoven, *Tetrahedron Letters*, 1974, 2503.
[64] J. Ficini and A. M. Touzin, *Tetrahedron Letters*, 1974, 1447.

sydnone species (31) is postulated to involve the hitherto unknown acyclic valence tautomer (31a).[65]

The reaction of ethyl propiolate with N-formyl-L-proline in acetic acid has been presumed to proceed *via* the intermediate 1,3-dipole (32).[66] Further examples of 1,3-dipolar additions to acetylenes include the reaction of nitrile oxides with silylacetylenes[67] and of phenyl azide with 1,1,4,4-tetraethoxybut-2-yne.[68] Diazocyclopentadienes add to acetylenes to give diaza[4,4] spirenes, which undergo 1,5-sigmatropic shifts to give aza-indolines (33) or indazoles (34).[69] Photolysis of the precursors also gives rise to the (1+2) cycloadducts such as (35).[70]

[65] G. V. Boyd and T. Norris, *J.C.S. Chem. Comm.*, 1974, 639.
[66] M. T. Pizzorno and S. M. Albonico, *J. Org. Chem.*, 1974, **39**, 731.
[67] L. Birkhofer and R. Stilke, *Chem. Ber.*, 1974, **107**, 3717.
[68] W. Winter and E. Müller, *Chem. Ber.*, 1974, **107**, 715.
[69] H. Dürr and W. Schmidt, *Annalen*, 1974, 1140.
[70] H. Dürr, B. Ruge, and B. Weiss, *Annalen*, 1974, 1150.

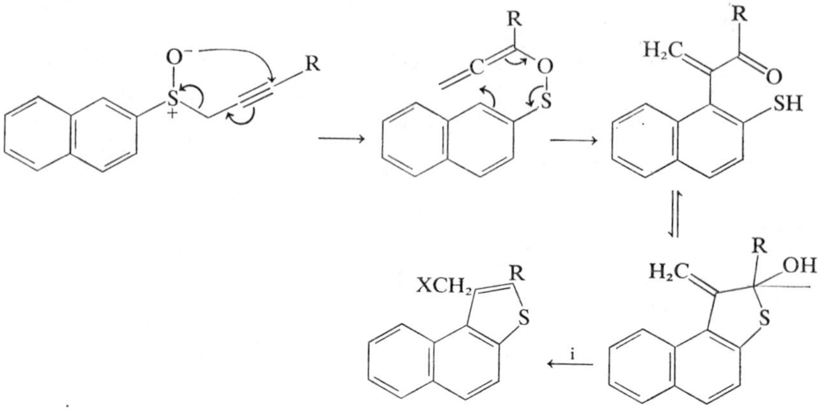

Scheme 9

Condensed thiophens are obtained *via* consecutive [2,3] and [3,3] sigmatropic rearrangements when arylprop-2-ynyl sulphoxides are heated in suitable protic solvents (Scheme 9).[71] An analogous type of reaction utilizes the addition of di(methoxycarbonyl)carbene to the phenyl propargyl sulphide (36) followed by a [2,3] sigmatropic rearrangement to the allene (37).[72]

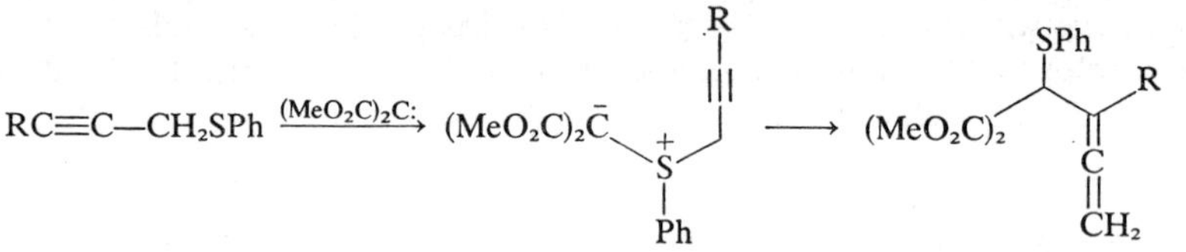

Thermolysis of homoallyl ethers proceeds in a concerted manner *via* an eight-centred transition state and follows a first-order rate law.[73] The thermal rearrangement of 7-propargyloxycycloheptatriene gives a mixture of bicyclo-[3,3,2]deca-3,7,9-trien-2-one (38) and the unstable 2,7-dihydrocyclohepta-[*b*]pyran (39).[74] The major photoproduct of 6-phenylhex-2-yne is the bicyclo-[6,3,0]undeca-1,3,5,7-tetraene (40), which is derived from the phenyl singlet excited state.[75] Thermal cyclization of *meta*-substituted-phenyl propargyl ethers yields a mixture of 5- and 7-substituted chromenes, and the factors which affect the product ratio have been investigated.[76]

[71] Y. Makisumi and S. Takada, *J.C.S. Chem. Comm.*, 1974, 848.
[72] P. A. Grieco, M. Meyers, and R. S. Finkelhor, *J. Org. Chem.*, 1974, **39**, 119.
[73] A. Viola, S. Madhowan, R. J. Proverb, B. L. Yates, and J. Larrahondo, *J.C.S. Chem. Comm.*, 1974, 842.
[74] A. Pryde, J. Zsindely, and H. Schmid, *Helv. Chim. Acta*, 1974, **57**, 1598.
[75] W. Lippke, W. Ferree, jun., and H. Morrison, *J. Amer. Chem. Soc.*, 1974, **96**, 2134.
[76] W. A. Anderson, E. J. LaVoie, and P. G. Whitkop, *J. Org. Chem.*, 1974, **39**, 881.

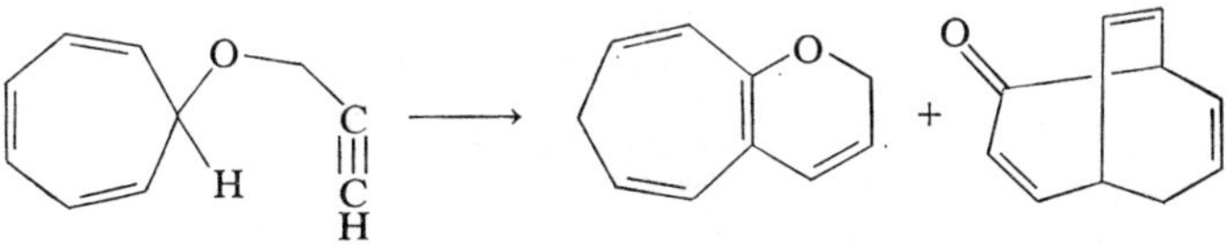

(39)   (38)

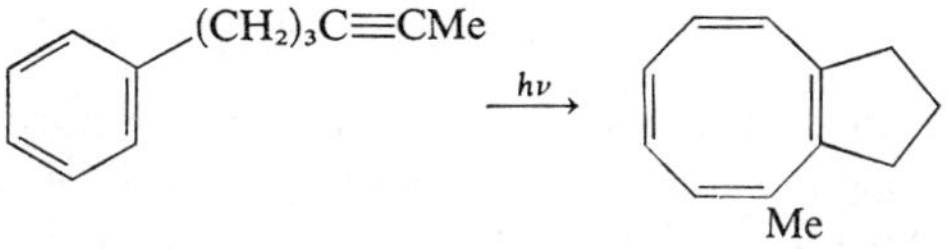

(40)

Other Additions to the Acetylenic Bond.—*Electrophilic Additions.* Studies of the initial reaction rate and product composition for the reaction of various acetylenes with hydrogen chloride give results which are consistent with reaction *via* competing Ad_E2 and *anti Ad*3 reaction mechanisms.[77] In a related study of the electrophilic addition of hydrogen chloride to phenyl [2-^{2}H]acetylene it was claimed that considerable participation by the forbidden *syn* [$2\pi + 2\sigma$] addition process occurred.[78] Kinetic studies of the electrophilic addition of 2,4-dinitrobenzenesulphenyl chloride with 1-phenylpropyne suggest that the episulphonium ion species (41) is an intermediate in the process.[79]

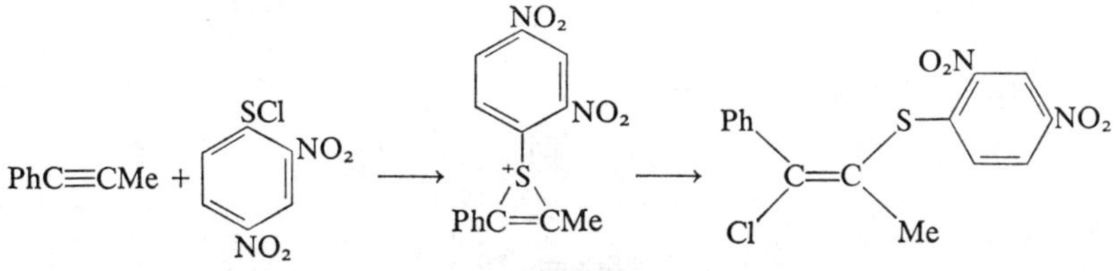

(41)

Although phenylacetylene polymerizes on treatment with xenon difluoride, alkylacetylenes react with the same reagent to give good yields of the tetrafluoro-adducts, and the process is catalysed by hydrogen fluoride.[80] Addition of BrCl and ICl to ethyl but-3-ynoate gives (42), but addition of sulphenyl halides results in the opposite orientation pattern and leads to (43).[81] Different

[77] R. C. Fahey, M. T. Payne, and D.-J. Lee, *J. Org. Chem.*, 1974, **39**, 1124.
[78] F. Marcuzzi, G. Melloni, and G. Modena, *Tetrahedron Letters*, 1974, 413.
[79] T. Okuyama, K. Isawa, and T. Fueno, *J. Org. Chem.*, 1974, **39**, 350.
[80] M. Zupan and A. Pollak, *J. Org. Chem.*, 1974, **39**, 2646.
[81] J. Tendil, M. Vemey, and R. Vessiere, *Tetrahedron*, 1974, **30**, 579.

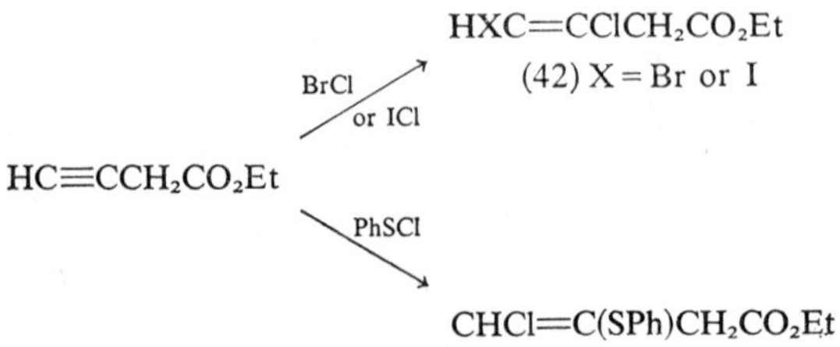

$$HC \equiv CCH_2CO_2Et \xrightarrow{\substack{BrCl \\ \text{or ICl}}} \begin{array}{c} HXC=CClCH_2CO_2Et \\ (42)\ X = Br\ \text{or}\ I \end{array}$$

$$\xrightarrow{PhSCl} \begin{array}{c} CHCl=C(SPh)CH_2CO_2Et \\ (43) \end{array}$$

mechanisms operate in each case, and these have been evaluated both by orientation and kinetic studies. Phenylsulphenyl chloride reacts by an Ad_E2 mechanism whereas ICl addition involves a combination of Ad_E3 and Ad_E4 mechanisms. The addition of phenylselenyl trifluoroacetate to acetylenes proceeds as expected, and hydrolysis of the trifluoroacetate moiety generates the corresponding ketone (Scheme 10).[82] Likewise, the addition of sulphonyl bromides RSO_2Br to acetylenes gives a mixture of *cis*- and *trans*-adducts in the presence of copper(II) bromide. In contrast, pure thermal addition gives only the *trans*-isomer (44). Both isomers readily lose hydrogen bromide on treatment with base and give the acetylenic sulphone.[83]

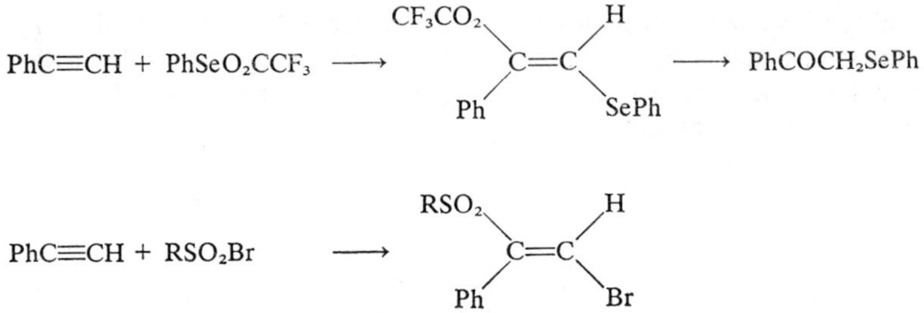

$$PhC \equiv CH + PhSeO_2CCF_3 \longrightarrow \underset{Ph}{\overset{CF_3CO_2}{>}}C=C\underset{SePh}{\overset{H}{<}} \longrightarrow PhCOCH_2SePh$$

$$PhC \equiv CH + RSO_2Br \longrightarrow \underset{Ph}{\overset{RSO_2}{>}}C=C\underset{Br}{\overset{H}{<}}$$

(44)

Scheme 10

Although vinyl cations are well-documented intermediates in the addition reactions of acetylenes, they have not been physically characterized to date. In trifluoroacetic acid solution ethynylferrocenes (45) bearing a 2-t-butyl substituent undergo protonation to give the corresponding vinyl cations. Although these are short-lived, they can be observed in this case by ^1H n.m.r. spectroscopy as a result of the stabilizing effect of the orbitals on the Fe atom.[84]

---

[82] H. J. Reich, *J. Org. Chem.*, 1974, **39**, 428.
[83] Y. Amiel, *J. Org. Chem.*, 1974, **39**, 3867.
[84] T. S. Abram and W. E. Watts, *J.C.S. Chem. Comm.*, 1974, 857.

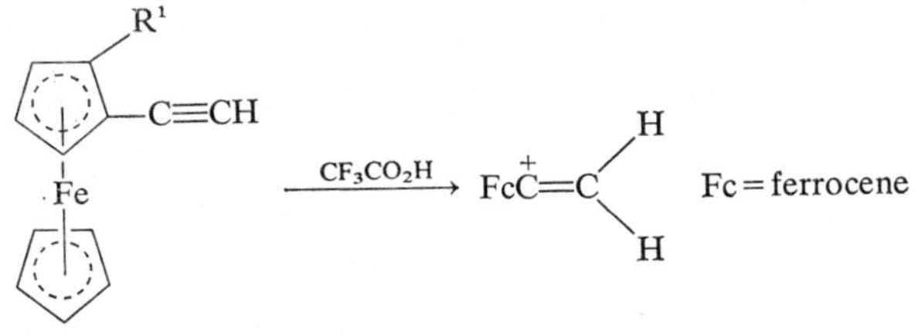

(45)

But-2-yne reacts with isobutyryl fluoroborate in methylene chloride–tetrachloroethane to give 2,3,5-trimethylcyclopent-2-enone as the major product, probably *via* a vinyl cation intermediate (Scheme 11).[85]

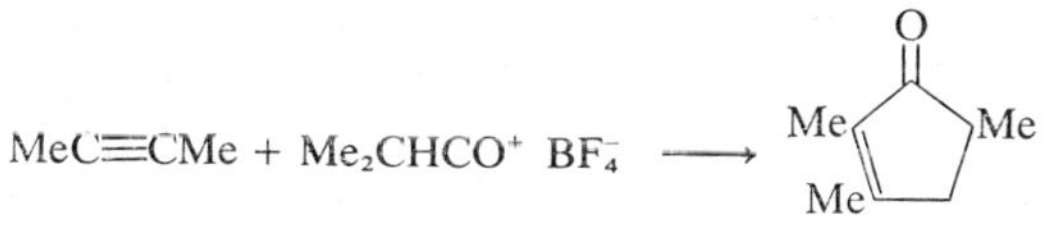

$$MeC{\equiv}CMe + Me_2CHCO^+\ BF_4^- \longrightarrow$$

Scheme 11

Nucleophilic Additions. Tetrahydropyran-2-thiol adds to acetylenes and olefins in the presence of base to give protected thiol derivatives in good yield. The thiol grouping can be liberated using silver nitrate followed by chloroform–HCl treatment. There was no evidence to show that these compounds existed in the tautomeric thioketone form (Scheme 12).[86]

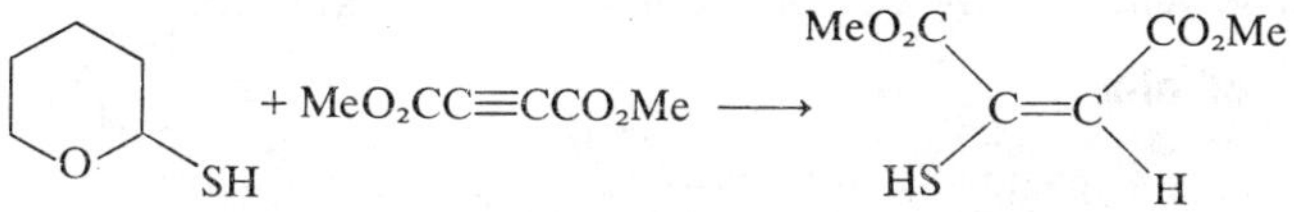

Scheme 12

Hydroxylamines add to toluene-*p*-sulphonylacetylenes by attack either *via* nitrogen or *via* oxygen, depending on the structure of the acetylene (Scheme 13).[87] The kinetics of the hydration of ynamines have been investigated in water–dioxan by stopped-flow spectrophotometry.[88] The very useful

---

[85] A. A. Schegolev, W. A. Smit, G. V. Roitburd, and V. F. Kucherov, *Tetrahedron Letters*, 1974, 3373.
[86] M. G. Missakin, R. Ketcham, and A. R. Martin, *J. Org. Chem.*, 1974, **39**, 2011.
[87] J. A. Sanders, K. Hovius, and J. F. N. Engberts, *J. Org. Chem.*, 1974, **39**, 2641.
[88] W. F. Verhelst and W. Drenth, *J. Amer. Chem. Soc.*, 1974, **96**, 6692.

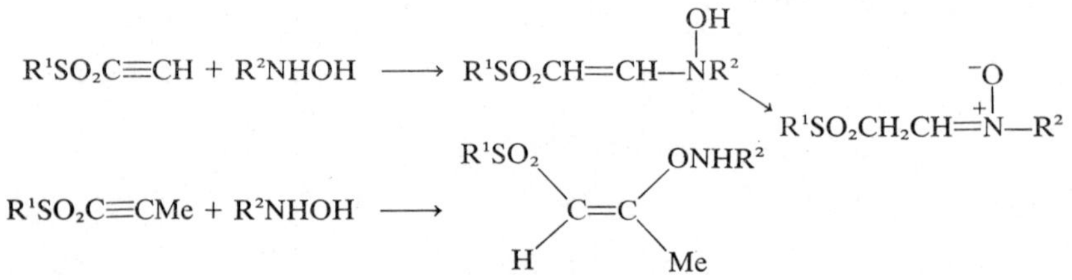

$R^1SO_2C\equiv CH + R^2NHOH \longrightarrow R^1SO_2CH=CH-NR^2(OH) \searrow$

$R^1SO_2CH_2CH=\overset{+}{N}(-O^-)-R^2$

$R^1SO_2C\equiv CMe + R^2NHOH \longrightarrow$ (C=C with R^1SO_2 and H on one carbon, $ONHR^2$ and Me on the other)

Scheme 13

addition of di-isobutylaluminium hydride to acetylenes to give *cis*-alanes has been studied kinetically, and the role of molecular association of the reagent investigated.[89] The reaction of the *cis*-alanes obtained by this procedure with allyl bromides in the presence of copper(I) chloride *via* a coupling reaction constitutes a stereoselective synthesis of *trans*-1,4-dienes (Scheme 14).[90] The lithium aluminium hydride reduction of alk-1-yn-3-ols has been

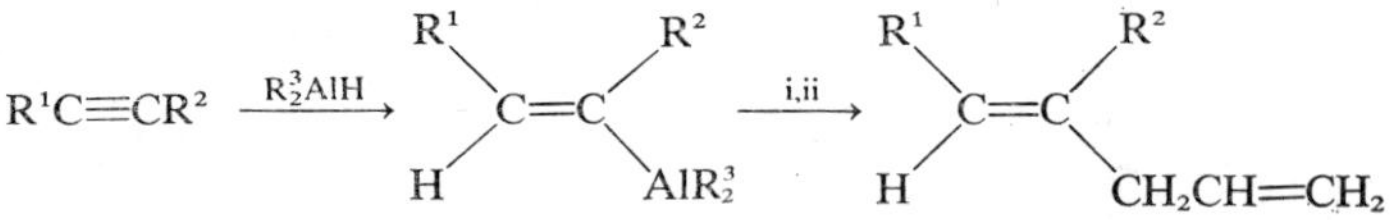

$R^1C\equiv CR^2 \xrightarrow{R^3_2AlH}$ (C=C with R^1,H / R^2,AlR^3_2) $\xrightarrow{i,ii}$ (C=C with R^1,H / R^2,$CH_2CH=CH_2$)

Reagents: i, CuCl; ii, $CH_2=CHCH_2Br$

Scheme 14

shown to proceed *via* a site-specific hydride transfer to C-2. The mechanism proposed (Scheme 15) rationalizes the observed reciprocal relationships between solvent basicity and the extent of *cis*-reduction for these systems.[91]

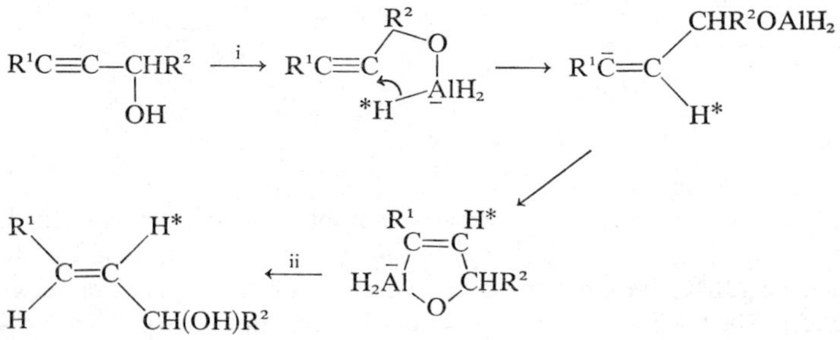

$R^1C\equiv C-CHR^2(OH) \xrightarrow{i} R^1C\equiv C(-R^2,O-AlH_2,*H) \longrightarrow R^1\bar{C}=C(CHR^2OAlH_2 / H^*)$

(C=C with R^1,H / H,CH(OH)R^2) $\xleftarrow{ii}$ (C=C with R^1,H^* / $H_2\bar{Al}$-O-CHR^2)

Reagents: i, LiAlH$^*_4$; ii, H_2O

Scheme 15

[89] J. G. Eisch and S.-G. Rhee, *J. Amer. Chem. Soc.*, 1974, **96**, 7276.
[90] R. A. Lynd and G. Zweifel, *Synthesis*, 1974, 658.
[91] B. Grant and C. Djerassi, *J. Org. Chem.*, 1974, **39**, 968.

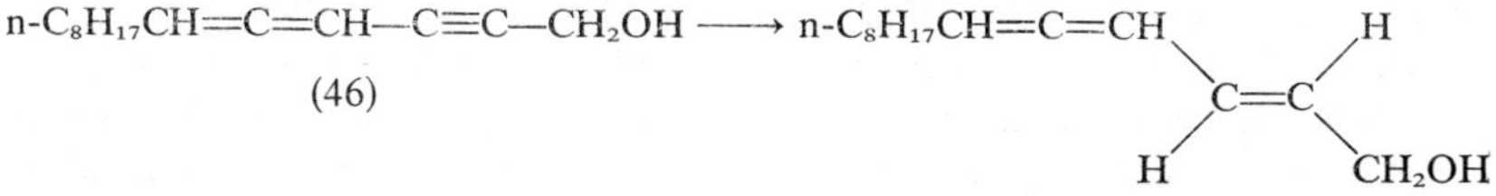

n-C$_8$H$_{17}$CH=C=CH—C≡C—CH$_2$OH ⟶ n-C$_8$H$_{17}$CH=C=CH\\C=C/H
(46)   H / \\ CH$_2$OH

(47)

Reduction of such systems where an alkoxy function is attached to the carbon atom adjacent to the triple bond leads to α-allenic alcohols instead.[92] A further example of this reaction is the reduction of the hydroxymethyl-acetylene (46) to the *trans*-olefin (47) with lithium tri-t-butoxyaluminium hydride.[93]

Metal-catalysed Reactions of Acetylenes.—The hydroformylation reaction of acetylenes is known to be more difficult to carry out than with alkenes. The results available indicate that, when this reaction is carried out with the catalyst HRh(CO)(PPh$_3$)$_3$, it occurs *via* the αβ-unsaturated aldehydes as intermediates followed by reduction, rather than *via* prior reduction of the acetylene to the alkene followed by hydroformylation.[94] The reaction of phenylacetylene with carbon monoxide catalysed by Rh$_2$(CO)$_4$Cl$_2$ gives appreciable yields of 3,6-diphenyl-1-oxabicyclo[3,3,0]octa-3,6-diene-2,4-dione (48).[95]

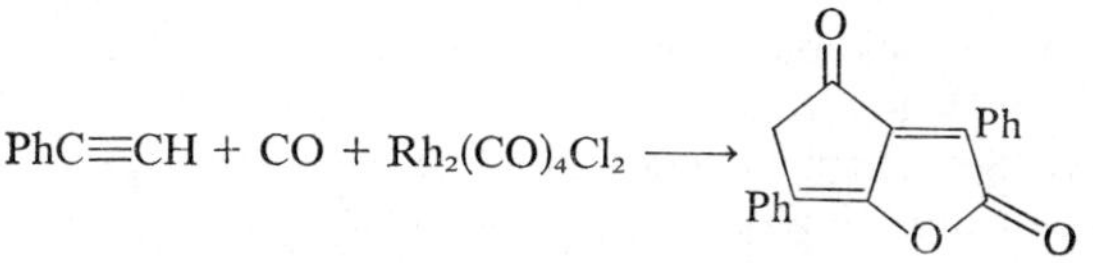

PhC≡CH + CO + Rh$_2$(CO)$_4$Cl$_2$ ⟶ (48)

The trimerization of acetylenes to aromatic compounds has been well studied, but linear terminal diacetylenes react with η^5-cyclopentadienylcobalt dicarbonyl, (η^5-C$_5$H$_5$)Co(CO)$_2$, to give trimers, the formation of which involves six acetylene functions (Scheme 16).[96] The complex (π-C$_5$H$_5$)Co(PPh$_3$)$_2$

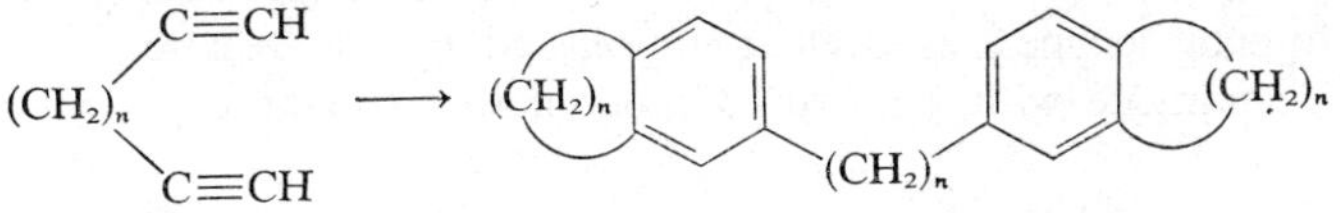

Scheme 16

[92] A. Claesson, L. I. Olsson, and C. Bogentoft, *Acta Chem. Scand.*, 1973, **27**, 2941.
[93] R. Baudouy and J. Gore, *Synthesis*, 1974, 573.
[94] C. Botteghi and C. Salomon, *Tetrahedron Letters*, 1974, 4285.
[95] J. Kiji, S. Yoshikawa, and J. Furukawa, *Bull. Chem. Soc. Japan*, 1974, **47**, 490.
[96] K. P. C. Volhardt and R. G. Bergman, *J. Amer. Chem. Soc.*, 1974, **96**, 4997.

reacts with one equivalent of an acetylene to give $(\pi\text{-}C_5H_5)Co(PPh_3)R^1C\equiv CR^2$, and further reaction with a second equivalent of an acetylene gives the air-stable species (49). These compounds can be decomposed by alkenes to give cyclohexadienes (50). Alternatively, their reaction with sulphur, selenium, or nitrosobenzene gives the corresponding thiophens, selenophens, or pyrroles, and this illustrates the versatility of (49) as a synthetic intermediate.[97]

$$(\pi\text{-}C_5H_5)Co(PPh_3)_2 \xrightarrow[\text{ii,}R^3C\equiv CR^4]{\text{i,}R^1C\equiv CR^2} (49) \xrightarrow{CH_2=CH_2} (50)$$

Compounds of the type $ArM(CO)_3$ (M = Cr, W, or Mo) catalyse the polymerization of phenylacetylene to compounds of high molecular weight *via* an intermediate which has been isolated and identified as a ladder compound composed of fused cyclobutane rings. The final product is a linear, polyconjugated polyethylene.[98] The molybdenum complex (51) can react with diphenylacetylene to give 6,6-dicyano-1,2,3,4-tetraphenylfulvene (52).[99]

The reaction of the oxoalkyne (53) with $Ni(CO)_4$ gives two products (54) and (55), and the possible involvement of metal-stabilized cyclobutadiene intermediates in their formation has been discussed.[100] Oxidation of cyclobutadieneiron tricarbonyl by ceric ion in the presence of dimethyl acetylenedicarboxylate or dibenzoylacetylene gives the interesting system (56).[101]

The codimerization of acetylenes and allyl halides in the presence of complexes such as $PdCl_2(PhCN)_2$ gives useful yields of penta-1,4-dienes.[102] Irradiation of solutions of cycloheptatrieneiron tricarbonyl and dimethyl acetylenedicarboxylate in THF at 0 °C gives (57). This is the first example of a $[2\pi + 6\pi]$ addition reaction which is forbidden by the Woodward–Hofmann rules, and it adds another example to the list of forbidden reactions that occur in the presence of transition metals.[103]

Other Reactions of Acetylenes.—The preparation of the anion radical (58) of cyclo-octatrienyne has been accomplished by the reaction of bromocyclo-octatetraene with a mirror of potassium metal at 100 °C. The e.s.r.

---

[97] Y. Wakatsuki, T. Kuramitsu, and H. Yamazaki, *Tetrahedron Letters*, 1974, 4549.
[98] M. F. Farona, P. A. Lofgren, and P. S. Woon, *J.C.S. Chem. Comm.*, 1974, 246.
[99] R. B. King and M. S. Saran, *J.C.S. Chem. Comm.*, 1974, 851.
[100] F. Wagner and H. Meier, *Tetrahedron*, 1974, **30**, 773.
[101] J. Meinwald and J. Mioduski, *Tetrahedron Letters*, 1974, 3839.
[102] K. Kaneda, F. Kawamoto, Y. Fujiwara, T. Imanaka, and S. Teranishi, *Tetrahedron Letters*, 1974, 1067.
[103] R. E. Davis, T. A. Dodds, F.-H. Hseu, J. C. Wagnon, T. Devon, J. Tancrede, J. S. McKennis, and R. Pettit, *J. Amer. Chem. Soc.*, 1974, **96**, 7562.

Acetylenes, Alkanes, Allenes, and Olefins

(51) → PhC≡CPh → (52)

(53) → (54) + (55)

(56)

(57)

(58)

spectrum of (58) is consistent with the assigned structure.[104] In the presence of alkali metals, 1-phenylbut-1-yne undergoes three types of reaction. These consist of (a) reductive metallation at the triple bond, which leads to metallated derivatives of trans-1-phenylbut-1-ene; allylic rearrangement of these latter compounds yields derivatives of cis- and trans-1-phenylbut-2-ene. (b) Propargylic rearrangement, which results in migration of the triple bond. (c) Addition of the metallated species from the acetylene to ethylene (generated by fragmentation of the solvent) to yield products of higher molecular weight. The relative importance of these processes depends on the metal used. Potassium gives evidence of all three but sodium gives mainly (a), owing to a rapid metal–hydrogen transfer step (Scheme 17). The initial step is probably

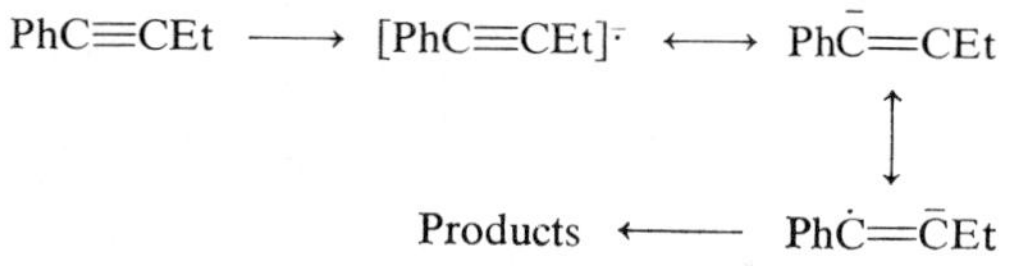

$$PhC{\equiv}CEt \longrightarrow [PhC{\equiv}CEt]^{-} \longleftrightarrow Ph\bar{C}{=}CEt$$
$$\updownarrow$$
$$Products \longleftarrow Ph\dot{C}{=}\bar{C}Et$$

Scheme 17

formation of the anion radical.[105] Similar observations have also been recorded in the reduction of hex-1- and -3-yne using sodium in hexamethylphosphortriamide–THF in the presence and absence of proton donors.[106]

The reaction of 2-chloromethylpyridine with sodium acetylide in liquid ammonia yields (59) and (60). Compound (59) is formed by a double alkylation of the dimer produced by direct metallation, whereas (60) is formed by double alkylation of the rearranged acetylene (Scheme 18).[107]

Contrary to earlier published results, the reaction of pyridine N-oxide and sodium acetylide does not yield 2-ethynylpyridine N-oxide but gives a mixture of (E,Z,E)-hepta-2,4-dien-6-ynal oxime (61) and (E,Z,E,E,Z,E)-dodeca-2,4,8,10-tetraen-6-ynedial dioxime (62).[108]

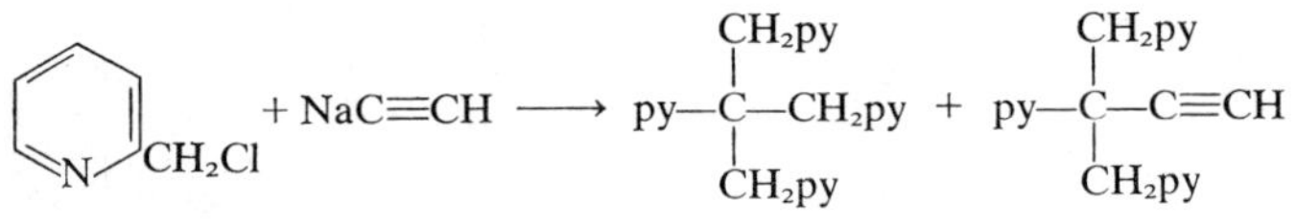

$$\underset{CH_2Cl}{\text{pyridine}} + NaC{\equiv}CH \longrightarrow py{-}\underset{CH_2py}{\overset{CH_2py}{C}}{-}CH_2py + py{-}\underset{CH_2py}{\overset{CH_2py}{C}}{-}C{\equiv}CH$$

py = 2-pyridyl   (59)   (60)

Scheme 18

[104] G. R. Stevenson, M. Colón, J. G. Concepción, and A. McB. Block, *J. Amer. Chem. Soc.*, 1974, **96**, 2283.
[105] J.-L. Derocque and F.-B. Sundermann, *J. Org. Chem.*, 1974, **39**, 1736.
[106] H. O. House and E. F. Kinloch, *J. Org. Chem.*, 1974, **39**, 747.
[107] A. E. Zune, U. Hollstein, and W. M. Litchman, *J. Org. Chem.*, 1974, **39**, 2461.
[108] U. Fritysche and S. Hünig, *Annalen*, 1974, 1407.

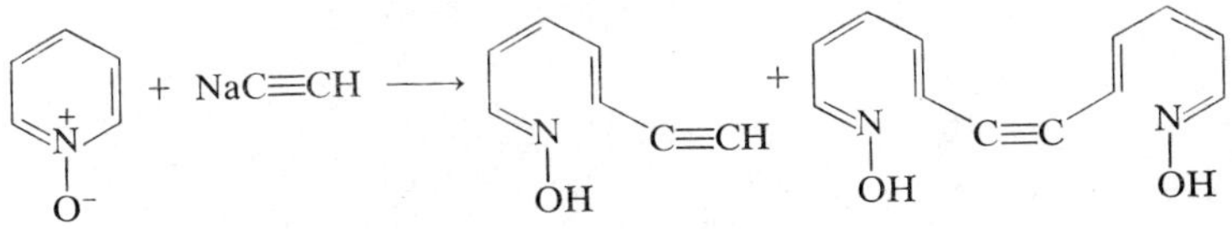

Acetylenes, Alkanes, Allenes, and Olefins

[Scheme showing pyridine N-oxide + NaC≡CH → (61) and (62)]

(61)   (62)

The synthesis of certain cyclo-octadecatetraenetetrayne-1,6-diones and -1,10-diones (tetradehydroannulenediones) has been accomplished by oxidative coupling of the required acetylenic precursors (Scheme 19).[109] The reaction of 1,12-diethynyldodecane dilithium salt with 1,12-dibromododecane in hexamethylphosphortriamide gives a modest yield of the symmetrical cyclized diacetylene.[110]

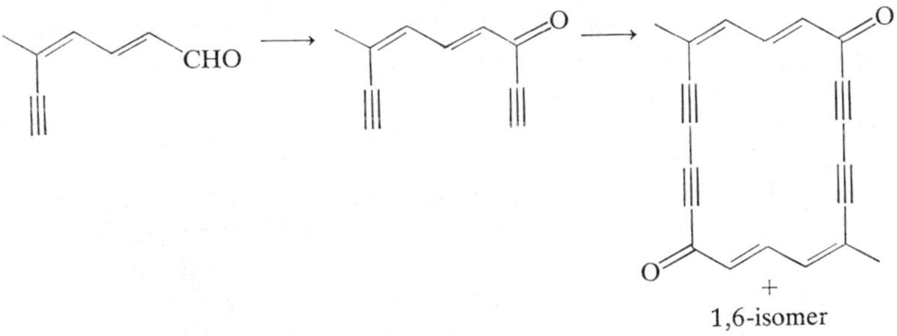

[Scheme 19 structures]

+
1,6-isomer

Scheme 19

The intriguing reaction of [10](9,10)anthracenophane-4,6-diyne (63) in sunlight gives the dimer (64). The multi-cycloaddition to (63) can be represented as a photochemically allowed $[_\pi 4_s + _\pi 2_a + _\pi 2_a + _\pi 4_s + _\pi 2_a + _\pi 2_a]$ cycloaddition. If this is concerted, then it would be the first case of a six-component cycloaddition.[111]

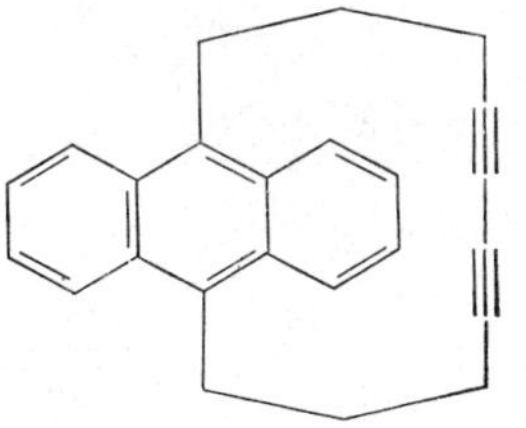

(63) 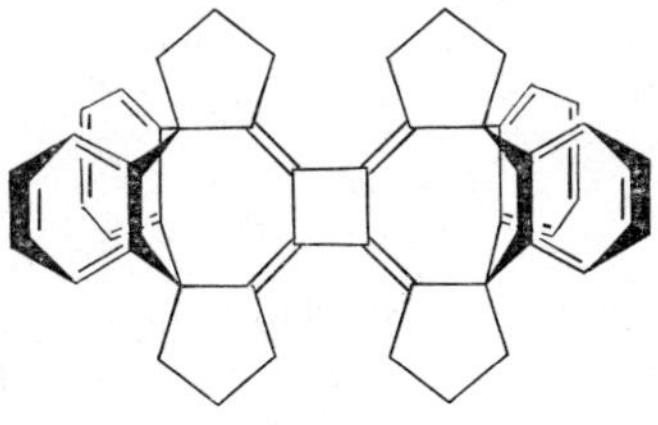   (64)

---

[109] N. Darby, K. Yamamoto, and F. Sondheimer, *J. Amer. Chem. Soc.*, 1974, **96**, 248.
[110] S. F. Karaev and M. M. Movsumzade, *Zhur. org. Khim.*, 1974, **10**, 880 (*Chem. Abs.*, 1974, **81**, 3422r).
[111] T. Inoue, T. Kaneda, and S. Misumi, *Tetrahedron Letters*, 1974, 2969.

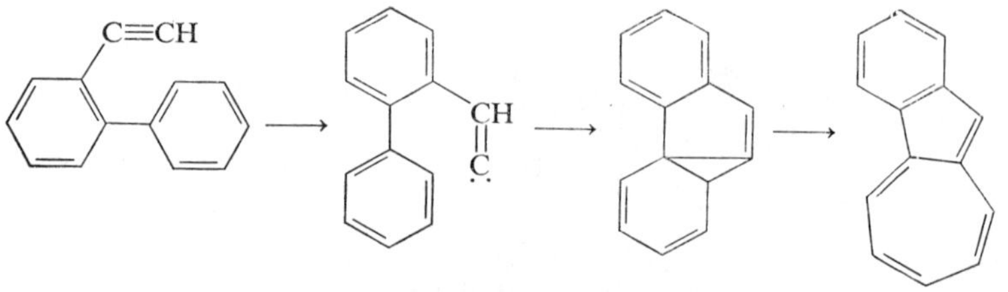

Scheme 20

Pyrolysis of labelled phenylacetylene, $Ph^{13}C\equiv CH$, at 550 °C under high vacuum over silica was shown by n.m.r. techniques to result in some rearrangement to $PhC\equiv^{13}CH$. This probably occurs *via* benzylidenecarbene $PhCH=C$: as an intermediate. Evidence in support of this hypothesis comes from a study of the pyrolysis of biphenyl-2-acetylene under similar conditions, which leads to formation of some 1,2-benzazulene (Scheme 20).[112]

Thermolysis of trimethylsilylethoxyacetylene at 120 °C results in smooth decomposition to give trimethylsilylketen, which is a useful acylating agent for hindered amines and alcohols.[113]

Triethylsilylbuta-1,3-diyne can be converted into 1-tri-n-butylstannyl-4-triethylsilylbuta-1,3-diyne and thence, by selective cleavage with bromine or iodine monochloride, into halogenobutadiynyl triethylsilanes. The latter compounds are useful intermediates for extending a terminal acetylene by two units *via* a Cadiot–Chodkiewicz coupling procedure.[114]

Ozonization of acetylenes in the presence of tetracyanoethylene (tcne) and pinacolone has been studied. In the presence of tcne, α-diketones are the major products whereas pinacolone yields anhydrides. The evidence suggests that tcne reacts with the intermediate zwitterion as an electron acceptor (Scheme 21).[115]

The mechanisms involved in the addition of dimethylsilylene to acetylenes have been further studied, and an overall reaction scheme has been proposed (Scheme 22).[116]

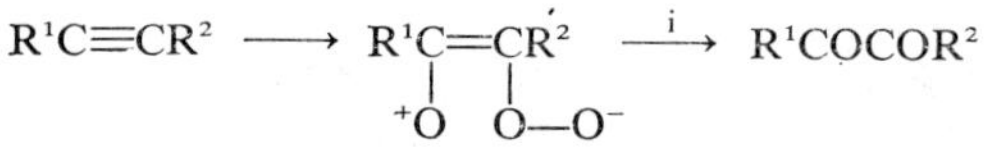

$$R^1C\equiv CR^2 \longrightarrow \underset{\underset{^+O}{|}\ \underset{O-O^-}{|}}{R^1C=CR^2} \xrightarrow{i} R^1COCOR^2$$

Reagent: i, tetracyanoethylene (tcne)

Scheme 21

---

[112] R. F. C. Brown, K. J. Harrington, and G. L. McMullen, *J.C.S. Chem. Comm.* 1974, 123.
[113] R. A. Ruden, *J. Org. Chem.*, 1974, **39**, 3607.
[114] B. N. Ghose and D. R. M. Walton, *Synthesis*, 1974, 890.
[115] N. C. Yang and J. Libman, *J. Org. Chem.*, 1974, **39**, 1782.
[116] T. J. Barton and J. A. Kilgour, *J. Amer. Chem. Soc.*, 1974, **96**, 7150.

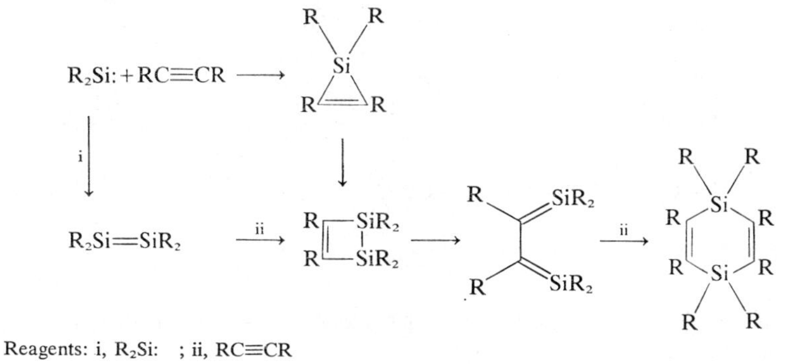
R_2Si: + RC≡CR
R
Si
$R_2Si=SiR_2$
i
ii
SiR_2
Reagents: i, R_2Si: ; ii, RC≡CR

Scheme 22

The alkynyl keten thioacetal (65) is generated by the ring cleavage of the thiophen (66) with ethyl-lithium and ethyl bromide.[117]

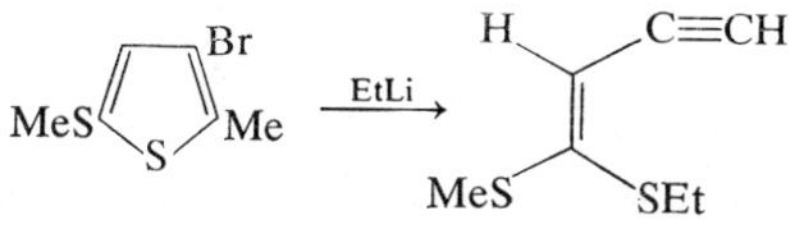
Br
MeS
Me
S
EtLi
H
C≡CH
MeS
SEt

(66) (65)

Physical Properties of Acetylenes.—The electronic spectra[118] of unsubstituted mono- to penta-acetylenes in the gas phase and in solution have been recorded. The photoelectron spectra of mono- and di-substituted silyl- and methyl-acetylenes have been assigned by comparison with the ionization potentials of acetylene, disilane, and ethane. The observed π splittings can be rationalized within a parametrized hyperconjugation model.[119] A series of substituted alkynylcarbenium ions and alkynoyl cations has been investigated by ^{13}C and ^{1}H n.m.r. spectroscopy. In the case of alkylcarbenium ions it has been concluded that the positive charge is extensively delocalized and that the mesomeric allenyl cations, which are vinyl cations, contribute extensively to the total ion structure. In contrast, the alkynoyl cations have the charge localized mainly on the oxygen atom as an oxonium ion (Scheme 23).[120]

The photoelectron spectra of several strained acetylenes (67)—(70) have been measured, and the splittings of the π-molecular orbitals predicted for

[117] S. Gronowitz and T. Frejd, *Acta Chem. Scand.*, 1973, **27**, 2242.
[118] E. Kloster-Jensen, H.-J. Haink, and H. Christen, *Helv. Chim. Acta*, 1974, **57**, 1731.
[119] W. Ensslin, H. Bock, and G. Becker, *J. Amer. Chem. Soc.*, 1974, **96**, 2757.
[120] G. A. Olah, R. J. Spear, P. W. Westerman, and J.-M. Dennis, *J. Amer. Chem. Soc.*, 1974, **96**, 5855.

$$\underset{X}{\overset{Y}{\rightthreetimes}}\overset{\oplus}{\underset{\ominus}{C}}-C\equiv C-Z \longleftrightarrow \underset{X}{\overset{Y}{\rightthreetimes}}C=C=\overset{\oplus}{\underset{\ominus}{C}}-Z$$

$$R-C\equiv C-\overset{+}{C}\equiv \overset{..}{O} \longleftrightarrow R-C\equiv C-\overset{+}{C}=\overset{..}{O} \longleftrightarrow R-\overset{+}{C}=C=C=O$$

Scheme 23

bent acetylenes using the theoretical MINDO/2 method have been found to be in close agreement.[121]

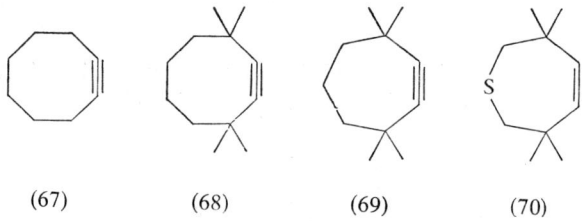

(67)   (68)   (69)   (70)

Acetylene–Allene Interconversions.—3-Chloro-3-methylbut-1-yne can be converted into the carbene Me$_2$C=C=C: using a two-phase system, and the allenic carbene adds to olefins in the organic phase to give useful yields of dimethylvinylidenecyclopropane derivatives.[122] At −40 °C the butynylenedisulphonium salts (71) give the butatrienesulphonium salts (72) on treatment with base, and these can be kept for several hours in solution at −10 °C. Tetra-arylbutatrienes add to cyclopentadienones across the central double bond. However, in (72) the terminal double bond acts as the dienophile, and reacts with cyclopentadiene to give a 50:50 mixture of the isomeric allenes (73) and (74).[123]

Allenic alcohols can be prepared in high yield by treatment of alkynyloxirans with Grignard reagents in the presence of powdered copper(I) iodide in THF followed by hydrolysis.[124] γ-Bromo-alkynol acetates can react with Grignard reagents to give reasonable yields of allenes (Scheme 24).[125] Terminal acetylenic Grignard reagents merely give the hydrolysis product (75), but propargyl Grignards give the alkynyl allenes (76) in reasonable yield. A similar type of transformation can be carried out using the organozinc derivatives.[126]

[121] H. Schmidt, A. Schweig, and A. Krebs, *Tetrahedron Letters*, 1974, 1471.
[122] T. Sasaki, S. Eguchi, and T. Ogawa, *J. Org. Chem.*, 1974, **39**, 1927.
[123] H. Braun, G. Strobl, and H. Gotzler, *Angew. Chem. Internat. Edn.*, 1974, **13**, 469, 470.
[124] P. Vermeer, J. Meijer, C. de Graaf, and H. Schreurs, *Rec. Trav. chim.*, 1974, **93**, 46.
[125] P. Perriot and M. Gaudemar, *Bull. Soc. chim. France*, 1974, 685.
[126] P. Perriot and M. Gaudemar, *Bull. Soc. chim. France*, 1974, 2603.

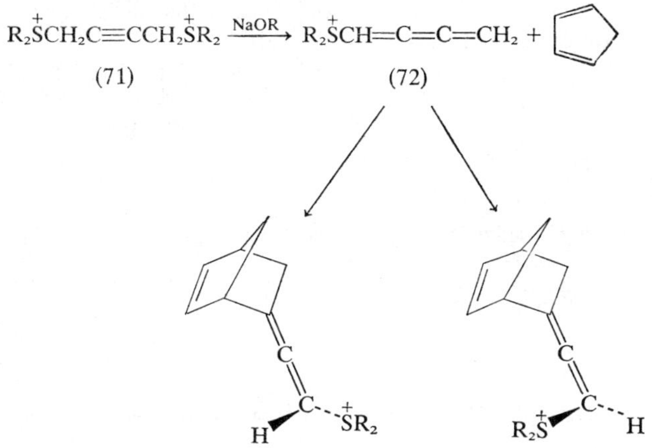

Scheme 24

$R_2\overset{+}{S}CH_2C\equiv CCH_2\overset{+}{S}R_2$ (71) $\xrightarrow{\text{NaOR}}$ $R_2\overset{+}{S}CH=C=C=CH_2$ (72) + ⬠

(73) / (74)

$MeCO_2CH_2C\equiv CCH_2Br + EtMgBr \longrightarrow MeCO_2CH_2\underset{Et}{C}=C=CH_2$

$\underset{Me}{MeCO_2CH}C\equiv CCH_2Br + C_5H_{11}C\equiv CMgBr \longrightarrow BrCH_2C\equiv C-\underset{Me}{CHOH}$ (75)

$+ HC\equiv CCH_2MgBr \longrightarrow MeCH(OH)C\equiv CCH_2CH=C=CH_2$ (76)

1,5-Di-t-butyl-3-bromopenta-1,4-diyne (77) undergoes normal displacement reactions with alcohols and cyanide ion, but the use of strong bases such as sodium ethoxide causes isomerization to the allene (78).[127] The base-catalysed isomerizations of the allyl and propargyl ethers (79) and (80) promoted by potassium t-butoxide proceed *via* the sequence of acetylene–allene isomerization, intramolecular [$_\pi 4 + {_\pi}2$] cycloaddition, and sigmatropic or prototropic hydrogen shift.[128]

3-Hydroxy-3-methylbut-1-yne reacts with sulphur dichloride at $-70\,^\circ$C to give the expected product (81). This compound cannot be isolated, and on warming decomposes *via* an acetylene–allene isomerization and migration from a C—O—S linkage to a C—S—O linkage. Further heating induces a facile cyclization and gives a quantitative yield of the thiophen 1,1-dioxide (82).[129]

---

[127] H. Hauptmann, *Tetrahedron Letters*, 1974, 3593.
[128] A. J. Bartlett, T. Laird, and W. D. Ollis, *J.C.S. Chem. Comm.*, 1974, 496.
[129] S. Braverman and D. Seger, *J. Amer. Chem. Soc.*, 1974, 96, 1245.

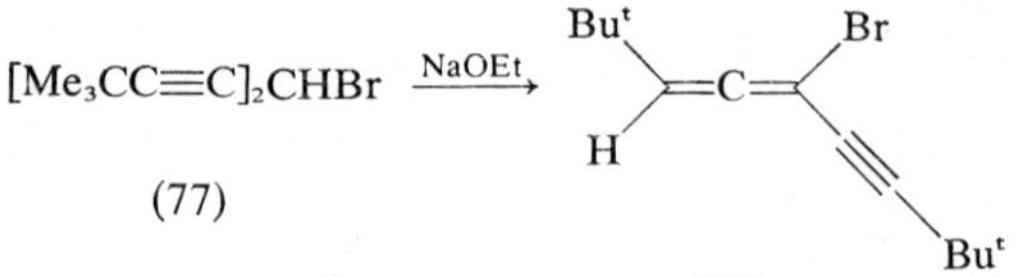

$[Me_3CC≡C]_2CHBr$ —NaOEt→ (78)

(77)

where (78) is Bu^t(H)C=C=C(Br)(Bu^t)... [structure shown]

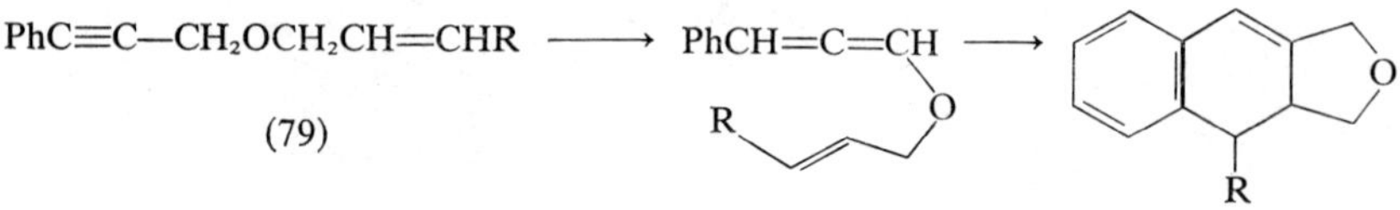

$PhC≡C-CH_2OCH_2CH=CHR$ ⟶ $PhCH=C=CH-O-CH_2-CH=CH-R$ ⟶ [bicyclic product with R]

(79)

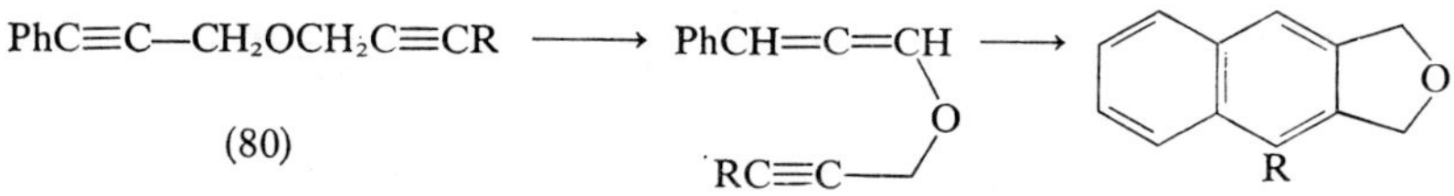

$PhC≡C-CH_2OCH_2C≡CR$ ⟶ $PhCH=C=CH-O-CH_2-C≡C-R$ ⟶ [bicyclic product with R]

(80)

$HC≡C-CMe_2OH + SCl_2$ ⟶ $[(HC≡C-CMe_2O)_2S]$ ⟶ [sulfone diene structure]

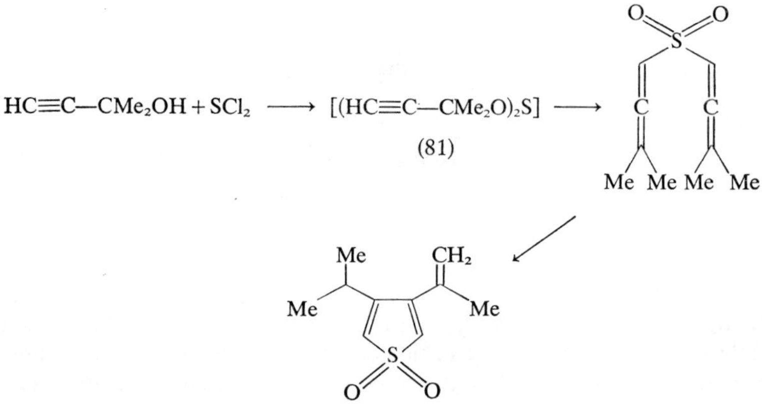

(81)

(82)

2 Alkanes

Treatment of alkylmercury(II) halides with iodo-(tri-n-butylphosphine)-copper(I) and three equivalents of t-butyl-lithium at −78 °C gives a reactive 'ate' complex which contains the alkyl group originally bonded to the mercury(II), and which can be alkylated to produce alkanes.[130] An extension of the titanocene system utilized by van Tamelen in dinitrogen fixation has been used to convert aldehydes, esters, and epoxides into alkanes in reasonable

---

[130] D. E. Bergbreiter and G. M. Whitesides, *J. Amer. Chem. Soc.*, 1974, **96**, 4937.

yield.[131] Reduction of acetoxybenzocyclobutane with lithium in liquid ammonia is known to give benzocyclobutane. Many benzyl esters have now been shown to be reduced[132] to the parent hydrocarbon under similar conditions. A fourfold excess of LiCuH(R), prepared from copper(I) hydride and butyl-lithium, is a very selective reducing agent. It converts bromides and tosylates into alkanes, and αβ-unsaturated ketones are reduced to ketones.[133]

The use of thionyl chloride in chlorination reactions of alkanes suggests that under photochemical or thermal conditions the reactive species are chlorine atoms.[134] Heptane, decane, and cyclohexane are oxidized by iron(II)–NN-diethylhydroxylamine or iron(II)–triethylamine N-oxide in trifluoroacetic acid to give trifluoroacetates, with no evidence of further oxidation. The overall yield is up to 50%, and the product distribution obtained with heptane is shown in Scheme 25.[135]

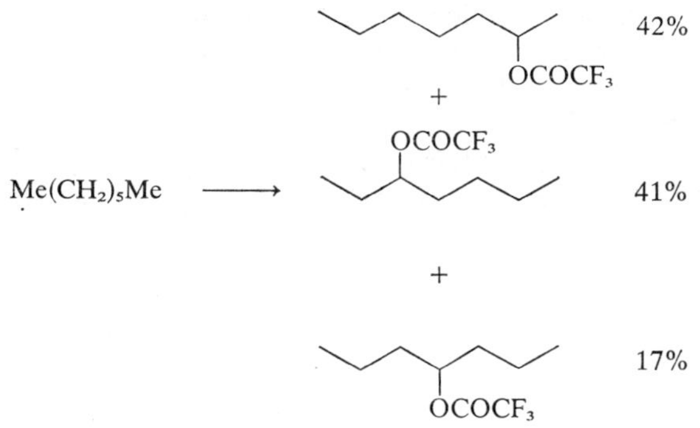

Scheme 25

The asymmetric cross-coupling reaction of racemic secondary alkyl Grignard reagents with alkyl halides has been achieved under very mild conditions by the use of nickel [(−)-2,3-O-isopropylidene-2,3-dihydroxy-1,4-bis(diphenylphosphino)butane] dichloride as catalyst, and gives optically active hydrocarbons.[136]

The first ionization potentials of n-alkanes are found to be a linear function of the inductive substituent constants.[137] The correlation is obtained by considering that a σ-bonding electron is ejected from the most central bond of the alkane. These values agree closely with those calculated. A computer

---

[131] E. E. Van Tamelen and J. A. Gladysz, *J. Amer. Chem. Soc.*, 1974, **96**, 5290.
[132] J. H. Markgraf, W. Hensley, and L. I. Shoer, *J. Org. Chem.*, 1974, **39**, 3168.
[133] S. Masamune, G. S. Bates, and P. E. Georghiou, *J. Amer. Chem. Soc.*, 1974, **96**, 3688.
[134] J. M. Krasniewski, jun., and M. W. Mosher, *J. Org. Chem.*, 1974, **39**, 1303.
[135] N. C. Deno and D. G. Pohl, *J. Amer. Chem. Soc.*, 1974, **96**, 6680.
[136] Y. Kiso, K. Tamao, N. Miyake, K. Yamamoto, and M. Kumada, *Tetrahedron Letters*, 1974, 3.
[137] H. F. Wilding and L. S. Levitt, *Tetrahedron*, 1974, **30**, 611.

model for flexible hydrocarbon chain behaviour in the mass spectrometer has been proposed.[138] Barriers to rotation about the central bond of 2,2-dimethyl-3-phenylbutanes are unexpectedly low, and the reasons for this have been discussed.[139]

3 Allenes

Synthesis of Allenes.—The reduction by lithium aluminium hydride of acetylenic alcohols to allenes continues to attract attention. Variations on this theme include the formation of α-allenic alcohols from 4-alkoxy-1-hydroxybut-2-ynes[140] and the asymmetric reduction of 2-en-4-ynols (83) to

$$RC\equiv C-CH=CHCH_2OH \longrightarrow RCH=C=CHCH_2CH_2OH$$
(83)

3,4-dienols using lithium aluminium hydride–3-*O*-benzyl-1,2-*O*-cyclohexylidene-α-D-glucofuranose complex.[141] Use of the Mannich reaction on α-acetylenic alcohols followed by reduction of the quaternary salt with lithium aluminium hydride constitutes a very useful process for acetylene–allene homologization, and this type of transformation is particularly useful in the steroid field, as shown in Scheme 26.[142] A synthesis of conjugated

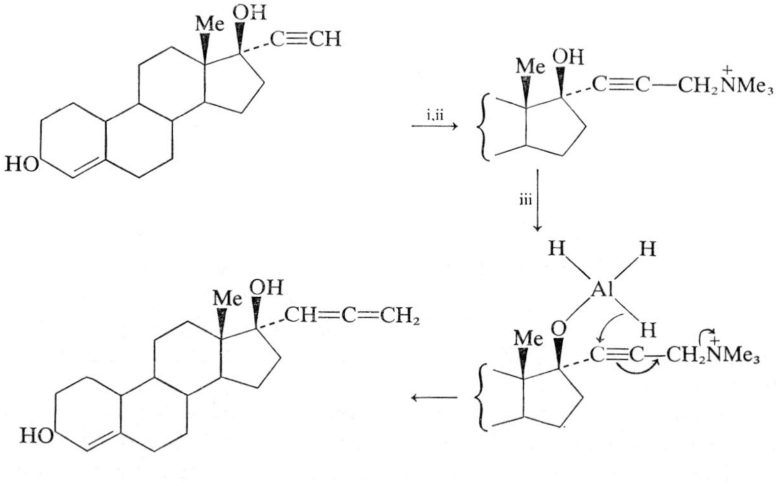

Reagents: i, CH₂O–NHMe₂; ii, MeI; iii, LiAlH₄

Scheme 26

[138] M. A. Winnik, D. Sanders, G. Jackowski, and R. E. Trueman, *J. Amer. Chem. Soc.*, 1974, **96**, 7510.
[139] J. E. Anderson and H. Pearson, *J.C.S. Perkin II*, 1974, 1779.
[140] A. Claesson, L.-I. Olsson, and C. Bogentoft, *Acta Chem. Scand.*, 1973, **27**, 2941.
[141] R. J. D. Evans, S. R. Landor, and J. P. Regan, *J.C.S. Perkin I*, 1974, 552.
[142] E. Galantay, I. Bacso, and R. V. Coombs, *Synthesis*, 1974, 344.

bis-allenes involves the coupling of an acetylenic allyl alcohol (84) with a propargyl chloride (85) and subsequent reduction of the product to the required allene (86) with lithium aluminium hydride.[143]

The lithium derivative of propargyl chloride reacts with trialkylboranes to give the intermediate (87), which can be cleaved with acetic acid to give the terminal allene in a one-pot procedure.[144]

A Wittig–Horner-type synthesis of allenes from the allylic alcohols (88), prepared from the acetylenic phosphonates (89), constitutes a new and mild method of preparation.[145]

---

[143] R. Baudouy and J. Gore, *Tetrahedron Letters*, 1974, 1593.
[144] T. Leung and G. Zweifel, *J. Amer. Chem. Soc.*, 1974, **96**, 5620.
[145] M. Barom Marszak, M. Simalty, and A. Seuleiman, *Tetrahedron Letters*, 1974, 1905.

The addition of dihalogenocarbenes to double bonds followed by treatment of the intermediate dihalogenocyclopropanes with strong base is a well-established route to allenes. Examples of this transformation reported during the year include the preparation of t-butylallene from t-butylethylene and dichlorocarbene,[146] and of 1,5-di-t-butyl-1,5-diphenylpentatetraene from the corresponding triene.[147] An interesting variation on the routes to the 15-membered ring ketones involves dichlorocarbene addition to the enol ether of cyclododecanone followed by treatment of the cyclopropane thus produced with methyl-lithium to give the allene (90). This undergoes hydrolysis with acid to give 3-methylcyclopentadec-2-en-1-one (91) (Scheme 27).[148]

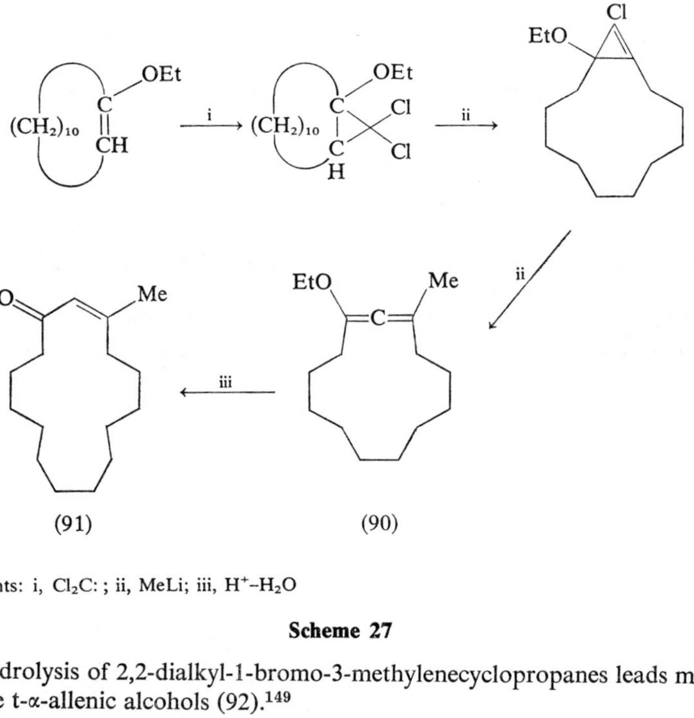

Reagents: i, Cl$_2$C: ; ii, MeLi; iii, H$^+$–H$_2$O

Scheme 27

Hydrolysis of 2,2-dialkyl-1-bromo-3-methylenecyclopropanes leads mainly to the t-α-allenic alcohols (92).[149]

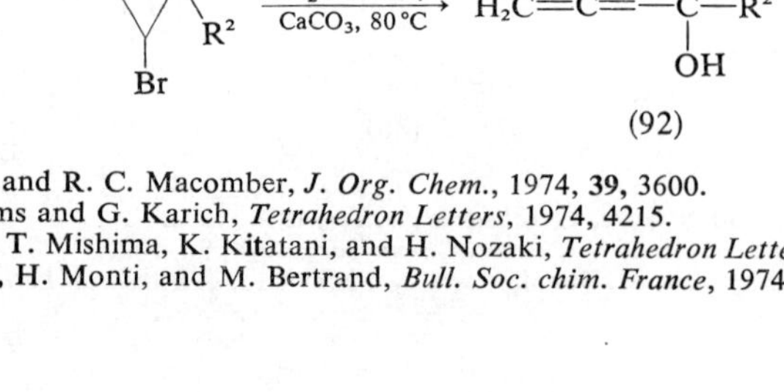

[146] K. C. Lilje and R. C. Macomber, *J. Org. Chem.*, 1974, **39**, 3600.
[147] J. C. Jochims and G. Karich, *Tetrahedron Letters*, 1974, 4215.
[148] T. Hiyama, T. Mishima, K. Kitatani, and H. Nozaki, *Tetrahedron Letters*, 1974, 3297.
[149] G. Leandri, H. Monti, and M. Bertrand, *Bull. Soc. chim. France*, 1974, 1919.

Reactions of Allenes.—Monosubstituted alkylallenes (93) give adducts of the type (94) on reaction with 2,4-dinitrobenzenesulphenyl chloride, and no detectable amounts of the isomeric adducts (95) are observed. With 1,3-dialkylallenes, mixtures of both (94) and (95) are formed, in which (95) predominates. The results for monoalkyl-substituted allenes contrast with those reported previously for phenylallene, in which (95) is by far the predominant product.[150] Electrophilic bromination, iodination, and chlorination of phenylallene have also been studied. Attack appears to occur primarily at the internal double bond, although earlier reports had shown that both addition of hydrogen chloride and sulphenylation took place at the terminal double bond to give adducts similar to (95).[151] The products obtained from the free-radical chlorination of allenes using iodobenzene dichloride have been compared with those obtained under ionic conditions. Thus cyclonona-1,2-diene gave the dichloride (96) with this reagent.[152] The stereochemistry of

$$R^1CH=C=CHR^2 \xrightarrow{ArSCl} R^1CH-C=CHR^2 + R^1CH-C=CHR^2$$
$$\phantom{R^1CH=C=CHR^2 \xrightarrow{ArSCl} R^1CH}\underset{Cl}{|}\underset{SAr}{|}\underset{SAr}{|}\underset{Cl}{|}$$

(93)   (94)   (95)

(96)

the oxythallation of cyclonona-1,2-diene has also been investigated. When optically active diene was used, the product which was obtained (97) was also optically active. The authors suggest that formation of the bridging intermediate π-complex is sufficient to prevent complete C—C bond rotation to form (98); the intermediacy of the thallinium ion (99) was also postulated.[153]

(99)   (98)   (97)

[150] T. L. Jacobs and R. C. Kammerer, *J. Amer. Chem. Soc.*, 1974, **96**, 6213.
[151] T. Okuyama, K. Ohashi, K. Izawa, and T. Fueno, *J. Org. Chem.*, 1974, **39**, 2255.
[152] M.-C. Lasne and A. Thuillier, *Bull Soc. chim. France*, 1974, 249.
[153] R. D. Bach and J. W. Holubka, *J. Amer. Chem. Soc.*, 1974, **96**, 7814.

Hydroboration of allenes with 9-borabicyclo[3,3,1]nonane gives a mixture of mono- and bis-adducts, the composition of which depends on the substitution pattern in the allene. Steric effects at either end of the allene system are predominant in determining which is formed.[154] Terminal allenes undergo rapid addition of alkyl or aryl sulphonyl iodides to give the products of addition across the terminal double bond.[155]

Oxidation of hindered allenes by peracids has given rise to many interesting products. Thus, 1,1-dimethyl-3-t-butylallene gives the 1,4-dioxaspiro[2,2] pentane, and other products from its decomposition. In contrast, 1,1-di-t-butylallene gives 2,2-di-t-butylcyclopropanone (100), which is formed by isomerization of (101). Tri-t-butylallene gives the expected monoadduct along with some of the rearranged bis-adduct (102). These results have been compared with those obtained using ozone as the oxidizing agent.[156] Oxidation of allylallenes with peracids leads to the formation of bicyclo[3,1,0]-hexanes, and a mechanism for their production has been proposed.[157]

t-Butylcyanoketen reacts with tetramethylallene in boiling benzene to give the [2 + 2] cycloadduct (103). Racemic 1,3-dimethylallene gives a mixture of two pairs of diastereoisomers of the (E) and the (Z) series. Earlier work has shown that the optically active allenes give racemic (Z)-adducts but optically active (E)-adducts. The present authors explain this observation by the

---

[154] L. Chevolot, J. Soulie, and P. Cadiot, *Tetrahedron Letters*, 1974, 3435.
[155] W. E. Truce, D. L. Heuring, and G. C. Wolf, *J. Org. Chem.*, 1974, **39**, 238.
[156] J. K. Crandall, W. W. Conover, J. B. Komin, and W. H. Machleder, *J. Org. Chem.*, 1974, **39**, 1723.
[157] J. Grimaldi, M. Malacria, and M. Bertrand, *Tetrahedron Letters*, 1974, 275.

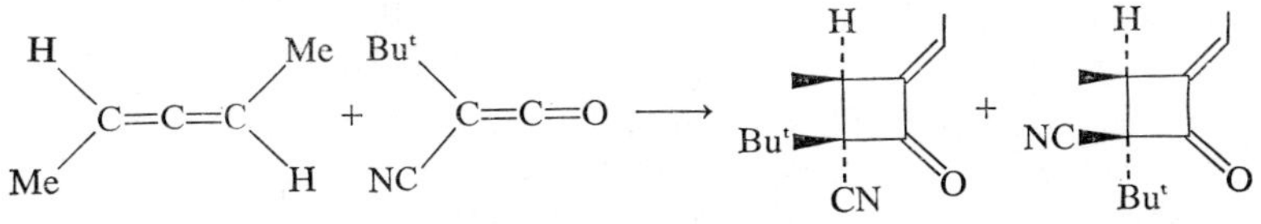

Scheme 28

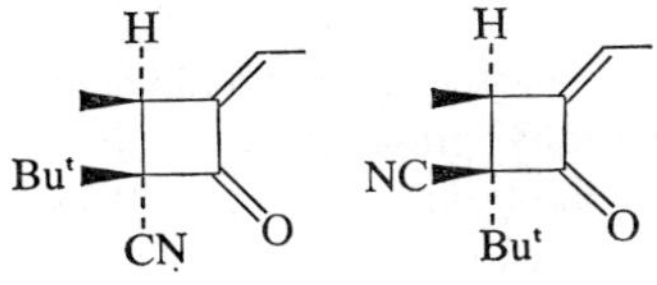

hindrance to rotation in the transition state by the antarafacial approach in the (Z) series and the consequent non-concertedness of the addition step (Scheme 28).[158]

The addition of hexafluorobutyne to tetramethylallene has also been investigated. Apart from the expected products (104) and (105) arising from cycloaddition and suprafacial addition, some unexpected isomer (106) was also formed. A possible explanation again is that perhaps both (105) and (106) are produced from a biradical intermediate such as (107).[159]

The photoaddition of allene to (108) is stereospecific, and is in accord with the theory of Wiesner, which postulates that the ground state of allene

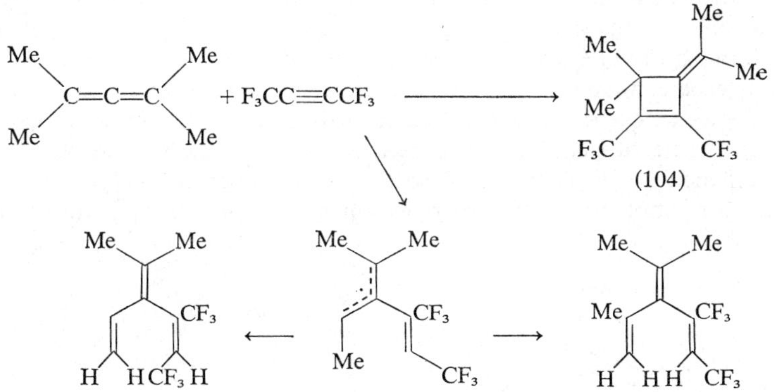

---

[158] H. A. Bampfield and P. R. Brook, *J.C.S. Chem. Comm.*, 1974, 171.
[159] H.-A. Chia, B. E. Kirk, and D. R. Taylor, *J.C.S. Perkin I*, 1974, 1209.

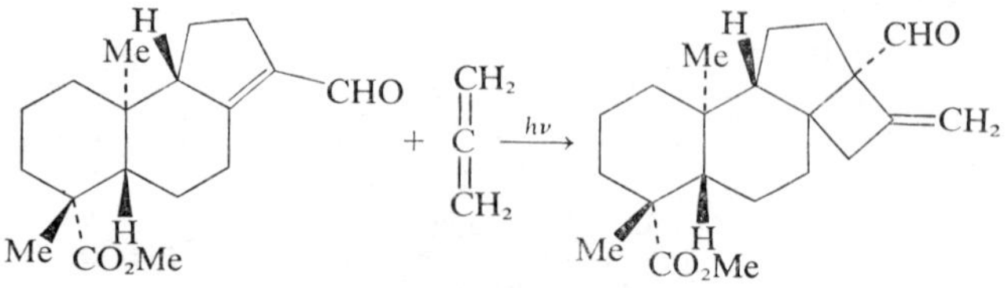

Scheme 29

reacts with the more stable configuration of the excited state of the conjugated carbonyl group (Scheme 29).[160] The [2,3] sigmatropic rearrangement of the allylic sulphonium ylide (109), followed by hydrolysis of the allenic thioether product, provides a very convenient route to artemisia ketone (110).[161]

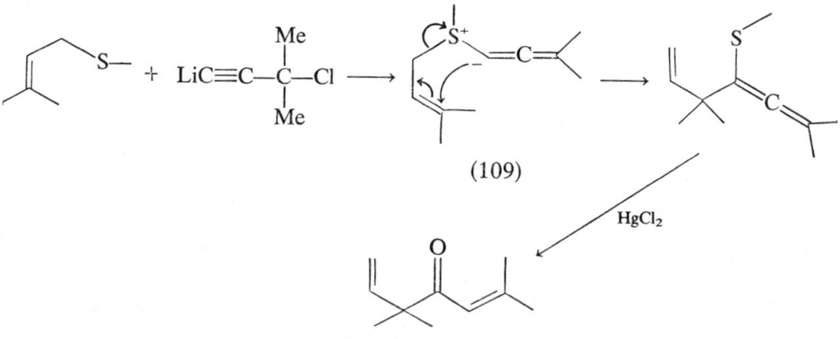

A very complete study has been conducted on the solvolysis of β-allenic tosylates. The results for acyclic allenic tosylates show that the allene group participates strongly, and cyclopropyl ketones and methylenecyclobutanols are formed. Use of kinetic and stereochemical data suggests the presence of a methylenebicyclobutonium ion type intermediate, which could capture water to give a methylenecyclobutanol before it isomerized to a cyclopropylvinyl cation. These observations have been extended to cyclic systems.[162] The corresponding γ-allenic tosylates give the vinylcyclopentenes (111) on acetolysis; the reaction proceeds with retention of optical activity and inversion

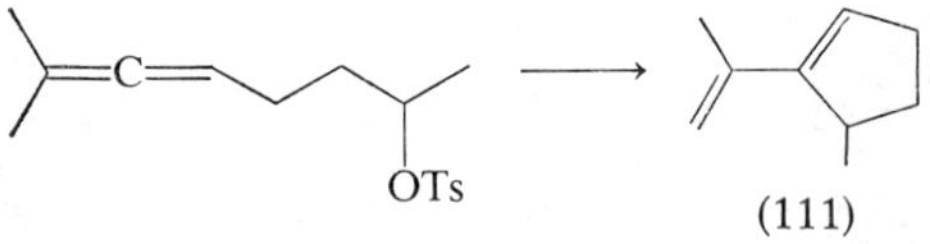

Scheme 30

[160] F. E. Ziegler and J. A. Kloek, *Tetrahedron Letters*, 1974, 315.
[161] D. Michelot, G. Linstrumelle, and S. Julia, *J.C.S. Chem. Comm.*, 1974, 10.
[162] M. Santelli and M. Bertrand, *Tetrahedron*, 1974, **30**, 235, 243, 251, 257.

Acetylenes, Alkanes, Allenes, and Olefins

of configuration, and several other products are also formed (Scheme 30).[163]

4 Olefins

Reviews within the olefin field for the year include the use of the phosphonate modification of the Wittig reaction (the Wittig–Horner reaction) in olefin synthesis,[164] the photochemistry of olefinic compounds,[165] and current thoughts on Bredt's rule for such systems.[166] Two reviews on organoboranes contain much relevant information on their reactions and synthetic transformations with olefins and acetylenes.[167,168] The sensitized photo-oxygenation of olefins has been discussed in detail.[169] The base-catalysed carbon–carbon addition reactions of olefins have been reviewed, and include the dimerization and oligomerization of important olefins such as propene and isoprene.[170] Thermal isomerism about double bonds has been surveyed,[171] as have all six-electron thermal pericyclic reactions, many of which involve olefinic systems.[172] Other articles include free-radical additions to multiple bonds,[173] aromatic substitution of olefins by palladium salts,[174] and the occurrence of vinyl and allenyl cations as reaction intermediates.[175]

Synthesis of Olefins.—Treatment of epoxides with lithium diphenylphosphide followed by oxidation with hydrogen peroxide gives, by a single inversion, β-hydroxydiphenylphosphine oxides. These can be fragmented stereospecifically to olefins of opposite configuration on treatment with base. Thus, *cis*-cyclo-octene has been converted into *trans*-cyclo-octene in 76% yield (Scheme 31).[176] The system ferric chloride–butyl-lithium deoxygenates

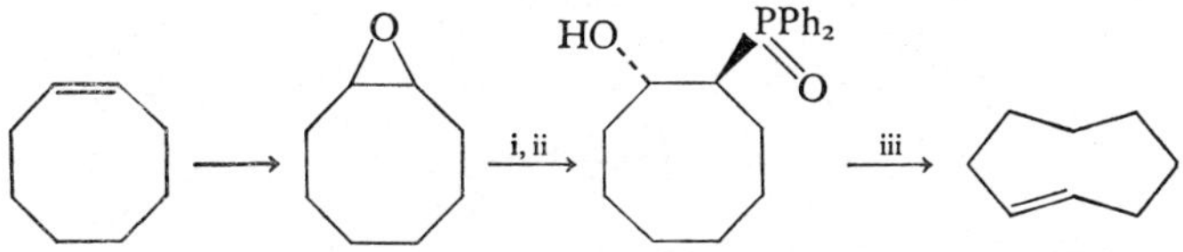

Reagents: i, LiPPh$_2$; ii, H$_2$O$_2$; iii, NaH

Scheme 31

---

[163] B. Ragonnet, M. Santelli, and M. Bertrand, *Helv. Chim. Acta*, 1974, **57**, 557.
[164] J. Boutagy and R. Thomas, *Chem. Rev.*, 1974, **74**, 87.
[165] J. D. Coyle, *Chem. Soc. Rev.*, 1974, **3**, 329.
[166] G. L. Buchanan, *Chem. Soc. Rev.*, 1974, **3**, 41.
[167] K. Smith, *Chem. Soc. Rev.*, 1974, **3**, 443.
[168] E. Negishi and H. C. Brown, *Synthesis*, 1974, 77.
[169] R. W. Denny and A. Nickon, *Org. Reactions*, 1973, **20**, 133.
[170] H. Pines, *Synthesis*, 1974, 309.
[171] H. Kessler, *Tetrahedron*, 1974, **30**, 1861.
[172] J. B. Hendrickson, *Angew. Chem. Internat. Edn.*, 1974, **13**, 47.
[173] P. I. Abell, 'Free Radicals', Wiley-Interscience, New York, 1973, Vol. 2, pp. 63—71.
[174] I. Moritani and Y. Fujiwara, *Synthesis*, 1973, 524.
[175] P. J. Stang, *Progr. Phys. Org. Chem.*, 1973, **10**, 205.
[176] A. J. Bridges and G. H. Whitman, *J. C. S. Chem. Comm.*, 1974, 142.

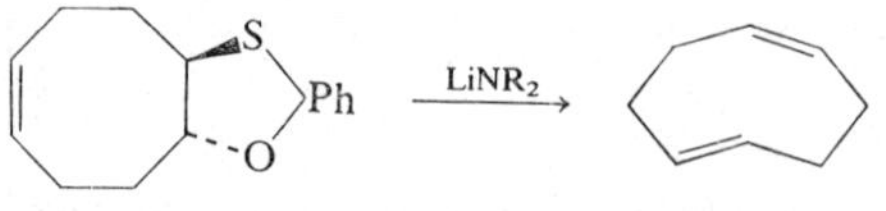

epoxides to olefins in high yield under mild conditions.[177] 2-Phenyl-1,3-oxathiolans, prepared from the epoxide *via* the mercaptoethanol, react with lithium dialkylamides to give the olefin stereospecifically. Thus (112), which can be readily prepared from cyclo-octa-1,5-diene, gives the strained *cis*, *trans*-cyclo-octa-1,5-diene.[178]

The pyrolysis of Δ^3-1,3,4-thiadiazolines under mild conditions is known to give episulphides which, on treatment with triphenylphosphine, give olefins in good yield. An alternative synthesis of the starting heterocycle has been discovered which utilizes the addition of diazo-compounds to thioketones and which can be used for the preparation of highly hindered olefins. Fortunately, the steric hindrance increases the stability of the required thioketones. Among examples of compounds prepared in this way is (113), as shown in Scheme 32. However, tetra-t-butylethylene remains an elusive molecule, even by this procedure.[179]

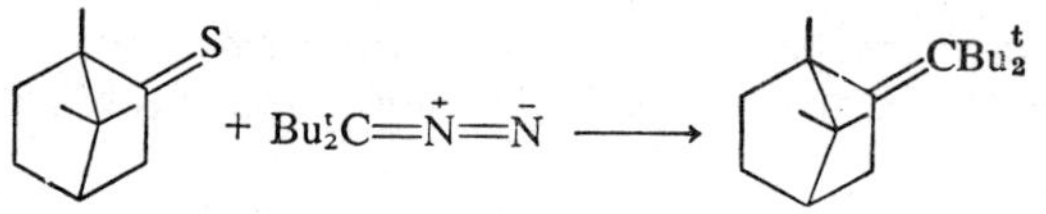

Scheme 32

1,4-Dienes can be prepared by the homo-1,4-elimination of $\alpha\alpha'$-dihydroxy-cyclopropanes with diphosphorus tetraiodide in pyridine.[180] Mild thermal treatment of α-chloroboronic esters, which are available from the base-induced reaction of borinic esters with dichloromethyl methyl ether, converts them into the chloroborate ester and the corresponding internal olefin (Scheme 33).[181]

Treatment of a β-hydroxy-ester with aluminium ethoxide or diethoxyaluminium chloride followed by warming with lithium di-isopropylamide gives the αβ-unsaturated ester regiospecifically, and, in certain cases, stereospecifically. The reaction is believed to proceed *via* a β-alanoxy-enolate similar to

---

[177] T. Fujisawa, K. Sugimoto, and H. Ohta, *Chem. Letters*, 1974, 883.
[178] M. Jones, P. Temple, E. J. Thomas, and G. H. Whitham, *J. C. S. Perkin I*, 1974, 433.
[179] D. H. R. Barton, F. S. Guziec, jun., and I. Shahak, *J. C. S. Perkin I*, 1974, 1794.
[180] T. Hanafusa, S. Imai, K. Ohkata, H. Suzuki, and Y. Suzuki, *J. C. S. Chem. Comm.* 1974, 73.
[181] J.-J. Katz, B. A. Carlson, and H. C. Brown, *J. Org. Chem.*, 1974, **39**, 2818.

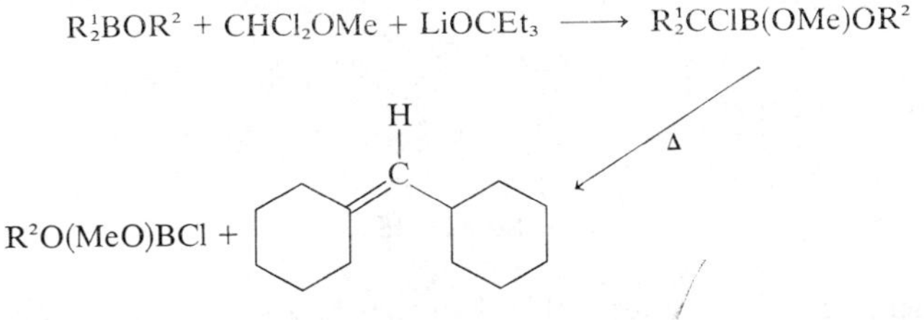

$R^1_2BOR^2 + CHCl_2OMe + LiOCEt_3 \longrightarrow R^1_2CClB(OMe)OR^2$

$R^2O(MeO)BCl +$ [structure: dicyclohexylidene methane with H] $\xrightarrow{\Delta}$

R^1 = Cyclohexyl

Scheme 33

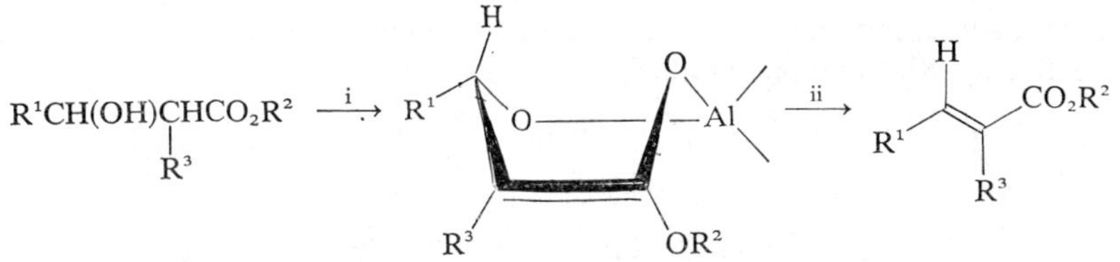

$R^1CH(OH)\underset{R^3}{C}HCO_2R^2 \xrightarrow{i} $ [aluminium chelate intermediate] $\xrightarrow{ii}$ [trans-alkene product with R^1, H, CO_2R^2, R^3]

Reagents: i, Al(OEt)$_3$; ii, LiNR$_2$

Scheme 34

that proposed in the reduction by lithium aluminium hydride of β-dicarbonyl enolates (Scheme 34).[182] A two-step synthesis of *trans*-alkenes from γ-substituted allyl phosphonates involves alkylation of the anion of this system with an alkyl halide followed by cleavage of the C—P bond with lithium aluminium hydride.[183]

The Wittig olefin synthesis starting from aldehydes can be performed in a two-phase system of benzene and aqueous alkali, with the alkylphosphonium halides acting as phase-transfer agents. Yields vary with the concentration of alkali, and competition occurs between olefin formation and salt decomposition. Only trace quantities of olefins are formed when ketones are used.[184]

The reaction of enolates with *trans*-1-butadienyltriphenylphosphonium bromide is an easy one-step synthesis of cyclohexa-1,3-dienes (Scheme 35).[185] The anion from α-phosphono-γ-butyrolactones reacts with aldehydes and ketones to give α-ylidene-γ-butyrolactones in high yield.[186]

Carbon–carbon double bonds are formed in good yield by the reaction of vicinal halides with sodium in liquid ammonia. The method is often more

---

[182] J. A. Katzenellenbogen and T. Utawanit, *J. Amer. Chem. Soc.*, 1974, **96**, 6153.
[183] K. Kondo, A. Negishi, and D. Tunemoto, *Angew. Chem. Internat. Edn.*, 1974, **13**, 407.
[184] W. Tagaki, I. Inoue, Y. Yano, and T. Okonogi, *Tetrahedron Letters*, 1974, 2587.
[185] P. L. Fuchs, *Tetrahedron Letters*, 1974, 4055.
[186] T. Minami, I. Niki, and T. Agawa, *J. Org. Chem.*, 1974, **39**, 3236.

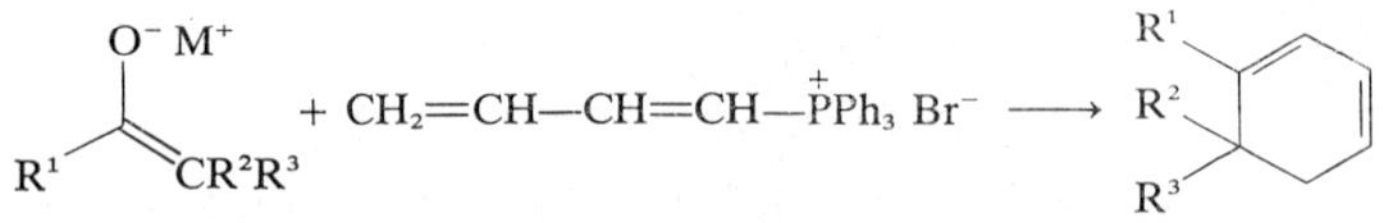

Scheme 35

convenient and gives better yields than that based on sodium–naphthalene. An example[187] is in the synthesis of (114) from cyclo-octatetraene (Scheme 36). This reaction can also be effected electrochemically at a stirred mercury cathode. Attempts to intercept a carbanionic intermediate were unsuccessful, indicating that reductive elimination is synchronous at both centres, or nearly so.[188] *gem*-Di-iodocyclopropanes are converted into *trans*-cycloalkenes on treatment with silver perchlorate in methanol.[189]

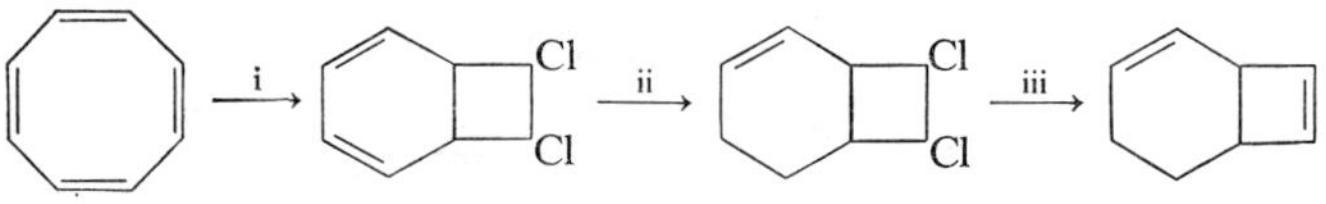

Reagents: i, Cl_2, $-30\,°C$; ii, Pd/H_2; iii, $Na-NH_3$

Scheme 36

Base-catalysed isomerization of 1,1,4,4-tetra-alkoxybut-2-ynes to the corresponding 1,3-dienes can readily be accomplished by slow addition of the alkyne to two equivalents of potassium amide in liquid ammonia.[190] The deconjugative isomerization of the dienone (115) is brought about by potassium t-butoxide in DMSO, followed by quenching of the anion with water (Scheme 37).[191] An analogue of the base-catalysed decomposition of *N*-nitroso-oxazolidones is provided by the decomposition[192] of sulphoximines of *N*-amino-2-oxazolidones in DMSO to give a stereospecific olefin synthesis. The starting compounds are readily prepared from the *N*-amino-compound by oxidation with lead tetra-acetate in DMSO (Scheme 38). Base elimination from *N*-aminotrialkylammonium salts provides an alternative to quaternization for the Hofmann olefin-elimination reaction.[193]

---

[187] E. L. Allred, B. R. Beck, and K. J. Voorhees, *J. Org. Chem.*, 1974, **39**, 1426.
[188] J. Casanova and H. R. Rogers, *J. Org. Chem.*, 1974, **39**, 2408.
[189] M. S. Baird, *J. C. S. Chem. Comm.*, 1974, 197.
[190] J. W. Scheeren and R. W. Aben, *Tetrahedron Letters*, 1974, 1019.
[191] A. van Wageningen, P. C. M. van Noort, F. W. M. van der Wielen, and H. Cerfontain, *Synthetic Comm.*, 1974, **4**, 325.
[192] J. D. White and M.-G. Kim, *Tetrahedron Letters*, 1974, 3361.
[193] H. Posvic and D. Rogers, *J. Org. Chem.*, 1974, **39**, 1588.

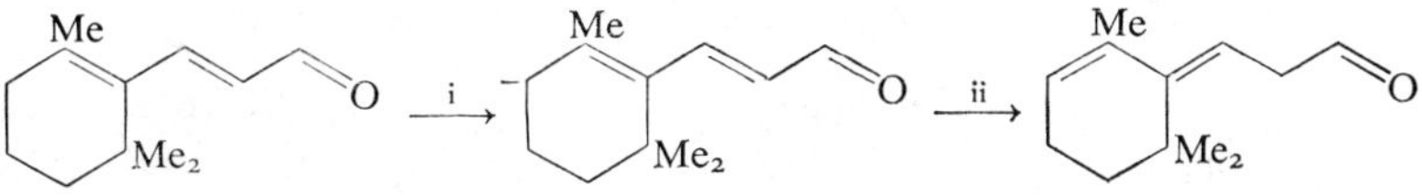

Acetylenes, Alkanes, Allenes, and Olefins

(115)

Reagents: i, KOBu^t-DMSO; ii, H$_2$O

Scheme 37

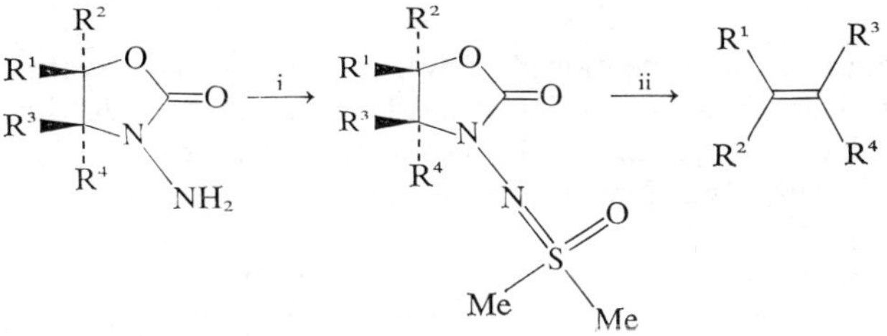

Reagents: i, Pb(OAc)$_4$-DMSO; ii, Δ

Scheme 38

Thiiran 1,1-dioxides react with certain anhydrous metal halides to give halogenosulphinates which, on treatment with water, undergo the equivalent of a decarboxylative elimination to give *cis*-olefins.[194] Oxidation of the dianion (116) with anhydrous cupric chloride gives a reasonable yield of diethyl fumarate. With analogous dianions, yields were lower, and generally *cis*–*trans* mixtures were obtained.[195]

A full report of the synthesis of alkenes from carbonyl compounds and carbanions α to silicon has been published. This type of transformation is analogous to the Wittig reaction. The carbanions were generated by the addition of alkyl-lithium compounds to vinyltriphenylsilane (117).[196] The allylic anion (118) derived from 3-methylbut-3-en-1-ol couples with 1-bromo-3-methylbut-2-ene to give the 1,5-diene (119).[197] In a similar reaction to the

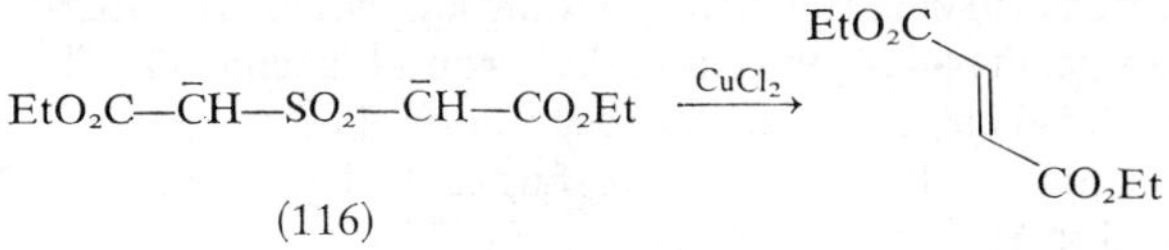

$$\text{EtO}_2\text{C}-\bar{\text{C}}\text{H}-\text{SO}_2-\bar{\text{C}}\text{H}-\text{CO}_2\text{Et} \xrightarrow{\text{CuCl}_2}$$

(116)

---

[194] E. Vilsmaier, R. Tropitzsch, and O. Vostrowsky, *Tetrahedron Letters*, 1974, 3275.
[195] J. S. Grossert, J. Buter, E. W. H. Asveld, and R. M. Kellogg, *Tetrahedron Letters*, 1974, 2805.
[196] T. H. Chan and E. Chang, *J. Org. Chem.*, 1974, **39**, 3264.
[197] G. Cardillo, M. Contento, and S. Sandri, *Tetrahedron Letters*, 1974, 2215.

Ph$_3$SiCH=CH$_2$ $\xrightarrow{\text{R}^1\text{Li}}$ Ph$_3$SiCHLiCH$_2$R^1 $\xrightarrow{\text{R}^2\text{R}^3\text{CO}}$ $\underset{\text{R}^3}{\overset{\text{R}^2}{\diagdown}}$C=CHCH$_2$R^1

(117)

[structure 118] + [structure with CH$_2$O$^-$Li$^+$] ⟶ [structure 119 with CH$_2$OH]

(118)   (119)

silicon transformation mentioned above, the stereospecific synthesis of *cis*- and *trans*-olefins from β-keto-silanes has been achieved. Reduction gives the alcohol, which, on treatment with boron trifluoride etherate in methylene chloride or with potassium hydride in THF, gives the *cis*- or *trans*-olefin respectively.[198]

Conventional procedures for the preparation of the cyclobutene (120) from the thionocarbonate ester, such as the use of triethyl phosphite, have proved

[structure] $\xrightarrow[\text{ii, Zn–EtOH}]{\text{i, RI}}$ [structure (120)]

(120)

to be unsuccessful. Alkylation of the thionocarbonate with 2-iodopropane followed by reduction with zinc in ethanol, however, gives (120) in a convenient mild procedure. This reaction sequence has been found to be generally applicable, but is not stereoselective, unlike the Corey procedure.[199]

The use of lithium divinylcuprates and their conjugate addition to enones continue to attract attention. A recent synthesis towards the sesquiterpene vernolepin involved such a reaction on (122). The resulting vinyl ketone could be transformed into (121) by three subsequent steps (Scheme 39).[200] *gem*-Dihalogeno-olefins react with lithium dialkylcuprates to give mixtures of *gem*-dialkylated olefins and dimers arising from cross-coupling reactions. With 1,2-dibromoalkenes the corresponding alkynes are formed first, followed by the stereospecific addition of an alkyl group (Scheme 40).[201]

A general 1,5-diene synthesis has been developed and used in a synthesis of all-*trans*-squalene from *trans*,*trans*-farnesol. This sequence involves the steps outlined in Scheme 41.[202]

[198] P. F. Hudrlik and D. Peterson, *Tetrahedron Letters*, 1974, 1133.
[199] E. Vedejs and E. S. C. Wu, *J. Org. Chem.*, 1974, 39, 3641.
[200] R. D. Clark and C. H. Heathcock, *Tetrahedron Letters*, 1974, 1713.
[201] J. Klein and R. Levene, *Tetrahedron Letters*, 1974, 2935.
[202] P. A. Grieco and Y. Masaki, *J. Org. Chem.*, 1974, 39, 2135.

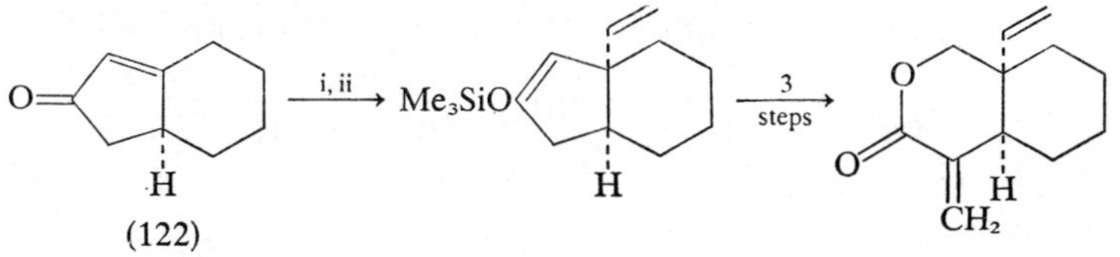

Scheme 39

Reagents: i, (=\\)$_2$CuLi; ii, Me$_3$SiCl

Scheme 40

$$PhCBr=CBrCO_2H \xrightarrow{i} PhC{\equiv}CCO_2H \longrightarrow \text{(Ph, Me / CO}_2\text{H, H)}$$

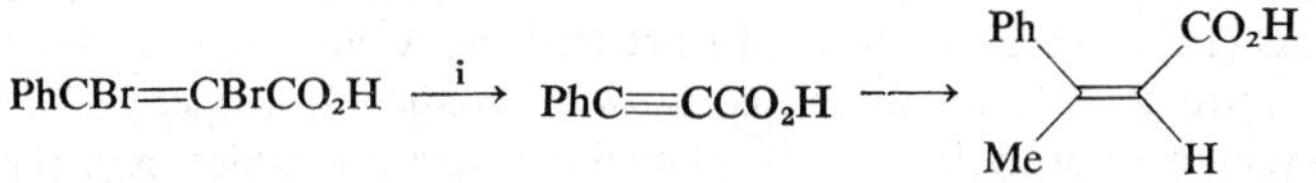

Reagent: i, Me$_2$CuLi

Scheme 41

$$RCH=CHCH_2OH \xrightarrow{i-iii} RCH=CH\bar{C}HSO_2Ph \xrightarrow{iv} RCH=CH-CHSO_2Ph$$
                                                                        |
                                                                $CH_2CH=CHR$

R = Farnesyl $RCH=CHCH_2CH_2CH=CHR$ (via v)

Reagents: i, PBr$_3$; ii, NaSO$_2$Ph; iii, BuLi–THF–HMPT; iv, RCH=CHCH$_2$Br; v, Li–EtNH$_2$

A synthesis of ωω'-diarylalkenes involves the disproportionation of alkenes with the catalyst system (Ph$_3$P)$_2$MoCl$_2$(NO)$_2$–EtAlCl$_2$. The reaction fails with styrene and tri- or tetra-substituted olefins.[203] Tetraphenylcyclobutane is cleaved by platinum(II) to give *trans*-stilbene.[204]

Introduction of a methylene function α to a carbonyl group is possible by alkylation of the starting carbonyl compound with iodomethyl phenyl sulphide, oxidation to the sulphoxide, and [2,3] sigmatropic rearrangement.[205] The thiophenyl group in the olefin (123) can be displaced by butylmagnesium bromide in the presence of cuprous iodide to give the corresponding butyl derivative with complete retention of configuration.[206]

---

[203] G. Descotes, P. Chevalier, and D. Sinou, *Synthesis*, 1974, 364.
[204] I. J. Harvie and F. J. McQuillin, *J.C.S. Chem. Comm.*, 1974, 806.
[205] B. M. Trost and R. A. Kunz, *J. Org. Chem.*, 1974, **39**, 2648.
[206] S. Kobayashi, H. Takei, and T. Mukaiyama, *Chem. Letters*, 1973, 1097.

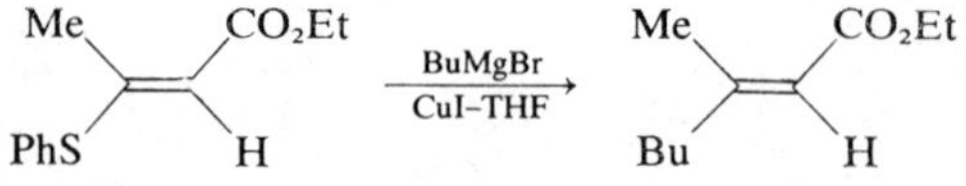

(123)

Isomeric oxetans have been prepared by photoreaction of benzaldehyde with pent-1-ene and separated by preparative g.l.p.c. On thermolysis the *trans*-isomer gives exclusively *trans*-1-phenylpent-1-ene, but the *cis*-isomer decomposes more slowly to give a mixture of *cis*- and *trans*-isomers.[207] The thermal decomposition of the 2,5-dimethyl-3,4-diazabicyclo[4,2,0]oct-3-enes (124) and (125) has been shown to proceed *via* a concerted cycloreversion to give pure 2,6-*cis,cis*- and 2,6-*cis,trans*-octadienes, respectively.[208] The thermolysis or photolysis of some phenyl cyclopropyl azides, *e.g.* (126), has been shown to give phenylpropenes in high yield.[209]

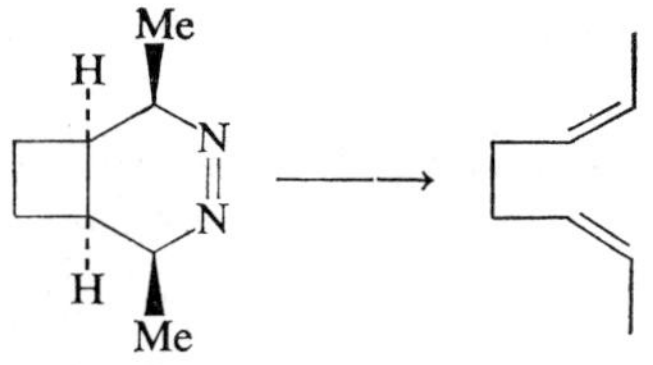

(124)

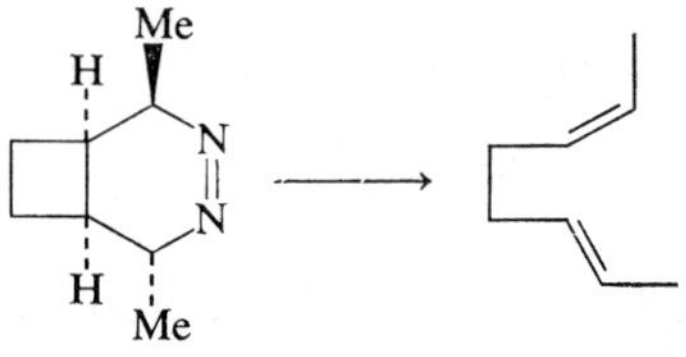

(125)

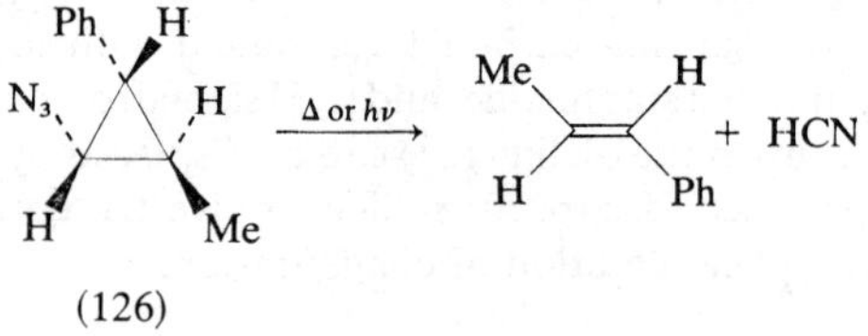

(126)

[207] N. Shimizu and S. Nishida, *J.C.S. Chem. Comm.*, 1974, 734.
[208] J. A. Berson, E. W. Petrillo, and P. Bickart, *J. Amer. Chem. Soc.*, 1974, **96**, 636.
[209] A. Hassner, A. B. Levy, E. E. McEntire, and J. E. Galle, *J. Org. Chem.*, 1974, **39**, 586.

Acetylenes, Alkanes, Allenes, and Olefins

The radical addition of anhydrous hydrogen bromide to terminal trimethylsilylalkynes gives good yields of 2-bromoalkenes. The method is claimed to be a significant improvement over other syntheses for this class of compound.[210] Arylsulphonyl thiocyanates react with olefins to give β-thiocyanato-sulphones, which can readily be converted into the sulphonyl olefins by elimination using triethylamine.[211]

A new synthesis of the interesting olefin tetrathiafulvalene utilizes the route shown in Scheme 42.[212]

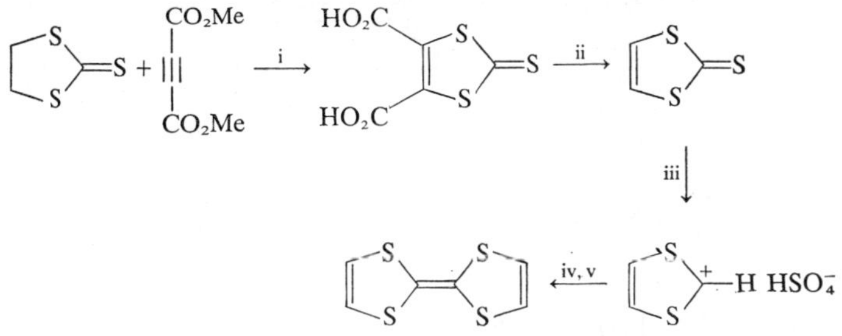

Reagents: i, H⁺aq.; ii, Δ, pyridine; iii, MeCO₃H; iv, NaPF₆; v, NEt₃

Scheme 42

Cycloaddition Reactions of Olefins.—*Diels–Alder Cycloadditions.* Only a few selected examples of the numerous Diels–Alder cycloadditions are reported in this section. The use of very high pressures (10 000 atm, 10 kbar) enables the reaction of enamines with dienes to take place at room temperature. A net volume contraction is expected for the conversion of double bonds into single bonds as the reactants approach the transition state, and the use of very high pressures increases the driving force for this change.[213] The self condensation of isoprene through a Diels–Alder reaction has been studied at four temperatures under pressures from 1 to 840 bar. Kinetic study gave the activation volume, which placed the transition state close to the final state. Thermodynamic studies showed the transition state to be highly ordered, and gave support to a one-step concerted mechanism.[214]

A new convenient route to benzocyclobutene and its derivatives utilizes

---

[210] R. K. Boeckman, jun., and D. M. Blum, *J. Org. Chem.*, 1974, **39**, 3307.
[211] G. C. Wolf, *J. Org. Chem.*, 1974, **39**, 3454.
[212] L. R. Melby, H. D. Hartzler, and W. A. Sheppard, *J. Org. Chem.*, 1974, **39**, 2456.
[213] W. G. Dauben and A. P. Kozikowski, *J. Amer. Chem. Soc.*, 1974, **96**, 3665.
[214] J. Rimmelin and G. Jenner, *Tetrahedron*, 1974, **30**, 3081.

the cycloaddition of dimethyl cyclobut-1-ene-1,2-dicarboxylate to butadienes followed by decarboxylation and aromatization.[215] 2-Hydroxy-5-oxo-5,6-dihydro-2H-pyran (127) is a novel, synthetically useful, asymmetric dienophile which contains a variety of centres for further functionalization.[216]

In contrast to other keten equivalents, such as 2-acetoxyacrylonitrile and 2-chloroacrylonitrile, nitroethylene undergoes cycloadditions even at $-100\,°C$, and this capacity has been exploited in the first step of a prostaglandin synthesis.[217] Vinyltriphenylphosphonium bromide can be used as an allene synthon in Diels–Alder additions. The intermediate phosphonium salts (128) can be converted via the ylide into alkenes (Scheme 43).[218]

The tetramethylketenimmonium ion (129) reacts with cis-dienes to give adducts such as (130).[219] The 1:1 addition product of tetraethyl ethylenetetracarboxylate and anthracene can be reduced with lithium aluminium hydride and cyclized with acid to the derivative (131). On heating to $600\,°C$ under vacuum, a retro-Diels–Alder reaction occurs to give $\Delta^{1,5}$-3,7-dioxabicyclo[3,3,0]octene, which cannot be synthesized by direct means. The precursor of (131) can also be transformed into 3,4-dialkylfurans.[220] 2-Acetoxyacrylonitrile reacts with 2,4-dihydroanisole and 2,4-dihydrotoluene, prepared by thermal isomerization of the 2,5-isomers, to give adducts which are useful starting materials for the synthesis of a range of bicyclic systems.[221]

---

[215] R. P. Thummel, J.C.S. Chem. Comm., 1974, 899.
[216] G. Jones, Tetrahedron Letters, 1974, 2231.
[217] S. Ranganathan, D. Ranganathan, and A. K. Mehrotra, J. Amer. Chem. Soc., 1974, 96, 5261.
[218] R. A. Ruden and R. Bonjouklian, Tetrahedron Letters, 1974, 2095.
[219] J. Marchand-Brynaert and L. Ghosez, Tetrahedron Letters, 1974, 377.
[220] J.-L. Ripoll, Tetrahedron Letters, 1974, 1665.
[221] I. Alfaro, W. Ashton, K. L. Rabone, and N. A. J. Rogers, Tetrahedron, 1974, 30, 559.

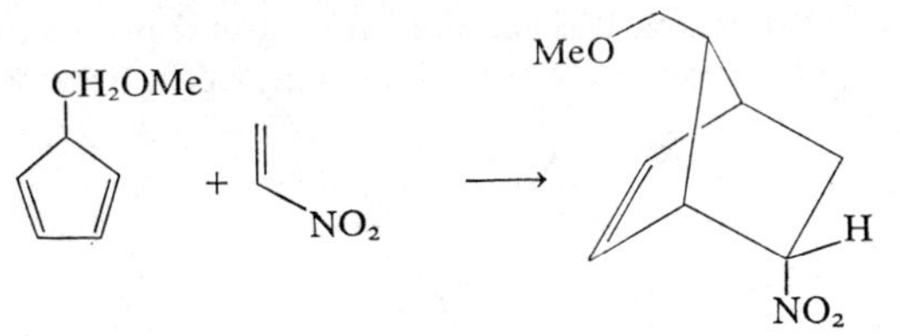

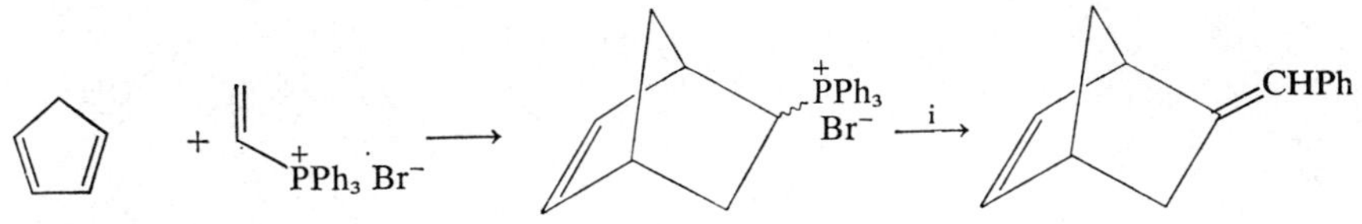

(128)

Reagent: i, PhCHO–LiNR2

Scheme 43

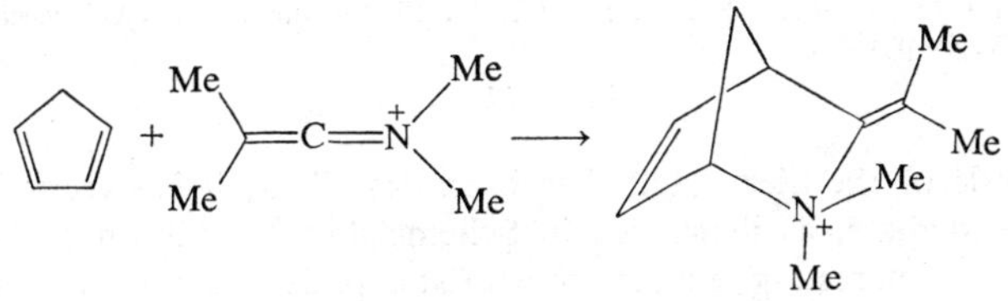

(129) (130)

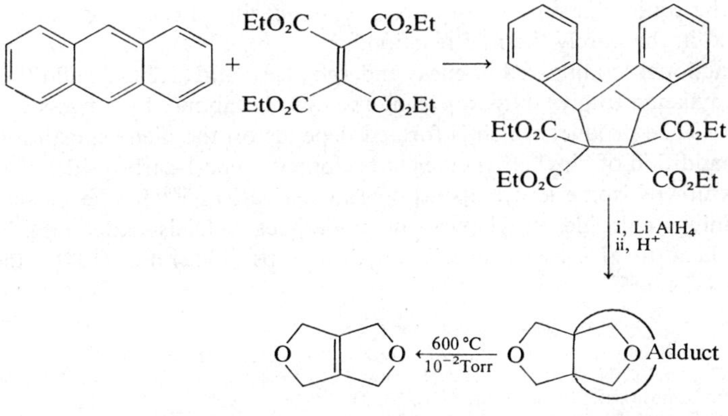

(131)

The butadiene (132), which was required for natural product synthesis, has been prepared by the sequence of reactions outlined in Scheme 44.[222]

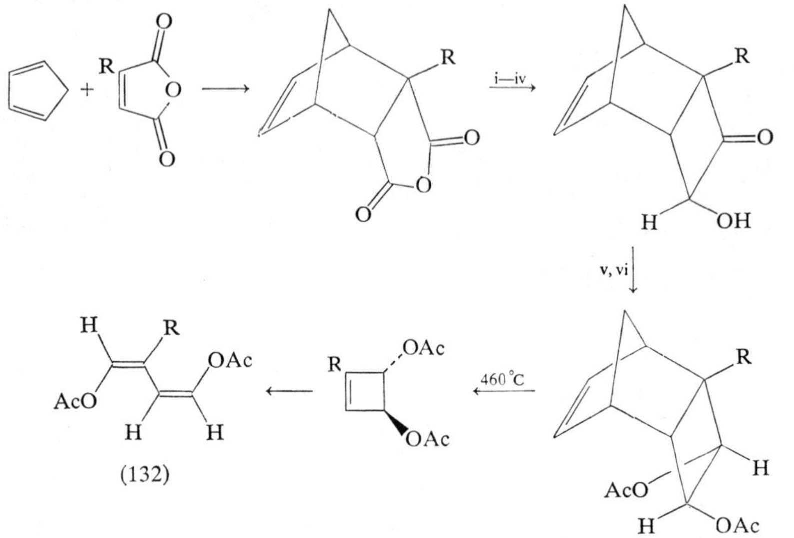

Reagents: i, NaOMe–MeOH; ii, Me_2SO_4; iii, Na–Me_3SiCl–toluene; iv, H_3O^+; v, $NaBH_4$; vi, Ac_2O–pyridine

Scheme 44

The synthesis of the diene (133) has been described; it is a very useful synthetic intermediate, as illustrated in Scheme 45.[223] Addition of Lewis-acids induces an interesting enhancement of stereoselectivity in the reaction of 1,4-diphenylbutadiene with β-nitrostyrene. The nitro endo-adduct is formed exclusively in the presence of Lewis acids, a result which contrasts with the equimolar mixture of nitro endo- and phenyl endo-adducts which is formed in the purely thermal reaction.[224]

Functional conjugated dienes undergo 1,4- and 1,2-cycloadditions to diphenylketen to give dihydropyrans and cyclobutanones by a two-step ionic process. The product which is formed depends on the diene substitution.[225] Cycloaddition of vinyl ethyl ether to cyclopent-1-ene-1-carboxaldehyde gives a mixture of isomeric cyclopent[c]pyran derivatives.[226] In the presence of aluminium chloride, ethyl pyruvate undergoes a Diels–Alder addition to pentadiene to give two isomeric dihydropyrans (134a) and (134b) and the pyran (134c).[227]

[222] M. E. Jung, *J.C.S. Chem. Comm.*, 1974, 956.
[223] S. Danishefsky and T. Kitahara, *J. Amer. Chem. Soc.*, 1974, **96**, 7807.
[224] P. C. Jain, Y. N. Mukerjee, and N. Anand, *J. Amer. Chem. Soc.*, 1974, **96**, 2996.
[225] J. P. Gonesnard, *Tetrahedron*, 1974, **30**, 3113.
[226] L.-F. Tietze, *Chem. Ber.*, 1974, **107**, 2491.
[227] K. Jankowski and R. Luce, *Tetrahedron Letters*, 1974, 2069.

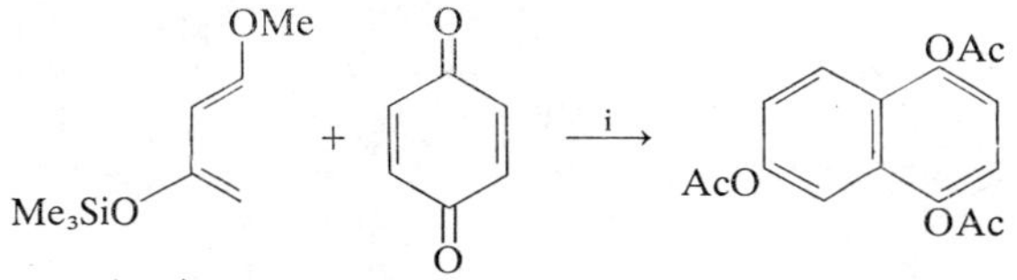

Acetylenes, Alkanes, Allenes, and Olefins

Scheme 45

Reagent: i, Ac₂O

(133)

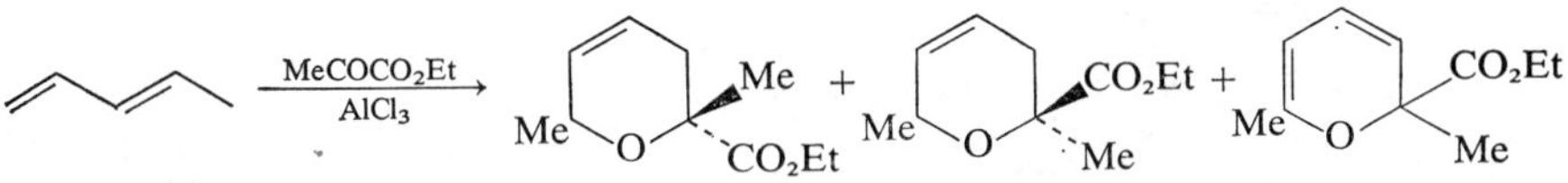

(134a) (134b) (134c)

1,3-Dipolar Additions. Benzonitrile N-oxides add across the double bond of α- and β-azidostyrenes to give dihydroisoxazoles. The product obtained from the α-isomer decomposes spontaneously to give the isoxazole, whereas decomposition of that obtained from the β-isomer requires the application of heat or treatment with triethylamine.[228] Similar types of addition to *cis*-3,4-dichlorocyclobutene have been reported.[229] The frontier-orbital method has been used to predict that certain heterocyclic zwitterions should 1,3-cycloadd to dienes.[230] The addition of several dienes to 1-methylquinolinium-3-olate to give adducts of the type (135) is the first observed reaction of this type. The reaction between isobutene and hexafluoroacetone azine does not give the [4 + 2] adduct but the azomethineimine (136), along with the bis-adduct (136b). This constitutes the first isolation and characterization of the 1,3-dipolar intermediate of criss-cross cycloaddition as postulated by Huisgen.[231]

Other Olefin Cycloadditions. The [2 + 2] cycloaddition reactions of tetracyanoethylene are known to be non-concerted processes. Addition of this olefin to *cis*-bicyclo[6,1,0]nona-2,4,6-triene is postulated to proceed *via* the *exo*-3-*trans*-1,3-bishomotropylium zwitterion (137) as an intermediate.[232] The previously reported [2 + 5] cycloaddition of tetracyanoethylene to the exomethylene-spirocyclopropane (138) has been found to be a two-stage process consisting of a [2 + 2] followed by a [3 + 4] cycloaddition reaction.

---

[228] G. L'abbé and G. Mathys, *J. Org. Chem.*, 1974, **39**, 1221.
[229] G. Bianchi, C. De Micheli, A. Gamba, and R. Gandolphi, *J.C.S. Perkin I*, 1974, 137.
[230] K.-L. Mok and M. J. Nye, *J.C.S. Chem. Comm.*, 1974, 608.
[231] K. Burger, W. Thenn, and A. Gieren, *Angew. Chem. Internat. Edn.*, 1974, **13**, 474.
[232] G. Boche, H. Weber, and J. Benz, *Angew. Chem. Internat. Edn.*, 1974, **13**, 207.

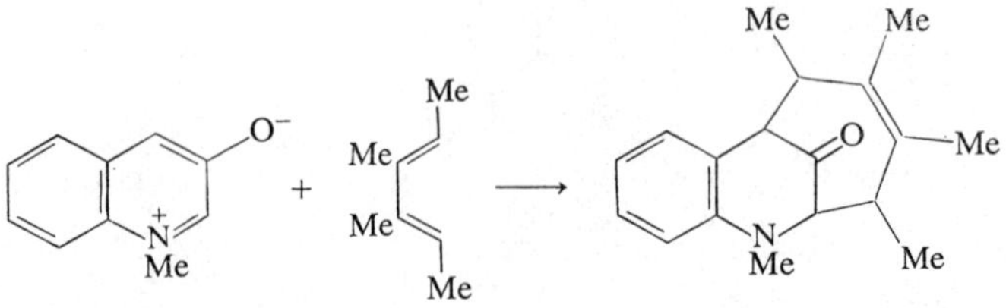

(135)

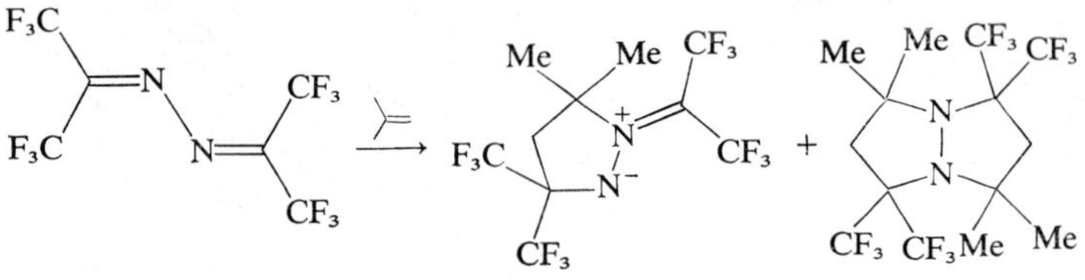

(136a) (136b)

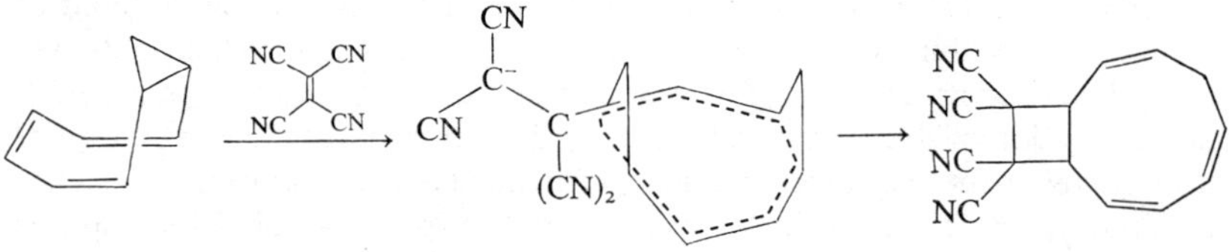

(137)

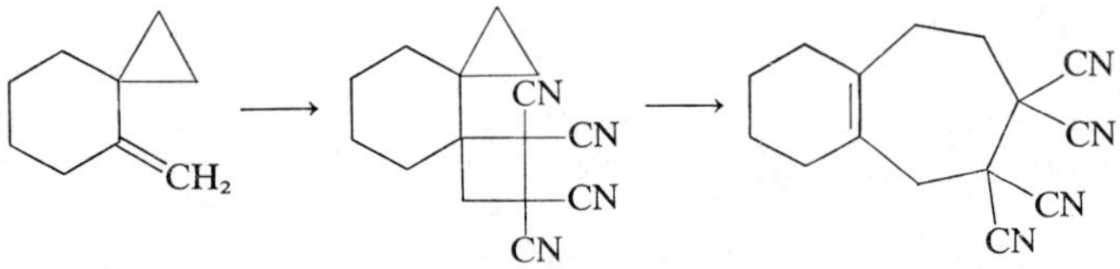

(138)

The spirocyclobutane adds to the spirocyclopentane to give a seven-membered ring.[233] Cyclopropyl olefins and enol ethers give cyclobutanes on reaction with tcne, and the reactivity of each type of substrate is strongly dependent on steric factors.[234] Vinylferrocenes also readily add tcne to give cyclobutanes. Both *cis*- and *trans*-isomers of 1-ferrocenylpropene react with tcne to give a single product, which indicates that the dipolar intermediate is sufficiently stable to allow rotation to occur before ring closure.[235]

[233] S. Sarel, A. Felzenstein, and J. Yovell, *J.C.S. Chem. Comm.*, 1974, 753.
[234] F. Effenberger and O. Gerlach, *Chem. Ber.*, 1974, **107**, 278.
[235] K. R. Berger, E. R. Biehl, and P. C. Reeves, *J. Org. Chem.*, 1974, **39**, 477.

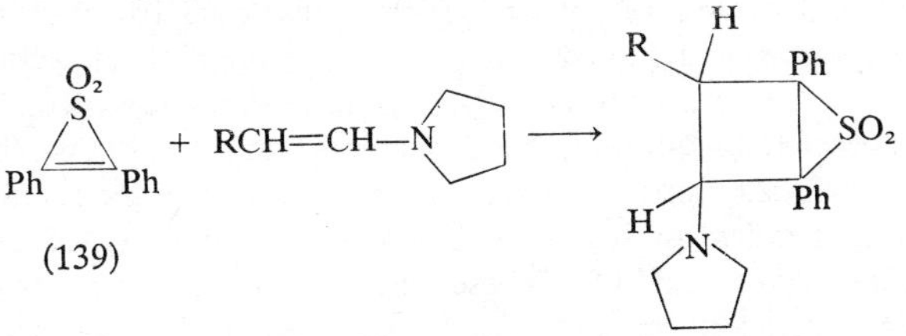

Scheme 46

2,3-Diphenylthiiren 1,1-dioxide (139) reacts with enamines in a thermal [2 + 2] cycloaddition to give the intermediate episulphone (140). This can decompose in several ways, and in some instances medium-sized sulphur-containing heterocycles can be formed in reasonable yields (Scheme 46).[236]

Observations of the polar and stereochemical aspects of the 1,2-photoaddition of olefins to benzene suggest that olefins which possess marked electron-donor properties give rise to 1,2-*endo*-adducts (141) whereas electron acceptors give 1,2-*exo*-products (142). Olefins which are weak donors and acceptors give 1,3-adducts (143).[237]

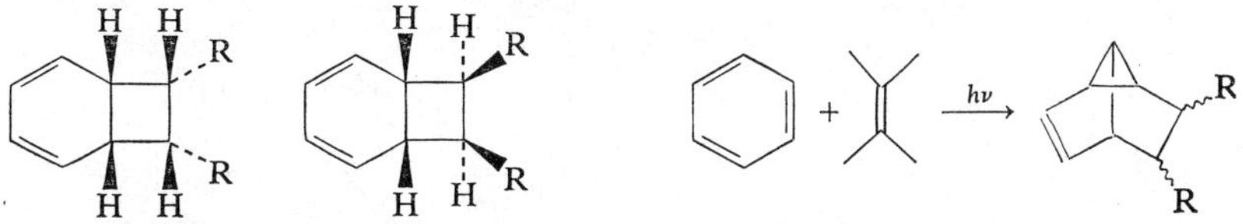

(141) R = —(CH$_2$)$_6$— (142) R = —CONHCO—   (143)

Photoreactions of 2-acetylthiophen and 2-acetylfuran with olefins give large amounts of the [4 + 2] product (144) as well as the [2 + 2] products (145) and (146). This suggests that the reactive photostate is predominantly $\pi\pi^*$, rather than $n\pi^*$, in nature, which is contrary to that observed with phenyl alkyl ketones.[238]

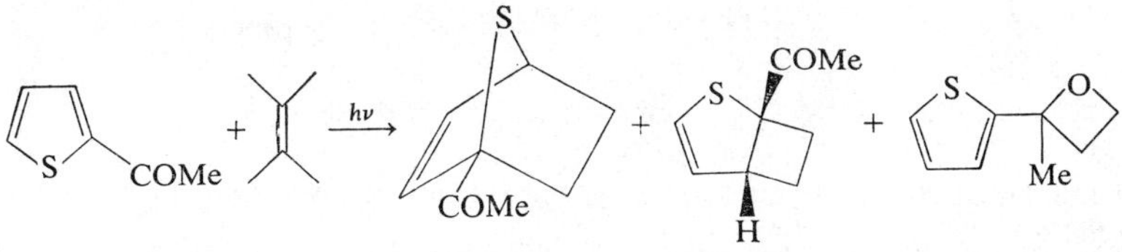

[236] M. H. Rosen and G. Bonet, *J. Org. Chem.*, 1974, **39**, 3805.
[237] D. Bryce-Smith, A. Gilbert, B. Orger, and H. Tyrrell, *J.C.S. Chem. Comm.*, 1974, 334.
[238] T. S. Cantrell, *J. Org. Chem.*, 1974, **39**, 2242.

Photochemical cycloadditions of 1,3-dimethyluracil to vinyl acetate and keten diethyl acetal give good yields of *cis*-fused 2,4-diazabicyclo[4,2,0]octa-3,5-dienes as well as the *trans*-isomers.[239] Similarly, phenoxypropenes readily photo-add across the 1,2-double bond of 1-naphthonitrile to give the expected adducts.[240] Cyclohexenone and cyclopentenone undergo moderately efficient [2 + 2] photochemical additions with a variety of conjugated dienes when diene concentrations are high. These observations account for the erratic results which are sometimes obtained when certain 1,3-dienes are used as triplet quenchers for such enones.[241]

The formation of $\beta\beta$-diphenylpropiophenone (147) as the sole product from the photoreaction of NN-dimethylbenzamide with 1,1-diphenylethylene has been shown to proceed by attack of the olefin in its excited triplet state on the amide in its ground state.[242]

Under $n \rightarrow \pi^*$ excitation, diaryl thioketones combine in dilute solution with the electron-rich double bond in ethyl vinyl ether to give thietans (148). In concentrated solution these react with a second molecule of thioketone to give the 1,4-dithian (149).[243] In a similar fashion, 1,3-diphenyl-2-thioparabanate and ethyl vinyl ether give the spiro-thietan (150) in good yield.[244] Likewise, a similar mode of addition has been established for the reaction of tetramethylethylene with thiophosgene and its derivatives.[245]

$$PhCONMe_2 + PhC=CH_2 \xrightarrow{h\nu} PhCOCH_2CHPh_2$$

(147)

$$Ph_2C=S + EtOCH=CH_2 \longrightarrow \underset{EtO}{\overset{Ph_2 \text{—} S}{\square}} + \underset{EtO}{\overset{S \text{—} Ph_2}{\underset{S \text{—} Ph_2}{\square}}}$$

(148)   (149)

(150)

[239] J. S. Swenton, J. A. Hyatt, J. M. Lisy, and J. Clady, *J. Amer. Chem. Soc.*, 1974, **96**, 4885.
[240] K. Mizuno, C. Pac, and H. Sakurai, *J.C.S. Chem. Comm.*, 1974, 648.
[241] T. S. Cantrell, *J. Org. Chem.*, 1974, **39**, 3063.
[242] Y. Katsuhara, R. Tsuji, K. Hara, Y. Shigemitsu, and Y. Odaira, *Tetrahedron Letters*, 1974, 453.
[243] H. Gotthardt, *Chem. Ber.*, 1974, **107**, 1856.
[244] H. Gotthardt and S. Nieberl, *Tetrahedron Letters*, 1974, 3397.
[245] H. Gotthardt, *Chem. Ber.*, 1974, **107**, 2544.

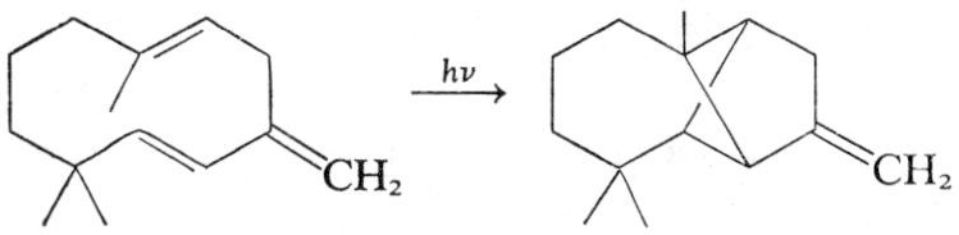

Acetylenes, Alkanes, Allenes, and Olefins

(151) → hv → (152)

Scheme 47

A key step in the synthesis of racemic α- and β-longipinenes was the photocyclization of the cyclodeca-1,5-diene (151) to the tricyclic system (152) (Scheme 47).[246] Irradiation of 1,3,6-triphenylhexa-1,3,5,-triene in the presence of iodine unexpectedly gives 1,2,4-triphenylbenzene.[247] 2- and 4-Allylanisoles have been cyclized photochemically to the corresponding cyclopropylanisoles.[248] The cycloaddition of diacylaminonitrenes to electrophilic or nucleophilic olefins often yields the corresponding aziridines stereospecifically (Scheme 48). The reactivity of these olefins can be interpreted on the theory of chemical reactivity resulting from the interaction of frontier orbitals and the direction of approach of the nitrene to the olefin.[249]

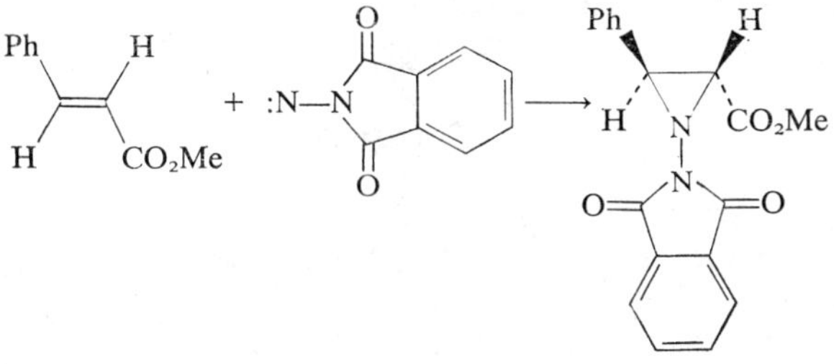

Scheme 48

Addition of a biradical species to a vinylcyclopropane derivative will generate a biradical with a cyclopropylcarbinyl site. Since the rearrangement of the cyclopropinylcarbinyl radical to an allylcarbinyl radical is known to be rapid, investigation of the products of such cycloaddition reactions gives information about the intermediates involved. In two papers the addition of carbenes to (153) and the photoaddition of aromatic carbonyl compounds to (154) are discussed in this light.[250] The phase-transfer method for generating

---

[246] M. Miyashita and A. Yoshikoshi, *J. Amer. Chem. Soc.*, 1974, **96**, 1917.
[247] W. Carruthers, N. Evans, and D. Whitmarsh, *J.C.S. Chem. Comm.*, 1974, 526.
[248] T. Clark and D. A. M. Watkins, *J.C.S. Perkin I*, 1974, 2124.
[249] H. Person, C. Fayat, F. Tonnard, and A. Foucaud, *Bull. Soc. chim. France*, 1974, 635.
[250] N. Shimizu and S. Nishida, *J. Amer. Chem. Soc.*, 1974, **96**, 6451; N. Shimizu M. Ishikawa, K. Ishikura, and S. Nishida, *ibid.*, p. 6456.

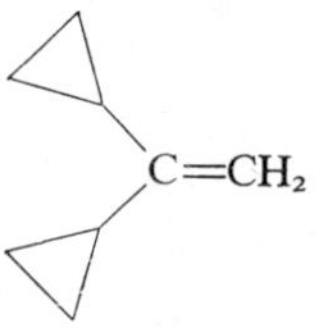

(153) 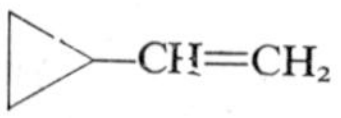   (154)

dichlorocarbene has been found to be superior to previous methods in reactions involving addition of CCl_2 to vinysilanes.[251] The relative reactivities of difluorocarbene with a variety of alkenes have been investigated with a view to possible selective synthesis of fluoro-steroids.[252]

Photolysis of 1,4-diphenylpent-4-en-1-ol (155) gives, amongst other products, up to 40% of acetophenone and α-methylstyrene. This represents the first example of a hydrocarbon analogue of the photoelimination reaction of ketones (Scheme 49).[253] The formation of (156) has been previously reported

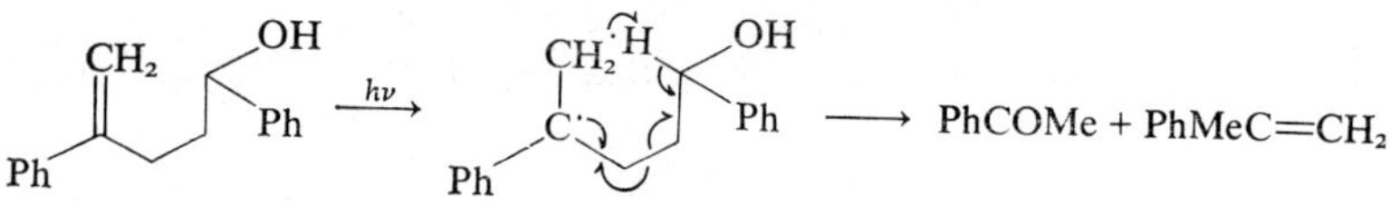

(155) $\xrightarrow{h\nu}$ ⟶ PhCOMe + PhMeC=CH_2

Scheme 49

and is thought to proceed through the biradical (157). If sulphur dioxide is used as a radical trap, the spiro-sulphone first formed undergoes a dienone–phenol rearrangement extremely easily to give (158).[254] A unique phenyl migration process from the excited singlet state of (159) to give (160) and (161) gives quantitative support for the mechanistic interpretation of the second π-bond involvement in the di-π-methane rearrangement.[255] This rearrangement has also been studied with various cyano-substituted derivatives of the parent 1,1,5,5-tetraphenyl-3,3-dimethylpentadiene studied previously. The 1,1-di-p-cyanophenyl derivative gives (162) exclusively, indicating a high degree of regiospecificity.[256]

The Ene Reaction. Ethyl *N*-phenylcarbamoylazoformate (163) adds to olefins by an ene-type reaction and is intermediate in reactivity between diethyl azodicarboxylate and the cyclic triazolinediones.[257] The configuration of the

[251] R. B. Miller, *Synthetic Comm.*, 1974, **4**, 341.
[252] R. A. Moss and D. J. Smudin, *Tetrahedron Letters*, 1974, 1829.
[253] J. M. Hornback, *J. Amer. Chem. Soc.*, 1974, **96**, 6773.
[254] R. M. Wilson and S. W. Wunderly, *J. Amer. Chem. Soc.*, 1974, **96**, 7350.
[255] H. E. Zimmerman and R. D. Little, *J. Amer. Chem. Soc.*, 1974, **96**, 5143.
[256] H. E. Zimmerman and B. R. Cotter, *J. Amer. Chem. Soc.*, 1974, **96**, 7445.
[257] G. T. Knight, M. J. R. Loadman, B. Saville, and J. Wildgoose, *J.C.S. Chem. Comm.* 1974, 193.

allylic hydrogen transferred in the ene reaction between β-pinene and maleic anhydride, and the preference for *endo* versus *exo* orientation, have been studied. Determination of the absolute configuration of the product, using stereospecifically C-3-labelled β-pinene, by degradation shows that the transition state is best depicted as (164).[258] The Lewis acid catalysis of ene reactions is

[258] R. K. Hill, J. W. Morgan, R. V. Shetty, and M. E. Synerholm, *J. Amer. Chem. Soc.* 1974, **96**, 4201.

exemplified by the reaction of β-pinene with methyl acrylate, which proceeds at room temperature in high yield in the presence of certain catalysts.[259] Benzyne reacts with *cis*-2-methylbut-1-en-1-yl acetate in the ene reaction to give (165). The *trans*-isomer, on the other hand, gives both (165) and (166). This is evidence that ene reactions with benzyne occur by approach from the least-hindered side of the double bond.[260] It has, however, been demonstrated

[259] B. M. Snider, *J. Org. Chem.*, 1974, **39**, 255.
[260] H. H. Wasserman and L. S. Keller, *Tetrahedron Letters*, 1974, 4355.

Acetylenes, Alkanes, Allenes, and Olefins

that the product of the reaction of benzyne with camphene arises *via* a biradical mechanism rather than by an ene reaction.[261]

Sigmatropic Rearrangements. Acetonyl allyl ethers undergo a [2,3] sigmatropic rearrangement with base to give 3-hydroxy-5-en-2-ones (Scheme 50).[262] Allylic alcohols are converted into homologous NN-dimethylamides on treatment with NN-dimethylformamide acetals. The reaction probably occurs by acetal exchange followed by imidate formation, carbene generation, and a [2,3] sigmatropic rearrangement, as shown in Scheme 51.[263]

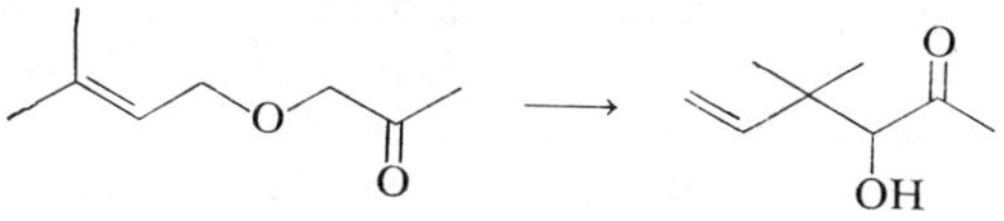

Scheme 50

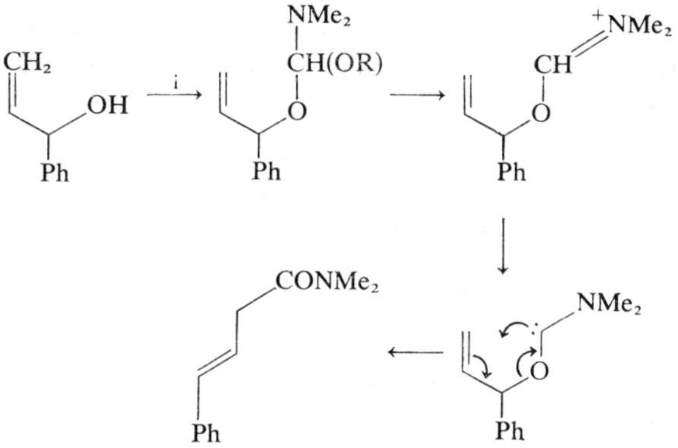

Reagent: i, $Me_2NCH(OR)_2$

Scheme 51

Recently, dyotropic rearrangements have been defined as pericyclic processes in which two σ-bonds interchange positions. The thermal rearrangement of (167) could be an example of this, but alternative mechanisms are possible as there is only 80% allyl inversion. In the presence of a radical inhibitor such as quinone, 97% allyl inversion occurs.[264]

Electrophilic Additions.—The interaction of ethylene, 2,3-dimethylbut-2-ene, and adamantylideneadamantane with a series of electrophiles in media of low acidity has been examined by ^1H and ^{13}C n.m.r. spectroscopies. Chemical

---

[261] G. Mehta and B. P. Singh, *Tetrahedron Letters*, 1974, 4297.
[262] A. F. Thomas and R. Dubini, *Helv. Chim. Acta*, 1974, **57**, 2084.
[263] G. Puchi, M. Cushman, and H. Wüest, *J. Amer. Chem. Soc.*, 1974, **96**, 5563.
[264] M. T. Reetz, *Angew. Chem. Internat. Edn.*, 1974, **13**, 402.

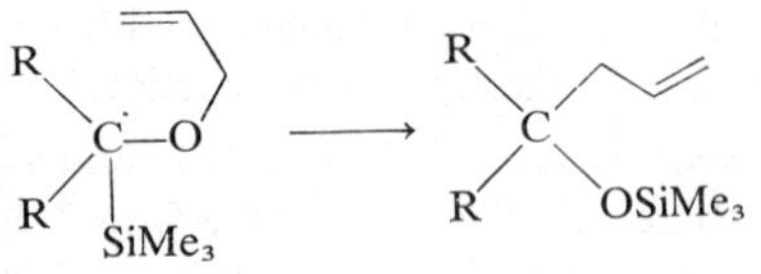

(167)

and spectral data indicate that adamantylideneadamantane alkonium-type π-complexes are formed preferentially as stable species which can be differentiated from open-chain β-substituted alkenium ions or three-membered-ring onium ions formed with less-hindered alkenes.[265]

Cyclohexene added dropwise to silver oxide in 98% sulphuric acid under an atmosphere of carbon monoxide gives 1-methylcyclopentanecarboxylic acid.[266] Reactions of olefins at low temperatures with benzeneselenyl bromide and silver trifluoroacetate, followed by sodium bicarbonate treatment, produce β-hydroxy-selenides, These are precursors to allylic alcohols by the well-established route of oxidation to the selenoxide and sigmatropic rearrangement (Scheme 52).[267]

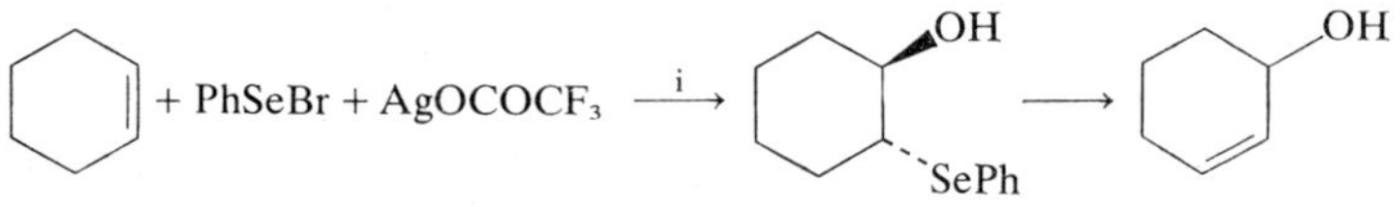

Reagent: i, H$_2$O

Scheme 52

Lithium perchlorate catalyses the addition of dichloroacetic acid to olefins. The specific salt effects observed have been discussed in terms of a general acid catalysis mechanism involving carbocation formation as the slow step.[268] The rates of addition of 4-chlorobenzenesulphenyl chloride to olefins have been investigated. All products are formed by stereospecific *anti*-addition across the double bond. Episulphonium ion intermediates are the most probable explanation for this, and the rate differences are primarily due to steric hindrance to the approach of the electrophile in the rate-determining transition state.[269] A kinetic study of the effects of the cyclopropyl group on electrophilic additions to the double bond reveals large accelerations relative to phenyl substitution when the transition state leads to a relatively open

---

[265] G. A. Olah, P. Schilling, P. W. Westerman, and H. C. Lin, *J. Amer. Chem. Soc.*, 1974, **96**, 3581.
[266] Y. Souma and H. Sano, *Bull. Chem. Soc. Japan*, 1974, **47**, 1717.
[267] D. L. J. Clive, *J. C. S. Chem. Comm.*, 1974, 100.
[268] J. Guenzet, M. Toumi, and A. Toumi, *Tetrahedron*, 1974, **30**, 159.
[269] C. L. Dean, D. G. Garratt, T. T. Tidwell, and G. H. Schmid, *J. Amer. Chem. Soc.* 1974, **96**, 4958.

positive charge on the adjacent carbon susceptible to resonance stabilization by the substituent.[270]

The addition of arylsulphonyl azides to olefins gives the arylsulphonyl imines (168). These can be reduced directly by sodium borohydride to give arylsulphonamides (169) or hydrolysed to the corresponding ketones (170).[271]

Addition of sulphur trioxide to substituted butenes at −78 °C in methylene chloride proceeds in high yield to give the sultone or sulphonic acid, depending on the substitution pattern. These conditions are milder than those normally necessary for the sulphonation of olefins.[272]

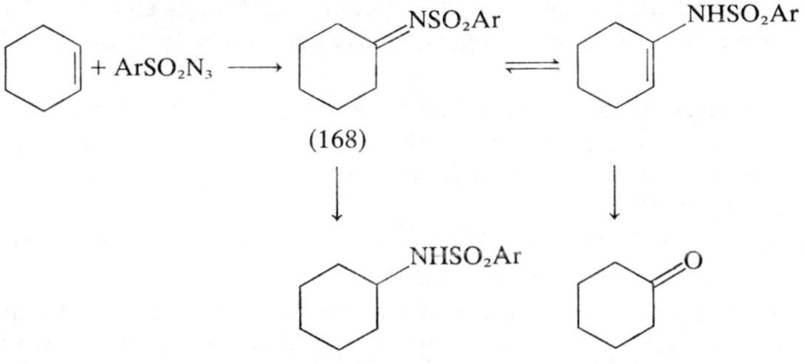

Halogenation of carbonyl compounds *via* the silyl enol ethers which are formed regiospecifically from the ketones, followed by β-cleavage of bromotrimethylsilane from the intermediate, provides a good route to the often unstable α-bromo-aldehydes and -ketones (Scheme 53). γ-Butyrolactone has been similarly converted into α-bromo-γ-butyrolactone.[273]

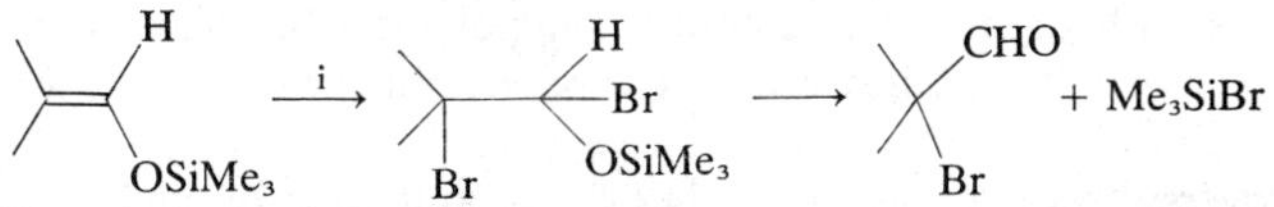

Reagent: i, Br_2

Scheme 53

In 80% methanol–water the olefin (171) undergoes addition of methanol by an A-S_E2 mechanism. Both the intermediate carbenium ion (172) and the final product (173) can be detected independently, and thus a complete

---

[270] D. G. Garratt, A. Modro, K. Oyama, G. H. Schmid, T. T. Tidwell, and K. Yates, *J. Amer. Chem. Soc.*, 1974, **96**, 5295.
[271] R. A. Abramovitch, G. A. Knaus, M. Pavlin, and W. D. Holcomb, *J.C.S. Perkin I*, 1974, 2169.
[272] M. D. Robbins and C. D. Broaddus, *J. Org. Chem.*, 1974, **39**, 2459.
[273] R. H. Reuss and A. Hassner, *J. Org. Chem.*, 1974, **39**, 1785.

(171)   (172)   (173)

equilibrium and kinetic analysis of the component steps can be achieved. Normally only one of these steps can be measured as it is the rate-determining step.[274]

Nitrogen dioxide is known to catalyse the geometric isomerization of but-2-enes and pent-2-enes. The results of a detailed kinetic study suggest that this process is unimportant in the bulk consumption of atmospheric olefin pollutants.[275]

Contrary to another report, the uncatalysed reaction between propene or isobutene and hydrogen chloride does occur at moderate pressures and at room temperature to give only one product. Kinetic studies suggest that this reaction takes place by a low-energy pathway which is not available at higher temperatures and is not the reverse of the decomposition pathway.[276] Addition of hydrogen bromide in acetic acid to cyclohexene and hex-1-ene is consistent with an $Ad3$ mechanism involving hydrogen bromide catalysis of the acetic acid addition.[277]

The bromochlorination of 3-t-butylcyclohexene has been studied with bromine monochloride and monopyridinebromine chloride. The results with bromine monochloride are suggestive of an addition mechanism involving rate- and product-determining formation of epibromonium ion intermediates. The results with the pyridine complex suggest that the steric course of the addition is controlled mainly during the nucleophilic rather than the electrophilic step.[278]

Controversy continues as to whether the bromination of olefins necessarily proceeds *via* cyclic bromonium ions. Dubois has argued against the accepted bromonium ion model and has interpreted his results on substituted stilbenes in terms of the three species shown in Scheme 54. The data obtained suggest that carbonium-ion-like and bromonium-ion-like transition states differ in their charge distribution. Arguments in favour of strain conservation, as proposed by Yates, need not necessarily involve bromine bridging, but other

[274] C. F. Bernasconi and W. J. Boyle, jun., *J. Amer. Chem. Soc.*, 1974, **96**, 6070.
[275] J. L. Sprung, H. Akimoto, and J. N. Pitts, jun., *J. Amer. Chem. Soc.*, 1974, **96**, 6549.
[276] F. Amar, D. R. Dalton, G. Eisman, and M. J. Haugh, *Tetrahedron Letters*, 1974, 3033.
[277] R. C. Fahey, C. A. McPherson, and R. A. Smith, *J. Amer. Chem. Soc.*, 1974, **96**, 4534.
[278] G. Bellucci, G. Ingrosso, F. Marioni, E. Mastrorilli, and I. Morelli, *J. Org. Chem.*, 1974, **39**, 2562.

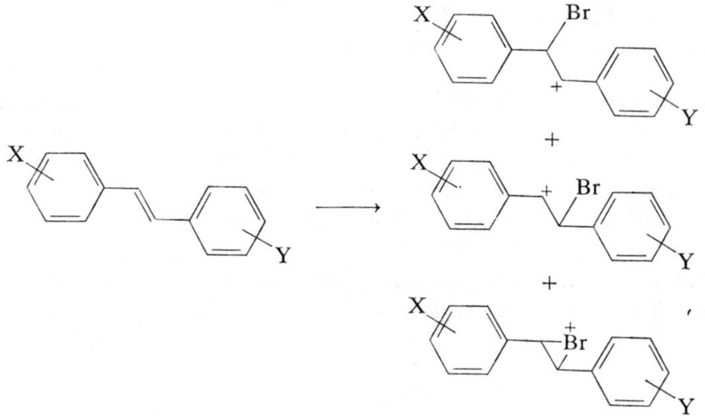

Scheme 54

possible effects such as counter-ions, or interactions between the carbon–bromine bond and the *p*-orbitals of the adjacent carbonium ion, should be considered.[279] Dubois has also investigated the bromination of allylbenzenes, which gives (174) and (175). The intermediacy of a phenonium ion is suggested.[280]

Competitive rate data for the addition of bromine to a series of alkenes in 1,1,2-trichlorotrifluoroethane at −35 °C show a considerably smaller range in relative rates to previous data for the bromination of similar alkenes in a polar solvent. The data indicate that the earlier reactant-like transition states are π-complex in character. Subsequent opening of the molecularly bonded π-complexes leads *via* participation of the non-bonded electron pairs of bromine to the σ-complexes or bromonium ions. Similar processes also operate for alkynes at 0 °C.[281]

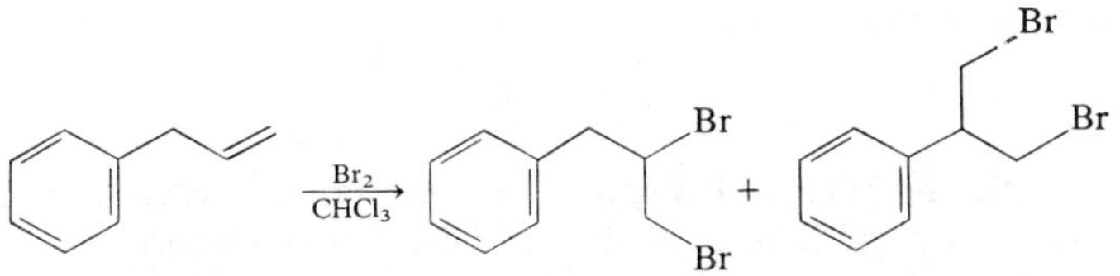

The relative nucleophilicity of DMSO and methanol has been measured

---

[279] M.-F. Ruasse and J. E. Dubois, *J. Org. Chem.*, 1974, **39**, 2441.
[280] D. Fain, J. Toullec, and J. E. Dubois, *Tetrahedron Letters*, 1974, 1725.
[281] G. A. Olah and T. R. Hockswender, jun., *J. Amer. Chem. Soc.*, 1974, **96**, 3574.

by product analysis of the bromination of olefins, and DMSO was found to be 2—7 times more effective.[282]

Hydroboration.—The full details have been published of the preparation and properties of 9-borabicyclo[3,3,1]nonane (9-BBN), with particular accent on its unique reactivity with olefins and on synthetic transformations of the products.[283] A simple new hydroboration procedure involves dissolving the olefin in dry THF, followed by addition of sodium borohydride and then an equal amount of acetic acid in THF. Oxidation with alkaline hydrogen peroxide after two hours gives high yields of the alcohols in the normal manner. Thus α-pinene (176) gives isopinocampheol (177) in good yield.[284]

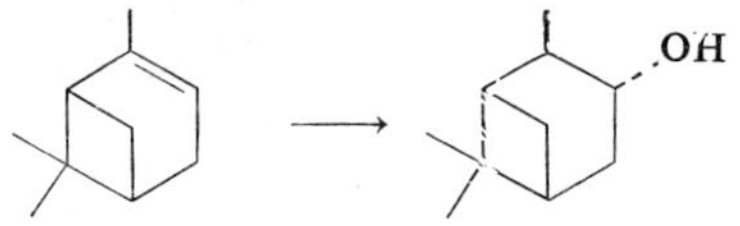

(176)   (177)

O-Mestiylenesulphonylhydroxylamine has been shown to be a superior reagent to hydroxylamine *O*-sulphonic acid and chloramine in terms of solubility and reactivity as an electrophilic aminating agent for the conversion of dialkylboranes into amines.[285] Excellent yields for the anti-Markovnikov esterification of alkenes are achieved using a hydroboration–mercuration–iodination sequence (Scheme 55).[286]

$$RCH=CH_2 \xrightarrow{i-iii} RCH_2CH_2OAc$$

Reagents: i, $\frac{1}{3}BH_3$; ii, $Hg(OAc)_2$; iii, I_2

Scheme 55

Oxymercuration and Oxythallation.—Aminomercuration reactions normally proceed sluggishly and in poor yield. If initial ethylene pressures of 600 p.s.i. are used, yields of up to 80% based on amine recovered are achieved in 1—2 hours.[287] The replacement of mercury by lithium in THF and subsequent hydrolysis by methanol constitutes an alternative method for the cleavage of the carbon–mercury bond which can be useful where borohydride reduction

---

[282] V. L. Heasley, R. A. Skidgel, G. E. Heasley, and D. Strickland, *J. Org. Chem.*, 1974, **39**, 3953.
[283] H. C. Brown, E. F. Knights, and C. G. Scouten, *J. Amer. Chem. Soc.*, 1974, **96**, 7765.
[284] V. Hach, *Synthesis*, 1974, 340.
[285] Y. Tamura, J. Minamikawa, S. Fujii and M. Ikeda, *Synthesis*, 1974, 196.
[286] R. C. Larock, *J. Org. Chem.*, 1974, **39**, 834.
[287] R. F. DeBrule and G. G. Hess, *Synthesis*, 1974, 197.

would cause complications.[288] The thermal decomposition of organomercurials in alkaline media is clearly different from that in acidic media. But-2-ene and mercuric sulphate give a 2:1 ratio of 2,3-epoxybutane to butan-2-one on base decomposition of the adduct.[289]

The oxidative coupling of *trans*-2-chloromercury-3-phenylpropenoic acid (178) using a Li_2PdCl_4–$CuCl_2$ reagent system gives a synthesis of the two unknown stereoisomers of the dibenzylidenesuccinic acids (179) and (180).[290]

A full comparative study of the mercury(II), thallium(III), and lead(IV) oxidations of oct-1-ene in methanol has been reported, and the conditions which affect the products formed by the decomposition of the intermediate organometallic adducts have been discussed.[291] The oxidation of alkenes with thallium(III) nitrate in methanol is known to lead to the formation of 1,2-dimethoxy-compounds and carbonyl compounds. Significant amounts of methoxy-nitrates and dinitrates are now shown also to be formed.[292] The oxythallation of 2-*endo*-norbornenecarboxylic acids (181) has been extended, and several intermediate organothallium species have been isolated in high yield. The adducts are remarkably stable and their structures have been confirmed by the large Tl–H coupling constants.[293]

[288] V. Aranda, J. Barluenga, A. Ara, and G. Asensio, *Synthesis*, 1974, 135.
[289] H. Arzoumanian, J. P. Aune, J. Guitard, and J. Metzger, *J. Org. Chem.*, 1974, **39**, 3445.
[290] H.-L. Elbe and G. Köbrich, *Chem. Ber.*, 1974, **107**, 1654.
[291] A. Lethbridge, R. O. C. Norman, and C. B. Thomas, *J.C.S. Perkin I*, 1974, 1929.
[292] R. J. Bertsch and R. J. Ouellette, *J. Org. Chem.*, 1974, **39**, 2755.
[293] A. McKillop, M. E. Ford, and E. C. Taylor, *J. Org. Chem.*, 1974, **39**, 2434.

Metal–Olefin Reactions.—Oxidation of cyclo-octa-1,5-diene with palladium chloride and lead tetra-acetate gives 2,6-diacetoxybicyclo[3,3,0]octane, which has been shown by X-ray analysis to be the di-*endo*-isomer (182).[294]

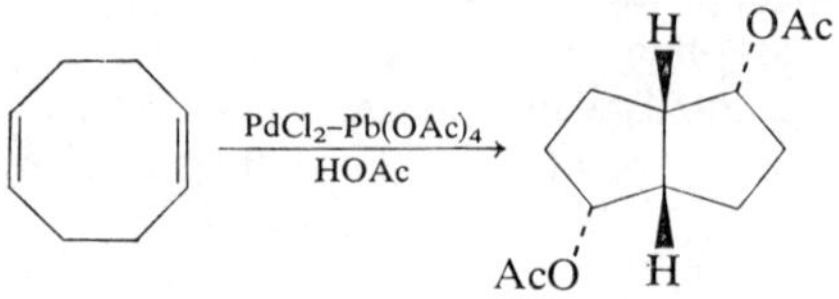

The oxidation of olefins by palladium(II) salts, promoted by cupric chloride, has been extended to other metal salts. Product distributions were found to vary widely with different metals, and this study provides evidence against a mechanism involving alkyl transfer from a noble metal to the copper(II).[295] The oxidation of long-chain and highly substituted α-olefins to methyl ketones, catalysed by palladium chloride, proceeds better in aqueous sulpholan than in previously used solvents.[296] 3,7-Dimethylocta-1,6-diene (183) can be transformed into a variety of products depending on the oxidant and conditions used (Scheme 56).[297]

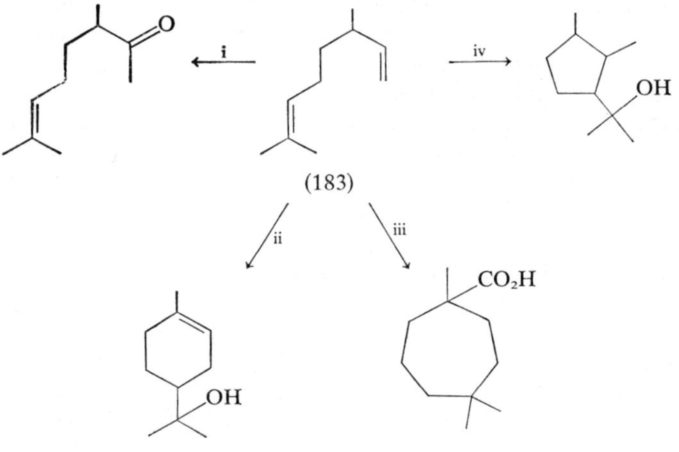

Reagents: i, $PdCl_2$–Cu_2Cl_2–DMF; ii, $PdCl_2$–aq.EtOAc; iii, HCO_2H_2–HSO_4; iv, $Hg(OAc)_2$; H_2O–THF–$NaBH_4$

Scheme 56

[294] P. M. Henry, M. Davies, G. Ferguson, S. Phillips, and R. Restivo, *J.C.S. Chem. Comm.*, 1974, 112.
[295] P. M. Henry, *J. Org. Chem.*, 1974, **39**, 3871.
[296] D. R. Fahey and E. A. Zuech, *J. Org. Chem.*, 1974, **39**, 3276.
[297] F. J. McQuillin and D. G. Parker, *J.C.S. Perkin I*, 1974, 809.

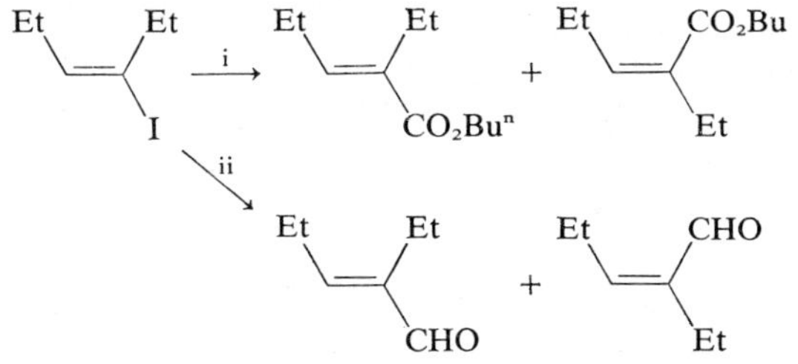

Scheme 57

Reagents: CO–Bu^{n}OH–Pd(PPh$_3$)$_2$I$_2$; ii, CO–H$_2$–NEt$_3$–Pd(PPh$_3$)$_2$I$_2$

Scheme 57

Vinylic bromides and iodides react with carbon monoxide and alcohols[298] or amines[299] in the presence of palladium triphenylphosphine complexes to give esters or amides (Scheme 57). The use of tertiary amines under these conditions gives aldehydes.[300]

A high-yielding general procedure for the formation of (π-allyl)palladium chloride dimers from olefins has been described and is useful in a new method of allylic alkylation.[301] The organometallic propellane (184) has been pre-

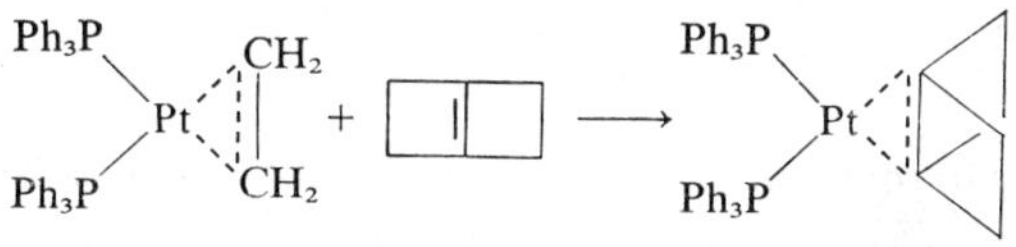

(184)

pared from $\Delta^{1,4}$-bicyclo[2,2,0]hexene and provides a convenient method for the separation and storage of this olefin. It can be regenerated using carbon disulphide.[302] The first example of a stable σ-bonded palladium complex (185) results from the addition of Pd—Cl to 5-vinyl-2-norbornene.[303]

Palladium compounds are active catalysts in the selective hydrosilylation of terminal olefins and conjugated dienes.[304] Butadiene reacts with nitroalkanes in the presence of triphenylphosphine-palladium complexes via oligomerization and displacement to give compounds such as (186). Reduction gives amines which have the primary amine group in the middle of the chain.[305]

---

[298] A. Schoenberg, I. Bartoletti, and R. F. Heck, *J. Org. Chem.*, 1974, **39**, 3319.
[299] A. Schoenberg and R. F. Heck, *J. Org. Chem.*, 1974, **39**, 3327.
[300] A. Schoenberg and R. F. Heck, *J. Amer. Chem. Soc.*, 1974, **96**, 7761.
[301] B. M. Trost and P. E. Strege, *Tetrahedron Letters*, 1974, 2603.
[302] M. E. Jason, J. A. McGinnety, and K. B. Wiberg, *J. Amer. Chem. Soc.*, 1974, **96**, 6531.
[303] W. T. Wipke and G. L. Goeke, *J. Amer. Chem. Soc.*, 1974, **96**, 4244.
[304] J. Tsuji, M. Hara, and K. Ohno, *Tetrahedron*, 1974, **30**, 2143.
[305] T. Mitsuyasu and J. Tsuji, *Tetrahedron*, 1974, **30**, 831.

(185)

$CH_2=CHCH=CH_2 + MeNO_2 \longrightarrow (CH_2=CHCH_2CH_2CH_2CH=CHCH_2)_3CNO_2$
(186)

The palladium nitrate–triphenylphosphine complex (1:3) catalyses the reaction of butadiene with Schiff bases to give 2,3,6-substituted piperidines (187).[306]

$CH_2=CHCH=CH_2 + R^1N=CHR^2 \longrightarrow$

(187)

A new route to muscone utilizes the addition of allene to the dodecatrienyl-nickel complex (188), followed by insertion of t-butylisocyanide and hydrolysis to 3-methylenecyclopentadeca-5,9,13-trienone. Hydrogenation of this latter intermediate gave muscone (Scheme 58).[307]

π-(2-Methoxyallyl)nickel bromide, prepared from 2-methoxyallyl bromide and nickel carbonyl, reacts with halides, ketones, and aldehydes in DMF to introduce the acetonyl function (Scheme 59).[308]

Reactions between 1,3-dienes and amines[309] and active-methylene compounds[310] in the presence of nickel catalysts give a mixture of four products, *e.g.* (189a—d). These reactions provide a synthetic route to a number of long-chain alicyclic derivatives, some of which could be used in terpene synthesis.

The Diels–Alder additions of butadiene and 2,3-dimethylbutadiene to methyl sorbate can be regiospecifically controlled by a nickel catalyst [Ni(acac)$_2$–Et$_3$Al–Ph$_3$P]. Thus, 2,3-dimethylbutadiene gives the three products

---

[306] J. Kiji, K. Yamamoto, H. Tomita, and K. Furukawa, *J.C.S. Chem. Comm.*, 1974, 506.
[307] R. Baker, R. C. Cookson, and J. R. Vinson, *J.C.S. Chem. Comm.*, 1974, 515.
[308] L. S. Hegedus and R. K. Stiverson, *J. Amer. Chem. Soc.*, 1974, 96, 3250.
[309] R. Baker, A. H. Cook, D. E. Halliday, and T. N. Smith, *J.C.S. Perkin II*, 1974, 1511.
[310] R. Baker, A. H. Cook, and T. N. Smith, *J.C.S. Perkin II*, 1974, 1517.

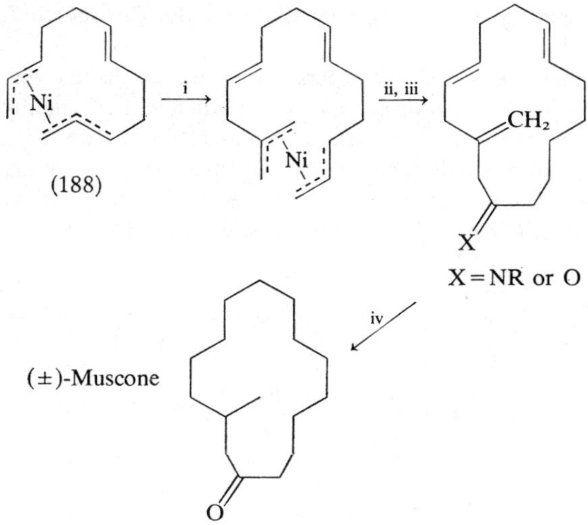

Acetylenes, Alkanes, Allenes, and Olefins

(188)

X = NR or O

(±)-Muscone

Reagents: i, H$_2$C=C=CH$_2$; ii, RNC; iii, H$^+$; iv, Pd/H$_2$

Scheme 58

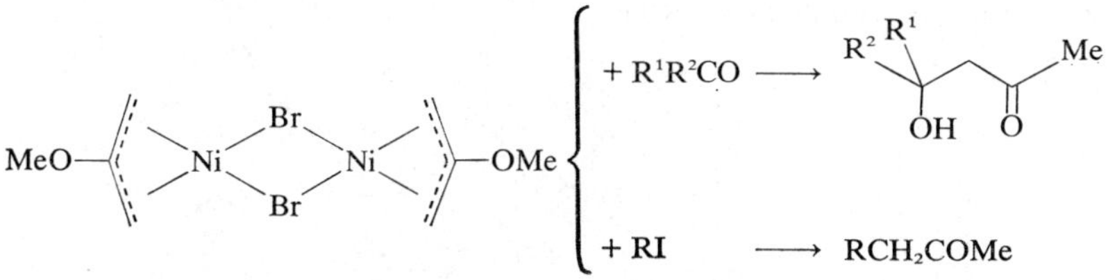

Scheme 59

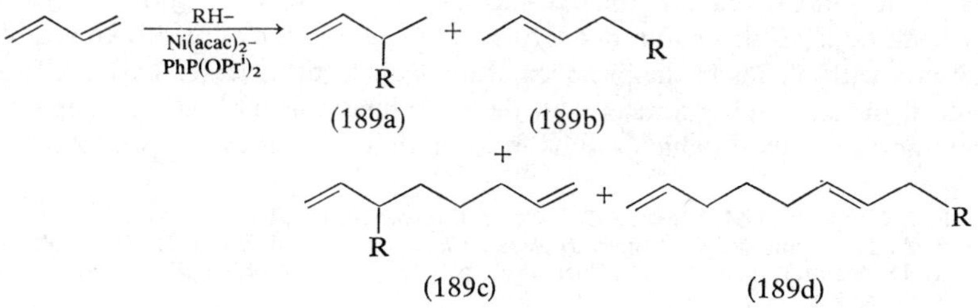

(189a) (189b)

(189c) (189d)

R = NR^{1}R^2 or activated —CH— group
e.g. PhCHCN, PhCHCOMe

(190a—c) thermally, but use of the catalyst at 40 °C results in formation of only (190c).[311]

A comprehensive series of papers has been published on the nickel-promoted skeletal rearrangements of 1,4-dienes and methylvinylcyclopropanes.[312] A cobaltacyclopentene complex (191) has been isolated as an intermediate in the cobalt-catalysed co-oligomerization of diphenylacetylene with cyano-olefins.[313] Acetone and oct-1-ene give the anti-Markovnikov addition product (192) in the presence of silver(II) oxide.[314]

$(\pi\text{-}C_5H_5)Co(PPh_3)(PhC{\equiv}CPh) \xrightarrow{CH_2{=}CHCN} $ (191)

$C_6H_{13}CH{=}CH_2 + MeCOMe \xrightarrow{AgO} C_6H_{13}CH_2CH_2COMe$
(192)

The development of a simple one-step synthesis of γ-lactones from olefins has been summarized and the free-radical nature of this oxidation confirmed (Scheme 60).[315] Dihydrofurans are formed by the reaction of readily enolizable ketones with olefins in the presence of manganese(III) acetate. This is a free-radical process and contrasts with the ionic mechanism observed when lead tetra-acetate is used (which results in formation of an isomeric product).[316]

---

[311] P. J. Garrett and M. Wyatt, *J.C.S. Chem. Comm.*, 1974, 251.
[312] P. A. Pinke and R. G. Miller, *J. Amer. Chem. Soc.*, 1974, **96**, 4221; P. A. Pinke, R. D. Stauffer, and R. G. Miller, *ibid.*, p. 4229; H. J. Golden, D. J. Baker, and R. G. Miller, *ibid.*, p. 4235.
[313] Y. Wakatsuki, K. Aoki, and H. Yamazaki, *J. Amer. Chem. Soc.*, 1974, **96**, 5284.
[314] M. Hajek, P. Silhavy, and J. Malek, *Tetrahedron Letters*, 1974, 3193.
[315] E. I. Heiba, R. M. Dessau, and P. G. Rodewald, *J. Amer. Chem. Soc.*, 1974, **96**, 7977.
[316] E. I. Heiba and R. M. Dessau, *J. Org. Chem.*, 1974, **39**, 3456.

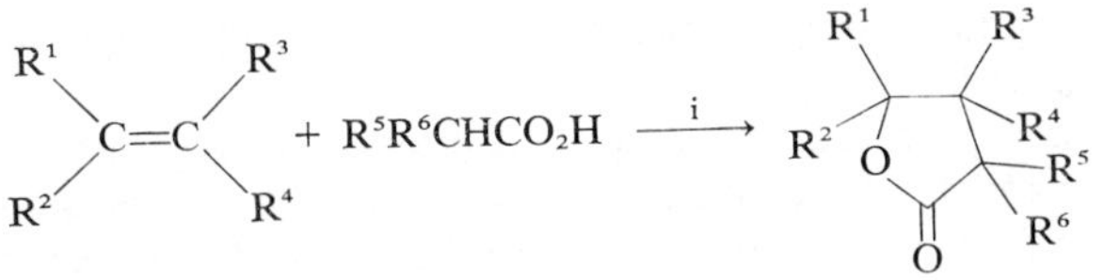

Reagent: i, Mn(OAc)$_3$

Scheme 60

Organozirconium chemistry appears to have an interesting future and many possible synthetic uses. All three isomeric octenes react with the zirconium complex (193) to give the same product (194), a process reminiscent of hydroboration. The product can be further transformed as shown in Scheme 61.[317]

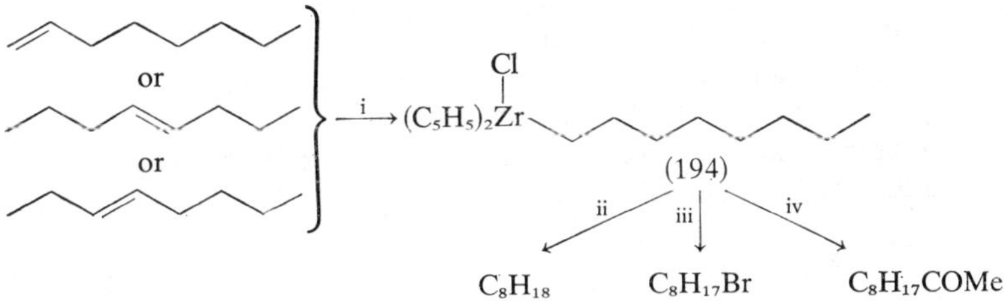

Reagents: i, (C$_5$H$_5$)$_2$ZrHCl (193); ii, H$^+$; iii, Br$_2$; iv, MeCOCl

Scheme 61

The thermodynamically less stable *trans*-fused tricyclododecanes and tricyclotetradecanes are formed by the photodimerizations, catalysed by copper(I) trifluoromethanesulphonate, of cyclohexene (195) and cycloheptene. Since cyclo-octene and acyclic olefins fail to react, a concerted orbital-symmetry-controlled process involving formation of copper(I)-trifluoromethanesulphonate-complexed *trans*-olefin intermediates is probable.[318]

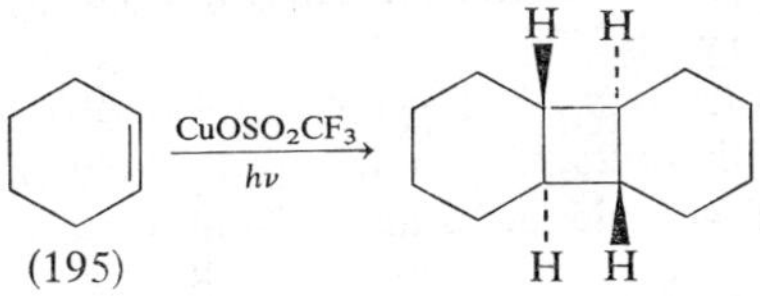

Oxidation of Olefins.—Conclusive evidence for formation of a π-complex between ozone and aromatic olefins comes from low-temperature studies. Ozone shows a characteristic band at 595—610 nm, but on complexation with mesitylphenylethylene at −195 °C this shifts to 460 nm. The ozone complexes react with olefins in a different way from ozone alone, and the steric

[317] D. W. Hart and J. Schwartz, *J. Amer. Chem. Soc.*, 1974, **96**, 8115.
[318] R. G. Salomon, K. Folting, W. E. Streib, and J. K. Kochi, *J. Amer. Chem. Soc.*, 1974, **96**, 1145.

size of the ozone–aromatic olefin complex is significant in determining product ratios.[319]

The two-step process of formation and cleavage of the kinetic silyl enolate with ozone provides a method for the oxidative cleavage of an asymmetric ketone away from the more highly substituted side. Moreover, the nucleophilicity of the silyloxyalkene bond allows this to be oxidized preferentially even in the presence of other olefinic bonds, such as in (196).[320]

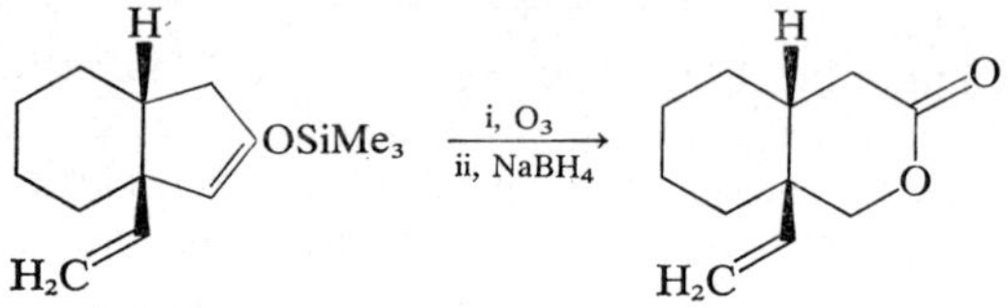

Oxidation of trimethylsilyl enol ethers with peracids in hexane followed by treatment with acid affords a general high-yield method for the preparation of acyloins (Scheme 62).[321] An improved work-up procedure for the oxidation of olefins to α-diketones using potassium permanganate in acetic anhydride has been described.[322]

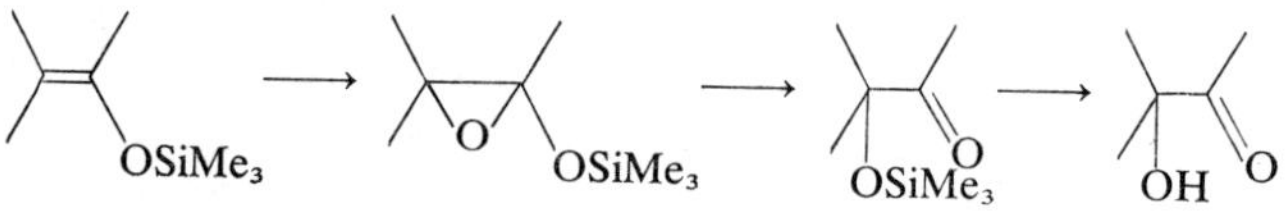

Scheme 62

A stable polymeric aromatic peracid has been prepared and used to oxidize alkenes to epoxides in good yield.[323] O-Benzylmonoperoxycarbonic acid is a new oxygenating agent for olefin epoxidation.[324] Epoxidation of the stereoisomeric hexa-2,4-dienes with m-chloroperbenzoic acid has been investigated. Of the six diastereoisomeric diepoxides formed, the attack gives a monoepoxide first in an s-trans-conformation, followed by preferential attack anti to the epoxide oxygen.[325]

Autoxidation of olefins with oxygen and iron(III) meso-tetraphenylporphyrin proceeds readily at 25 °C by a free-radical chain process to give mixtures of epoxides, allylic alcohols, and αβ-unsaturated ketones.[326] Atomic

[319] P. S. Bailey, J. W. Ward, T. P. Carter, jun., E. Nieh, C. M. Fischer, and A.-I. Y. Khashab, *J. Amer. Chem. Soc.*, 1974, **96**, 6136.
[320] R. D. Clark and C. H. Heathcock, *Tetrahedron Letters*, 1974, 2027.
[321] G. M. Rubottom, M. A. Vazquez, and D. R. Pelegrina, *Tetrahedron Letters*, 1974, 4319.
[322] H. P. Jensen and K. B. Sharpless, *J. Org. Chem.*, 1974, **39**, 2314.
[323] C. R. Harrison and P. Hodge *J.C.S. Chem. Comm.*, 1974, 1009.
[324] R. M. Coates and J. W. Williams, *J. Org. Chem.*, 1974, **39**, 3054.
[325] G. E. Heasley, R. V. Hodges, and V. L. Heasley, *J. Org. Chem.*, 1974, **39**, 1769.
[326] D. R. Paulson, R. Ullman, R. B. Sloane, and G. L. Closs, *J.C.S. Chem. Comm.*, 1974, 186.

oxygen in its ground state is produced from mercury vapour, an organic acceptor, and nitrous oxide. It reacts with butadiene to give radical intermediates, the decomposition pathways of which have been studied using [2,3-^{2}H$_2$]butadiene as starting material.[327] Anodic oxidation of tetraphenylethylene in methylene chloride containing trifluoroacetic acid gives the cation radical, which is very stable but which undergoes quantitative conversion into 9,10-diphenylphenanthrene (197).[328]

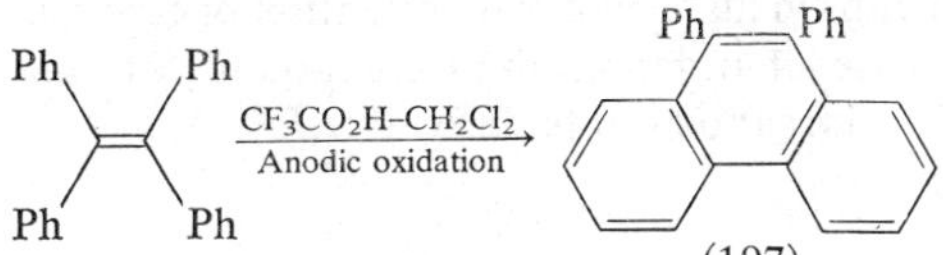

Stereoselective epoxidation of acyclic allylic alcohols is possible using t-butyl peroxide in the presence of vanadyl acetylacetonate. The product can be transformed stereospecifically, using known reactions, into an alkene (Scheme 63).[329]

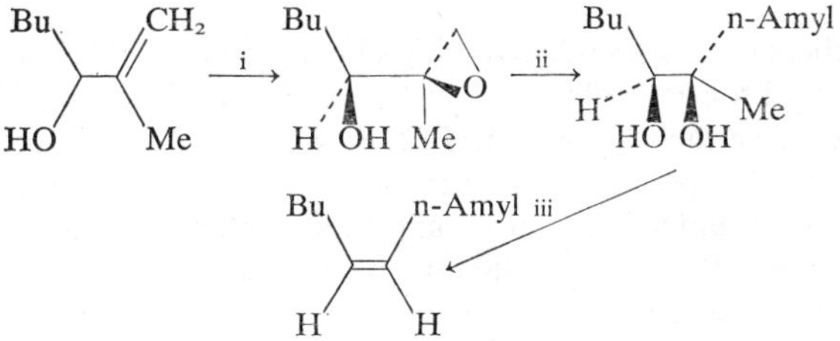

Reagents: i, Bu^{t}OOH–VO(acac)$_2$; ii, Bu$_2$CuLi–Et$_2$O; iii, Me$_2$NCH(OMe)$_2$–Δ–Ac$_2$O

Scheme 63

Other Reactions of Olefins.—α-Methoxyvinyl-lithium (198) is a conveniently prepared acyl anion equivalent that is of considerable synthetic potential.

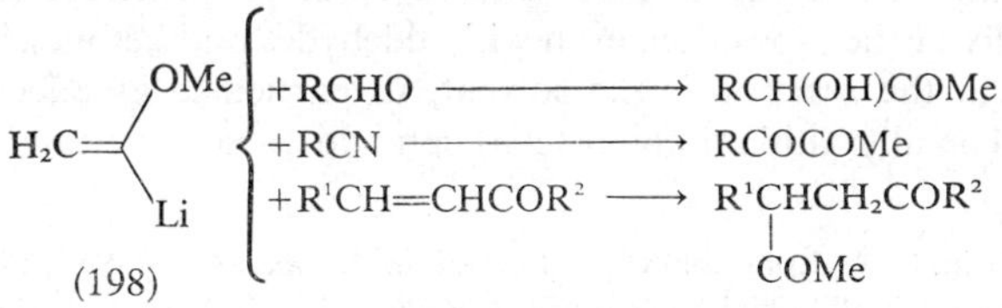

$$H_2C=C(OMe)(Li) \quad (198)$$

+RCHO ⟶ RCH(OH)COMe
+RCN ⟶ RCOCOMe
+R^{1}CH=CHCOR^2 ⟶ R^{1}CHCH$_2$COR^2
$\quad\quad\quad\quad\quad\quad\quad\quad\quad\quad\;$ |
$\quad\quad\quad\quad\quad\quad\quad\quad\quad\quad$ COMe

[327] J. J. Havel and K. H. Chan, *J. Org. Chem.*, 1974, **39**, 2439.
[328] U. Svanholm, A. Ronlán, and V. D. Parker, *J. Amer. Chem. Soc.*, 1974, **96**, 5108.
[329] H. Yamamoto, H. Nozaki, K. B. Sharpless, S. Tanaka, R. C. Michaelson, and J. D. Cutting, *J. Amer. Chem. Soc.*, 1974, **96**, 5254.

Its reaction with electrophiles and subsequent mild hydrolysis gives the corresponding carbonyl compounds. Other reactions with halides, aldehydes, ketones, and nitriles proceed as expected. It undergoes exclusive 1,2-addition to αβ-unsaturated ketones and can thus be used to prepare 1,4-diketones.[330] In a similar way, vinyl-lithium reagents from halogenosilanes will add to αβ-unsaturated ketones after initial conversion into the corresponding diorganocuprates. The addition compounds may be further transformed into ketones (199).[331] Vinyl-lithium reagents can be alkylated with primary halides to give olefins in high yield, with retention of configuration. This is thus a more efficient use of organolithium compound for this type of reaction than the use of a diorganocuprate derivative.[332]

(199)

Allylic ethers can be metallated with s-butyl-lithium in THF. The reaction of the anion with electrophiles to give (200) or (201) depends on the steric bulk of the substituent group R. For R = $SiEt_3$, (200) is formed almost exclusively.[333] The reaction with carbonyl compounds gives reverse regioselectivity. The counter-ion is also very important, as shown in the reactions of the anion of allyl methyl ether with cyclohexanone to give (202) and (203).[334]

(200)   (201)

In contrast, the allylic dianion derived from propenethiol (204) reacts preferentially at the γ-position, even with aldehydes and ketones.[335] Another key factor in the reaction is the solvent, in particular its effect on anion solvation. The allyl-lithium anion (205) can add to acetone to give (206) or

---

[330] J. E. Baldwin, G. A. Höfle, and O. W. Lever, jun., *J. Amer. Chem. Soc.*, 1974, **96**, 7125.
[331] R. K. Boeckman, jun., and K. J. Bruza, *Tetrahedron Letters*, 1974, 3365.
[332] G. Linstrumelle, *Tetrahedron Letters*, 1974, 3809.
[333] W. C. Still and T. L. Macdonald, *J. Amer. Chem. Soc.*, 1974, **96**, 5561.
[334] D. A. Evans, G. C. Andrews, and B. Buckwalter, *J. Amer. Chem. Soc.*, 1974, **96**, 5560.
[335] K. Geiss, B. Seuring, R. Pieter, and D. Seebach, *Angew. Chem. Internat. Edn.*, 1974, **13**, 479.

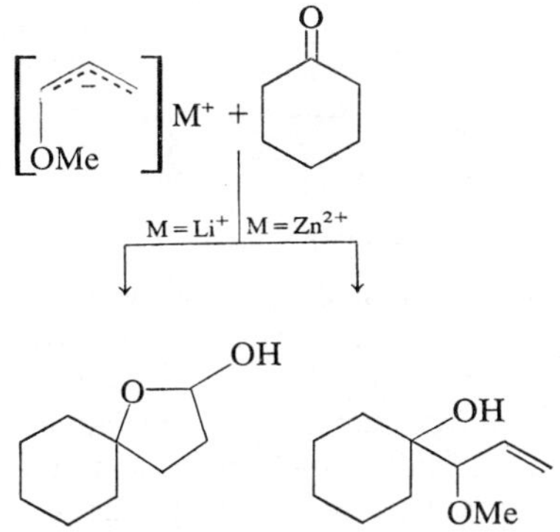

(207), depending on whether the reaction is carried out in diazabicyclo-octane (γ-substitution) or in a macrocyclic diamino-polyether (α-substitution), respectively.[336]

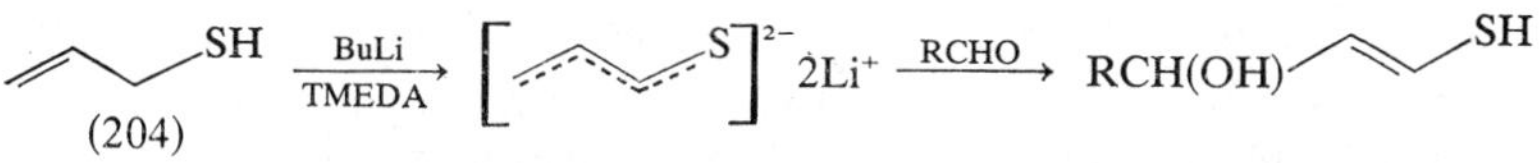

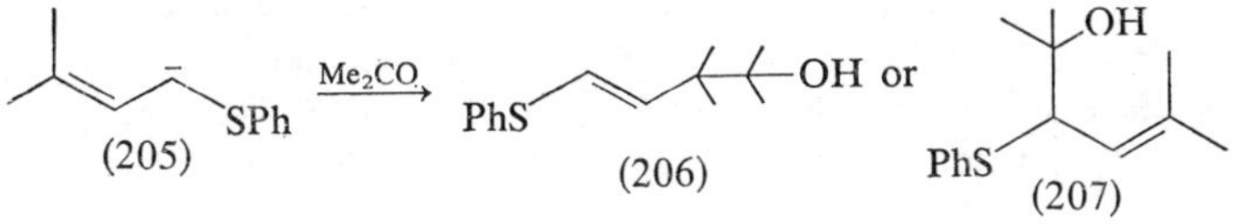

Some simple alkenes and unconjugated dienes can be di- and tri-metallated by the treatment with butyl-lithium–tetramethylethylenediamine for 1—4 days (Scheme 64).[337]

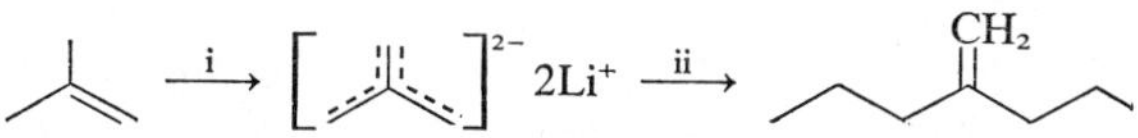

Reagents: i, BuLi–TMEDA; ii, EtI

Scheme 64

The preparation and some Michael reactions of the sulphonium ylide (208) have been described and some novel vinylcyclopropanes are formed, in reasonable yield.[338]

The kinetics of the gas-phase hydrogenation of olefins by di-imide suggest that the overall rate is controlled by the isomerization of *trans*-di-imide to

---

[336] P. M. Atlanti, J. F. Biellmann, S. Dube, and J. J. Vicens, *Tetrahedron Letters*, 1974, 2665.
[337] R. B. Bates, W. A. Beavers, M. G. Greene, and J. H. Klein, *J. Amer. Chem. Soc.*, 1974, **96**, 5641.
[338] J. R. Neff, R. R. Gruetzmacher, and J. E. Nordlander, *J. Org. Chem.*, 1974, **39**, 3814.

cis-di-imide. It has been estimated that the reaction of *cis*-di-imide with *trans*-di-imide to give nitrogen and hydrazine is 4—8 times faster than its reaction with ethylene.[339]

A mechanistic study of the mercury(II) oxide–iodine reaction with olefins, epoxides, and iodohydrins leads to the conclusion that a regiospecific mechanism involving attack by positive iodine on the double bond is involved.[340] Lewis-base catalysis by phosphorus tribromide or phosphorus oxychloride of the anti-Markovnikov addition of hydrogen bromide to olefins has been examined, and a mechanism proposed for the initiation of the homolytic cleavage of the H—Br bond.[341] Bis(trifluoromethyl) trioxide adds to olefins that are not prone to radical polymerization to give fluorocarbon peroxides.[342] The additions of silylene to butadiene and hexa-2,4-diene have been studied. The reaction is non-concerted, as judged by the stereochemistry of the hexa-2,4-diene adduct.[343] Butadiene gives the corresponding spiro-silane derivative (209) when treated with silicon tetrachloride and active magnesium.[344]

Many interesting papers have appeared on the use of silyl enol ethers and vinylsilanes as synthetic intermediates. Conia and his workers have shown that silyl enol ethers of $\alpha\beta$-unsaturated methyl ketones can be transformed into α-methylated ketones or alicyclic ketones depending on the conditions

---

[339] S. K. Vidyarthi, C. Willis, R. A. Back, and R. M. McKitrick, *J. Amer. Chem. Soc.* 1974, **96**, 7647.
[340] C. P. Forbes, A. Goosen, and H. A. H. Laue, *J.C.S. Perkin I*, 1974, 2346.
[341] G. A. Olah and T. R. Hockswender, jun., *J. Org. Chem.*, 1974, **39**, 3478.
[342] A. Hohorst, J. V. Paukstelis, and D. D. DesMarteau, *J. Org. Chem.*, 1974, **39**, 1298.
[343] O. F. Zeck, Y. Y. Su, G. P. Gennaro, and Y.-N. Tang, *J. Amer. Chem. Soc.*, 1974, **96**, 5967.
[344] R. G. Salomon, *J. Org. Chem.*, 1974, **39**, 3602.

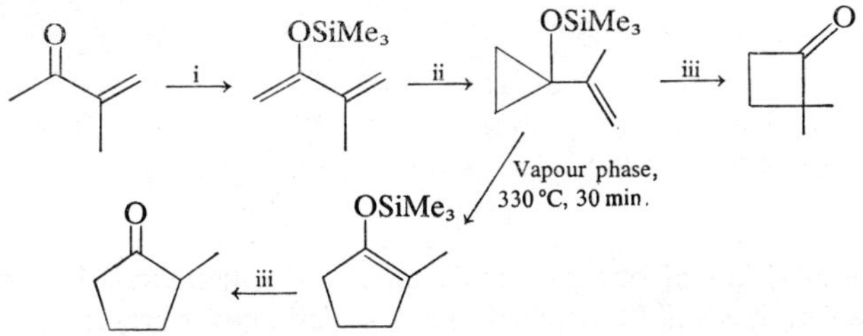

Acetylenes, Alkanes, Allenes, and Olefins

Reagents: i, LiNPr$^i_2$–Me$_3$SiCl–THF:H$_2$O(1:1); ii, CH$_2$I$_2$–Zn/Ag–Et$_2$O; iii, H$^+$

Scheme 65

used (Scheme 65).[345,346] The vinylsilanes (210)—(212) have been used as latent functionalities for HCOCH$_2$CH$_2$X, CH$_3$COCH$_2$X, and RCH$_2$COCH$_2$-CH$_2$X.[347] In contrast to most vinylsilanes, the β-hydroxyvinylsilanes are readily cleaved by fluoride ion in DMSO or acetonitrile.[348]

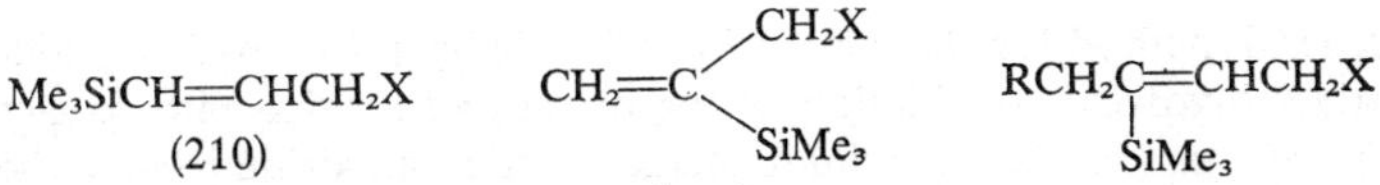

Me$_3$SiCH=CHCH$_2$X (210)

CH$_2$=C(CH$_2$X)(SiMe$_3$) (211)

RCH$_2$C(SiMe$_3$)=CHCH$_2$X (212)

Two variations on the Robinson annelation reaction using vinylsilanes have been described. The precursor (213) is employed in one, as shown in Scheme 66. The other utilizes α-trimethylsilyl vinyl ketones (214) as a Michael acceptor with regiospecifically generated enolates.[349,350]

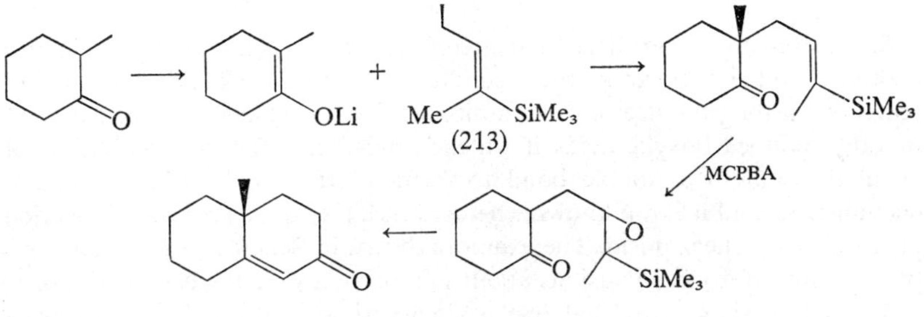

Scheme 66

---

[345] C. Girard and J. M. Conia, *Tetrahedron Letters*, 1974, 3327.
[346] C. Girard, P. Amice, J. P. Barnier, and J. M. Conia, *Tetrahedron Letters*, 1974, 3329.
[347] G. Stork, M. E. Jung, E. Colvin, and Y. Noel, *J. Amer. Chem. Soc.*, 1974, **96**, 3685.
[348] T. H. Chan and W. Mychajlowskij, *Tetrahedron Letters*, 1974, 3479.
[349] G. Stork and M. E. Jung, *J Amer. Chem. Soc.*, 1974, **96**, 3683.
[350] R. K. Boeckman, jun., *J. Amer. Chem. Soc.*, 1974, **96**, 6179; G. Stork and J. Singh, *ibid.*, p. 6181.

(214)

Michael addition of phenylselenide anion to α-methylene lactones leads to a convenient method for protection of the olefin group; regeneration of the double bond is *via* the selenoxide.[351] *trans*-1,4,9-Decatriene undergoes thermal and base cyclization to 5-methyl-3a,4,5,7a-tetrahydroindane (215). The cyclization probably occurs *via* formation of 1,6,8-decatriene as an intermediate.[352,353]

(215)

A general method for the conversion of α-olefins into α-hydroxy-aldehyde derivatives involves co-oxidation of the olefin and a thiol with oxygen and subsequent treatment with sodium acetate–acetic anhydride to give the protected aldehyde (216).[354]

(216)

Short-lived anion radicals of activated olefins can be generated at a mercury cathode, and under anhydrous conditions undergo electrocarboxylation reactions in the presence of carbon dioxide. The products are succinic acids or only monocarboxylic acids if water is present.[355] The cycloaddition of an allyl anion to a double bond to form a cyclopentyl anion is seldom encountered, and it is not known whether it is a $[_\pi 2_s + _\pi 4_s]$ concerted reaction or a two-stage mechanism. The example shown in Scheme 67 illustrates this process but no firm conclusions about the mechanism have been reached.[356]

The solvolysis of vinyl halides to vinyl cations still continues to attract attention. Amongst several such papers must be mentioned the first preparation of a stable vinyl cation (Scheme 68) and the stabilizing effect of the

---

[351] P. A. Grieco and M. Miyashita, *Tetrahedron Letters*, 1974, 1869.
[352] A. N. Sagredos, *Annalen*, 1974, 176.
[353] R. Moser and A. N. Sagredos, *Annalen*, 1974, 1028.
[354] S. Iriuchijima, K. Maniwa, T. Sakakibara, and G. Tsuchihashi, *J. Org. Chem.*, 1974, 39, 1170.
[355] D. A. Tyssee and M. M. Baizer, *J. Org. Chem.*, 1974, 39, 2819.
[356] G. W. Klumpp and R. F. Schmitz, *Tetrahedron Letters*, 1974, 2911.

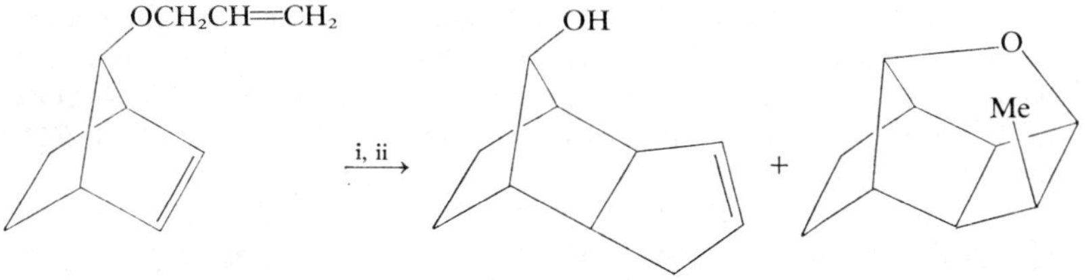

Scheme 67

Reagents: i, MeLi; ii, H$_2$O

Scheme 67

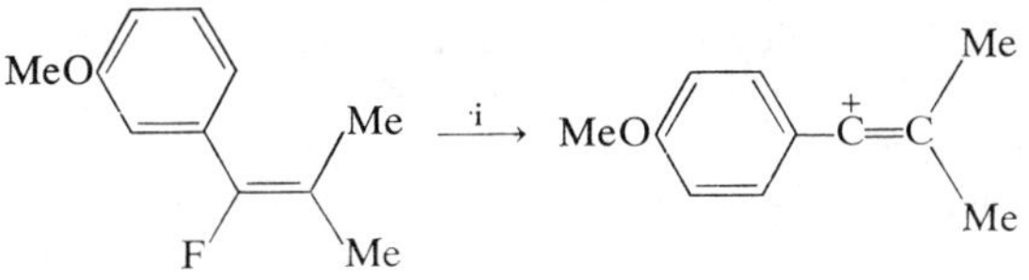

Scheme 68

Reagent: i, SbF$_5$–SO$_2$ClF

Scheme 68

ferrocene group on such cations.[357,358] *ortho*-Substituted arylvinyl bromides and tosylates solvolyse *via* vinyl cations up to 4 × 10^5 times faster than the parent compound in acetic acid. The twisting of the aryl ring means that much less energy is needed to break the conjugation and form the vinyl cation.[359] The stereochemistry of solvolytic displacements and intramolecular nucleophilic substitutions by a remote double bond at a vinyl cation centre has been investigated.[360]

Theoretical and Spectral Properties of Olefins.—MINDO/3 calculations on the Diels–Alder cycloaddition reaction between buta-1,3-diene and ethylene indicate that the transition state is very unsymmetrical, implying a high degree of radical character. Similar results have been calculated for the reactions of cyclobutadiene with ethylene and acetylene. It has been concluded that these typical Diels–Alder reactions cannot be classed as concerted pericyclic processes.[361] The minimum-energy conformations of all possible methyl-, isopropyl-, and t-butyl-substituted ethylenes, as well as of some mixed methyl-t-butyl-ethylenes, have been calculated using a consistent force field derived by least-squares fitting of a large set of observed conformational, vibrational, and thermochemical olefin properties.[362]

---

[357] H.-U. Siehl, J. C. Carnahan, jun., L. Eckes, and M. Hanack, *Angew. Chem. Internat. Edn.*, 1974, **13**, 675.
[358] D. Kaufmann and R. Kupper, *J. Org. Chem.*, 1974, **39**, 1438; see also ref. 84.
[359] K. Yates and J.-J. Périé, *J. Org. Chem.*, 1974, **39**, 1902.
[360] T. C. Clarke and R. G. Bergman, *J. Amer. Chem. Soc.*, 1974, **96**, 7934.
[361] M. J. S. Dewar, A. C. Griffin, and S. Kirschner, *J. Amer. Chem. Soc.*, 1974, **96**, 6225.
[362] O. Ermer and S. Lifson, *Tetrahedron*, 1974, **30**, 2425.

Photoelectron spectra of iodoethylenes show that spin–orbit interactions cannot be neglected in MO models for molecules with heavy atoms.[363] Spectra of mono-, di-, and tetra-substituted ethylenes with —SMe, —OMe, and —NMe$_2$ substituents furnish information on the n–π conjugation.[364] Similar results for halogen substituents have also been reported.[365]

The ionization potentials of alkenes, which can be measured precisely and rapidly, are often a linear function of other characteristic parameters. They can be used to predict such properties as stability constants of alkene–metal complexes, bond-localization and excitation energies of π-complexes, and stability constants of charge-transfer complexes.[366]

E.s.r. studies have included the investigation of the cyclization of the 5-hexenyl radical to the cyclopentylmethyl radical,[367] and the formation of [RH]$\bar{\cdot}$ ion radicals by electron transfer of unsaturated hydrocarbons on rare-earth zeolites.[368]

The c.d. spectra of olefins are not subject to notable solvent shifts due to solvent density, contrary to a recent report of effects in perfluoro-alcohols. This should be interpreted with great caution, as olefins are readily isomerized in these systems.[369] Experimental evidence is available for the role of π-bond torsion in the c.d. spectra of olefins.[370]

---

[363] K. Wittel, H. Bock, and R. Manne, *Tetrahedron*, 1974, **30**, 651.
[364] H. Bock, G. Wagner, K. Wittel, J. Sauer, and D. Seebach, *Chem. Ber.*, 1974, **107**, 1869.
[365] K. Wittel and H. Bock, *Chem. Ber.*, 1974, **107**, 317.
[366] D. Grosjean, P. Masclet, and G. Monvier, *Bull. Soc. chim. France*, 1974, 567.
[367] D. Lal, D. Griller, S. Husband, and K. U. Ingold, *J. Amer. Chem. Soc.*, 1974, **96**, 6355.
[368] S. Krzyzanowski, *J.C.S. Chem. Comm.*, 1974, 1036.
[369] N. H. Andersen and Y. Ohta, *J.C.S. Chem. Comm.*, 1974, 730.
[370] N. H. Andersen, C. R. Costin, and J. R. Shaw, *J. Amer. Chem. Soc.*, 1974, **96**, 3693

2
Functional Groups other than Alkanes, Acetylenes, Allenes, and Olefins

BY E. W. COLVIN

1 Carboxylic Acids

Preparation.—Meyers[1] has extended the utility of oxazines and oxazolines in the preparation of homologous acetic acids and derivatives. The chiral oxazoline (1), readily prepared[2] from the (−)-diol (2) of known absolute configuration, leads to (S)-2-methylalkanoic acids (Scheme 1); the optical yields are very high, and the diol can be recovered and recycled. Use of the enantiomeric (+)-diol yields (R)-acids.

This method has been modified to provide an asymmetric synthesis[3] of dialkylacetic acids (Scheme 2); initial introduction of the group R^1 with the lower configurational priority always gives the (S)-acid, and *vice versa*. Other variants include routes to chiral 3-alkyl-,[4] 2-chloro-,[5] 2-methoxy- and 2-hydroxy-alkanoic[6] acids, the last providing the first successful application of *aliphatic* aldehydes to such a purpose.

The chloromethyloxazine (3) functions[7] as a common precursor for 2-chloro-, arylacetic, and αβ-unsaturated acids.

Pursuing his studies on α-metallated isocyanides, Schöllkopf[8] has described the conversion of aldehydes into homologous acetic acids by reaction with the isocyanomethylphosphonate (4) (Scheme 3); the lithium salt of the phosphonate is required for the analogous reaction with ketones.

Substituted acetic acids can be obtained from ketones *via* glycidonitriles,[9] as shown in Scheme 4.

---

[1] A. I. Meyers, D. L. Temple, R. L. Nolan, and E. D. Mihelich, *J. Org. Chem.*, 1974, **39**, 2778; A. I. Meyers, E. D. Mihelich, and K. Kamata, *J.C.S. Chem. Comm.*, 1974, 768.
[2] A. I. Meyers, G. Knaus, and K. Kamata, *J. Amer. Chem. Soc.*, 1974, **96**, 268.
[3] A. I. Meyers and G. Knaus, *J. Amer. Chem. Soc.*, 1974, **96**, 6508.
[4] A. I. Meyers and K. Kamata, *J. Org. Chem.*, 1974, **39**, 1603.
[5] A. I. Meyers, G. Knaus, and P. M. Kendall, *Tetrahedron Letters*, 1974, 3495.
[6] A. I. Meyers and G. Knaus, *Tetrahedron Letters*, 1974, 1333.
[7] G. R. Malone and A. I. Meyers, *J. Org. Chem.*, 1974, **39**, 618, 623.
[8] U. Schöllkopf, R. Schröder, and D. Stafforst, *Annalen*, 1974, 44; *cf.* U. Schöllkopf, R. Schröder, and E. Blume, *ibid.*, 1972, **766**, 130.
[9] D. R. White and D. K. Wu, *J.C.S. Chem. Comm.*, 1974, 988.

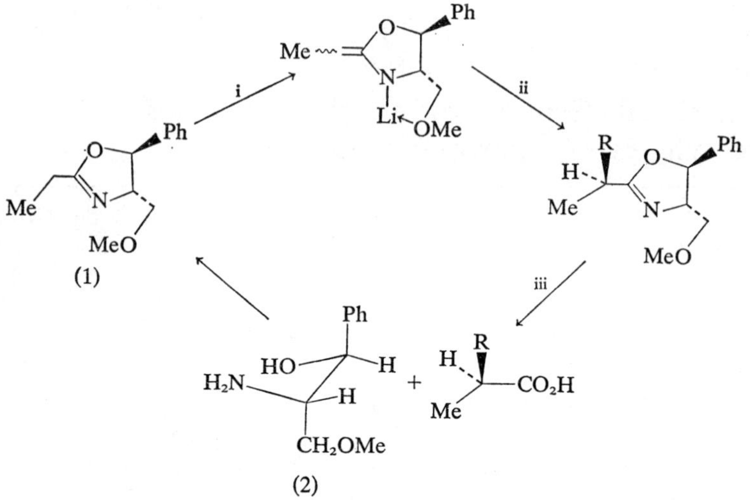

Reagents: i, LiNPr$^i_2$; ii, RI; iii, H$_3$O$^+$

Scheme 1

The anion of NN-di-isopropylformamide[10] reacts with aldehydes and ketones to provide a useful synthesis of α-hydroxy-acids (Scheme 5) which is of particular value in those cases where cyanohydrin formation is difficult; the use of esters as electrophiles leads to α-keto-acids.

Reaction of polymer-supported esters such as (5) with Grignard reagents affords[11] optically active α-hydroxy-acids (Scheme 6); the polymer can be used repeatedly.

An extremely simple yet effective method for the direct carboxylation of various active methylene compounds has been described,[12] and is illustrated in Scheme 7.

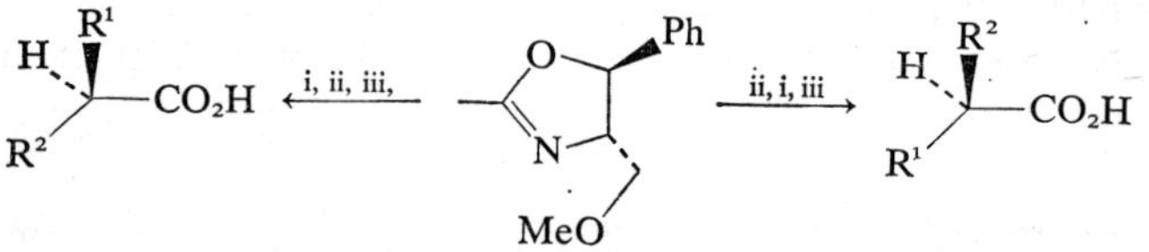

Reagents: i, LiNPr$^i_2$–R^{1}X; ii, LiNPr$^i_2$–R^{2}X; iii, H$_3$O$^+$

Scheme 2

[10] R. R. Fraser and P. R. Hubert, *Canad. J. Chem.*, 1974, **52**, 185.
[11] M. Kawana and S. Emoto, *Bull. Chem. Soc. Japan.*, 1974, **47**, 160.
[12] E. Haruki, M. Arakawa, N. Matsumura, Y. Otsuji, and E. Imoto, *Chem. Letters*, 1974, **427**; see also Y. Otsuji, M. Arakawa, N. Matsumura, and E. Haruki, *ibid.*, 1973, 1193; T. Ito and Y. Takami, *ibid.*, 1974, 1035.

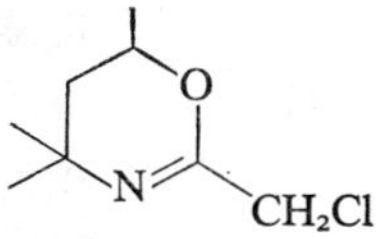

Functional Groups other than Alkanes, Acetylenes, Allenes, and Olefins 83

(3)

(EtO)$_2$P(O)CH$_2$N̈=C: + RCHO —i→ RCH—CH(N=CH—)—P(O)(OEt)$_2$

(4)

↓ ii, iii

RCH$_2$CO$_2$H ←iv— RCH=C(NHCHO)(P(O)(OEt)$_2$)

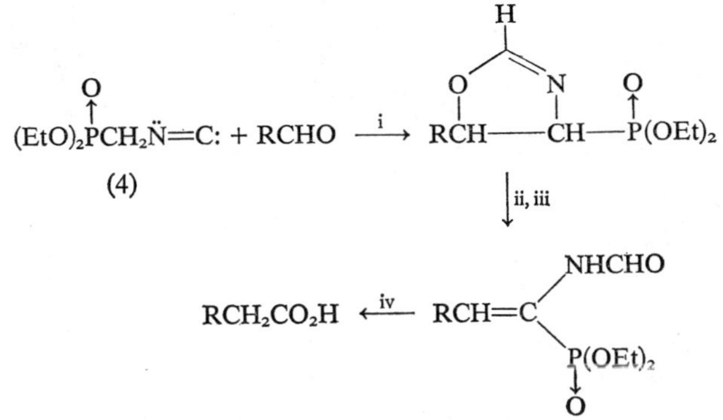

Reagents: i, NaCN(catalytic)–EtOH; ii, KOBu^t–THF; iii, H$^+$; iv, H$_3$O$^+$

Scheme 3

Alkylidenephosphoranes react[13] with CO_2 to give, after treatment with hydroxide to induce loss of triphenylphosphine oxide, carboxylic acids in excellent yields (Scheme 8).

α-Alkoxycarbonylation of carboxylic acid dianions has been presented[14] as a simple route to monoesters of malonic acids; this complements the previously reported α-carboxylation[15] of ester dianions. A new procedure[16]

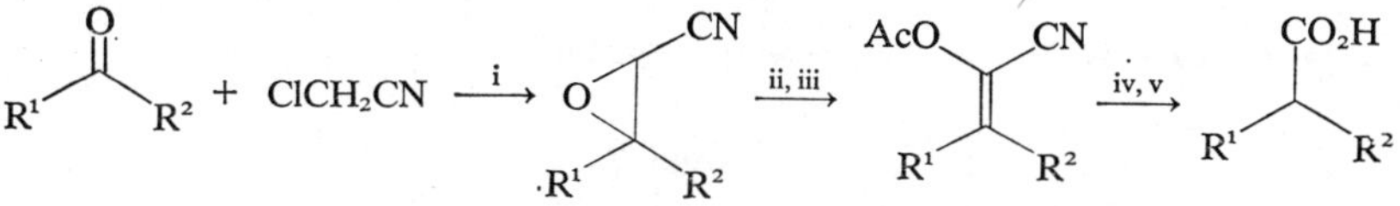

R^{1}C(O)R^2 + ClCH$_2$CN —i→ epoxide(CN, R^1, R^2) —ii, iii→ (AcO)(CN)C=C(R^1)(R^2) —iv, v→ R^{1}CH(CO$_2$H)R^2

Reagents: i, NaOAm^t; ii, HCl gas; iii, Ac$_2$O–py–Et$_3$N; iv, HO$^-$; v, H$_3$O$^+$

Scheme 4

[13] H. J. Bestmann, Th. Denzel, and H. Salbaum, *Tetrahedron Letters*, 1974, 1275.
[14] A. P. Krapcho, E. G. E. Jahngen, and D. S. Kashdan, *Tetrahedron Letters*, 1974, 2721.
[15] S. Reiffers, J. Strating, and H. Wynberg, *Tetrahedron Letters*, 1971, 2339.
[16] A. Szabolics, J. Szammer, and L. Noszkó, *Tetrahedron*, 1974, **30**, 3647.

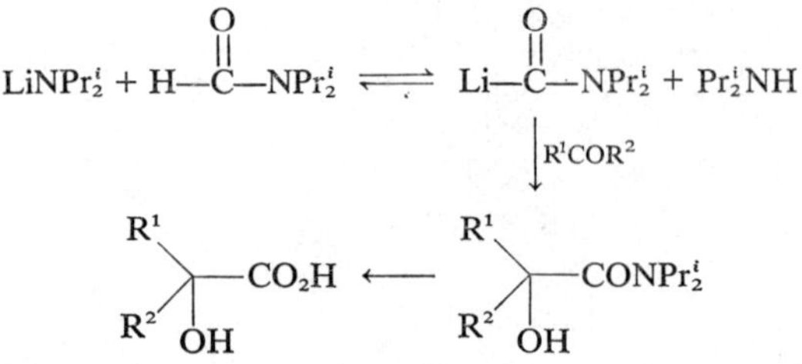

Scheme 5

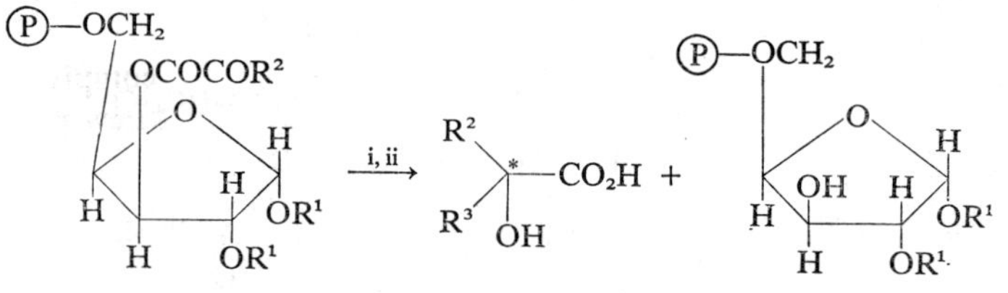

(5) $R^2 = Me$ or Ph

Reagents: i, R^3MgX; ii, HO^-

Scheme 6

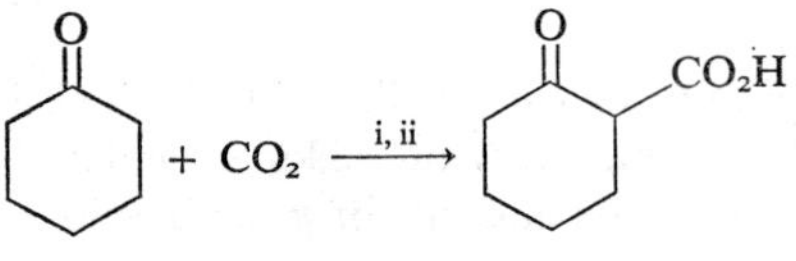

63%

Reagents: i, 1,8-diazabicyclo[5,4,0]undecene; ii, H_3O^+–0 °C

Scheme 7

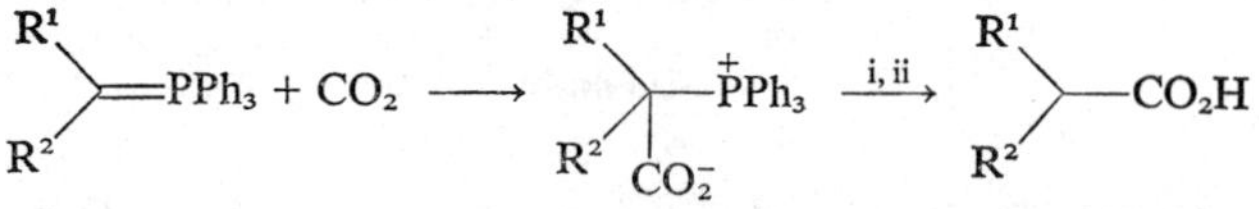

Reagents: i, HO^-–H_2O(–Ph_3PO); ii, H_3O^+

Scheme 8

Functional Groups other than Alkanes, Acetylenes, Allenes, and Olefins 85

$$R_2CHCO_2^-M^+ \xrightleftharpoons{^{14}CO_2} R_2CH^{14}CO_2^-M^+$$

Scheme 9

for the preparation of carboxyl-labelled aliphatic acids by an exchange reaction with labelled CO_2 (Scheme 9) presumably proceeds *via* a malonate intermediate. Zinc dust reduction of bromoacetic acid in D_2O affords [$^2H_1$]acetic acid with greater than 99% isotopic purity.[17]

A one-step synthesis[18] of $\alpha\beta$-unsaturated acids from ketones or aldehydes can be achieved (Scheme 10) *via* the phosphonate (6).

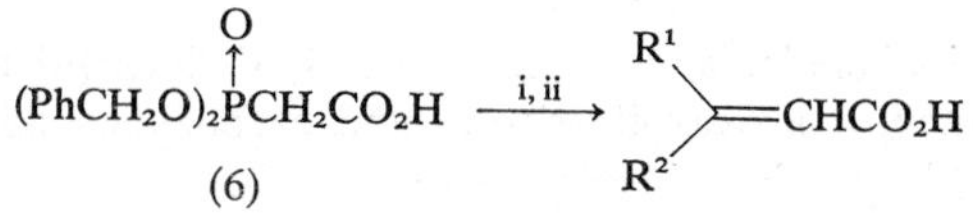

$$(PhCH_2O)_2\overset{\overset{O}{\uparrow}}{P}CH_2CO_2H \xrightarrow{i, ii} \underset{R^2}{\overset{R^1}{>}}{=}CHCO_2H$$

(6)

Reagents: i, 2LiNPr$^i_2$; ii, R^1COR^2

Scheme 10

Methods[19,20] have been described for the syntheses of dioxocarboxylic acids of the type (7); an improved route[21] to β-ketoadipic acid has been reported. The absolute configuration of a number of naturally occurring 2-alkylmalic acids has been shown to be (R), by asymmetric synthesis[22]

$$MeCOCH_2CO(CH_2)_n CO_2H$$

(7)

of the (S)-enantiomers. The sex pheromone of the female furniture carpe beetle, *Anthrenus flavipes* (Le Conte), has been synthesized[23] and identified as (Z)-dec-3-enoic acid; the beetle reputedly gained entry to the U.S.A. in 1911 on furniture stuffed with Russian horse hair.

Protection.—The 9-anthrylmethyl system (8) affords protection[24] to carboxylic acids under a wide variety of conditions; methanethiolate anion effects rapid cleavage[25] (Scheme 11). The method can also be used for the protection

---

[17] D. A. Robinson, *J.C.S. Chem. Comm.*, 1974, 345.
[18] G. A. Koppel and M. D. Kinnick, *Tetrahedron Letters*, 1974, 711.
[19] R. O. Pendarvis and K. G. Hampton, *J. Org. Chem.*, 1974, **39**, 2289.
[20] C. Kashima, N. Mukai, and Y. Tsuda, *Chem. Letters*, 1973, 539.
[21] B. J. Whitlock and H. W. Whitlock, *J. Org. Chem.*, 1974, **39**, 3144.
[22] S. Brandänge, S. Josephson, and S. Vallén, *Acta Chem. Scand.*, 1974, **B28**, 153.
[23] H. Fukui, F. Matsumura, M. C. Ma, and W. E. Burkholder, *Tetrahedron Letters*, 1974, 3563.
[24] N. Kornblum and A. Scott, *J. Amer. Chem. Soc.*, 1974, **96**, 590.
[25] C. W. Jaeger and N. Kornblum, *J. Amer. Chem. Soc.*, 1972, **94**, 2545.

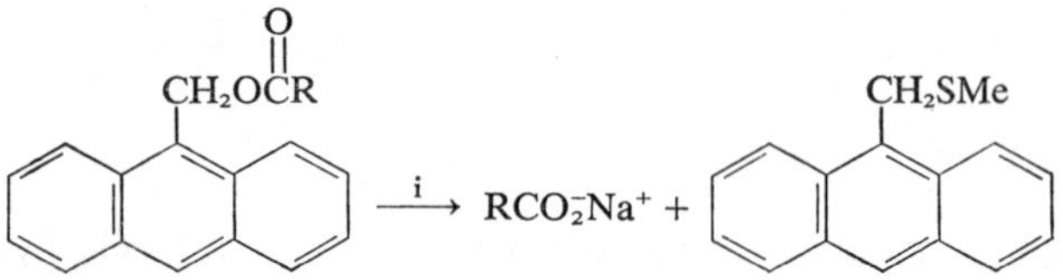

[Structure: 9-(acyloxymethyl)anthracene (CH₂OCR with C=O) → RCO₂⁻Na⁺ + 9-(methylthiomethyl)anthracene (CH₂SMe)]

(8)

Reagent: i, Na⁺ ⁻SMe–HMPA, 25 °C, 30 s, or Na⁺ ⁻SMe–DMF, −20 °C, 1 h

Scheme 11

of phenols, thiophenols, and alkanethiols. Full descriptive details[26] have been given of the use of oxazolines to protect carboxylic acids against attack by Grignard and hydride reducing agents.

Properties.—From measurements of gas-phase equilibria, the intrinsic acidities of α-, β-, and γ-chloro-substituted aliphatic acids have been evaluated;[27] the effects of the chloro substituent parallel those in solution, but they are much larger. The attenuation in solution is attributed to weaker hydrogen-bonding of the chloro-stabilized acid anions to water molecules. Related studies have been carried out on substituent effects on the intrinsic acidities of benzoic acids,[28] and on the intrinsic acidities of carbon acids.[29] Some of the techniques and results of gas-phase acidity and basicity determinations have been reviewed.[30]

Derivatives and Reactions.—Convenient methods[31] have been described for the α-chlorination and α-iodination of carboxylic acid chlorides. Propiolyl chloride is readily prepared[32] from propiolic acid and phosphorus pentachloride. Imidazole functions[33] as a recyclable catalyst in the phosgenation of lauric acid to lauroyl chloride. Diazotization of α-amino-acids in poly(hydrogen fluoride)–pyridine leads to α-fluorocarboxylic acids[34] in good yield; with alkyl carbamates as substrates, alkyl fluoroformates[35] are produced (Scheme 12).

---

[26] A. I. Meyers, D. L. Temple, D. Haidukewych, and E. D. Mihelich, *J. Org. Chem.*, 1974, **39**, 2787.
[27] R. Yamdagni and P. Kebarle, *Canad. J. Chem.*, 1974, **52**, 861.
[28] R. Yamdagni, T. B. McMahon, and P. Kebarle, *J. Amer. Chem. Soc.*, 1974, **96**, 4035.
[29] T. B. McMahon and P. Kebarle, *J. Amer. Chem. Soc.*, 1974, **96**, 5940.
[30] C. Agami, *Bull. Soc. chim. France*, 1974, 869.
[31] D. N. Harpp, L. Q. Bao, C. J. Black, R. A. Smith, and J. Gleason, *Tetrahedron Letters*, 1974, 3235.
[32] W. J. Balfour, C. C. Greig, and S. Visaisouk, *J. Org. Chem.*, 1974, **39**, 725.
[33] C. F. Hauser and L. F. Theiling, *J. Org. Chem.*, 1974, **39**, 1134.
[34] G. A. Olah and J. Welch, *Synthesis*, 1974, 652.
[35] G. A. Olah and J. Welch, *Synthesis*, 1974, 654.

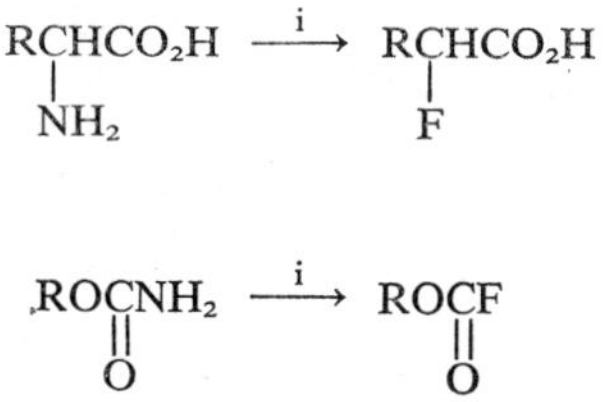

Functional Groups other than Alkanes, Acetylenes, Allenes and Olefins

$$\begin{array}{c} \text{RCHCO}_2\text{H} \\ | \\ \text{NH}_2 \end{array} \xrightarrow{i} \begin{array}{c} \text{RCHCO}_2\text{H} \\ | \\ \text{F} \end{array}$$

$$\begin{array}{c} \text{ROCNH}_2 \\ \| \\ \text{O} \end{array} \xrightarrow{i} \begin{array}{c} \text{ROCF} \\ \| \\ \text{O} \end{array}$$

Reagent: i, $NaNO_2-(HF)_xF^-$ pyH$^+$

Scheme 12

Allyl difluorochlorovinyl ethers (9) undergo a facile [3,3]sigmatropic rearrangement[36] to give $\gamma\delta$-unsaturated acid fluorides, and thence specifically fluorinated carboxylic acids (Scheme 13).

Acetyl hypobromite (10) has been isolated (Scheme 14) as a pale yellow deliquescent low-melting solid.[37]

The decarbonylation of acid chlorides with chlorotris(triphenylphosphine)-rhodium(I) has been subjected to further study;[38] α-chiral acid chlorides with no β-hydrogen atoms are decarbonylated to racemic alkyl chlorides, whereas substrates with β-hydrogen atoms afford Saytzeff-type olefins with one less carbon atom. Thermodynamic parameters suggest an intramolecular free radical reaction, analogous to the related decarbonylation of aldehydes, although the latter reaction proceeds with a high degree[39] of configurational retention.

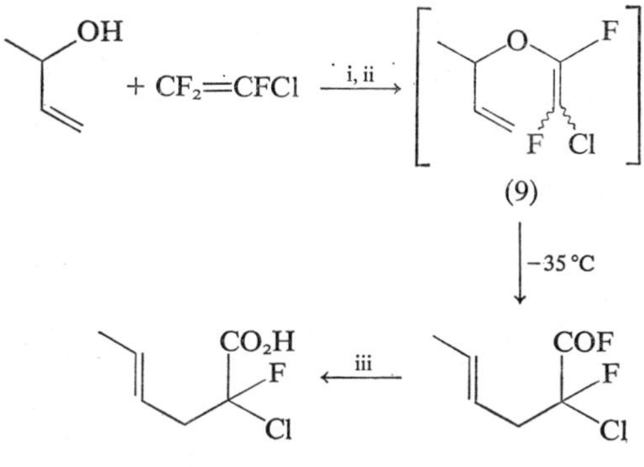

Reagents: i, Na(catalytic); ii, Bu^{n}Li, $-70\,°C$; iii, H$_2$O

Scheme 13

[36] J. F. Normant, O. Reboul, R. Sauvêtre, H. Deshayes, D. Masure, and J. Villieras, *Bull. Soc. chim. France*, 1974, 2072.
[37] J. J. Reilly, D. J. Duncan, T. P. Wunz, and R. A. Patsiga, *J. Org. Chem.*, 1974, **39**, 3291.
[38] J. K. Stille and M. T. Regan, *J. Amer. Chem. Soc.*, 1974, **96**, 1508; J. K. Stille and R. W. Fries, *ibid.*, p. 1514; J. K. Stille, F. Huang, and M. T. Regan, *ibid.*, p. 1518.
[39] H. M. Walborsky and L. E. Allen, *J. Amer. Chem. Soc.*, 1971, **93**, 5465.

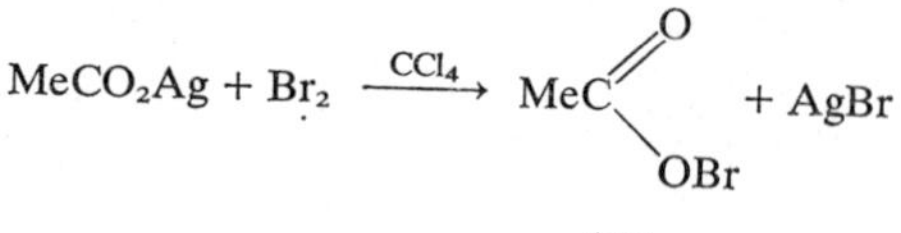

$$\text{MeCO}_2\text{Ag} + \text{Br}_2 \xrightarrow{\text{CCl}_4} \text{MeC(=O)OBr} + \text{AgBr}$$

(10)

Scheme 14

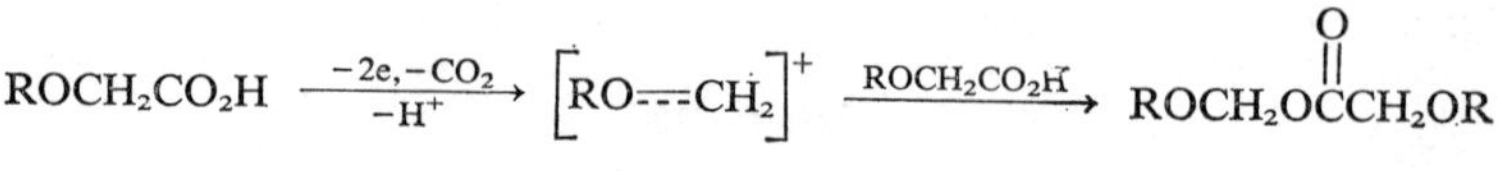

$$\text{ROCH}_2\text{CO}_2\text{H} \xrightarrow[-\text{H}^+]{-2e, -\text{CO}_2} [\text{RO}\text{---}\text{CH}_2]^+ \xrightarrow{\text{ROCH}_2\text{CO}_2\text{H}} \text{ROCH}_2\text{OCCH}_2\text{OR}$$

Scheme 15

Trimethylaluminium exhaustively methylates[40] carboxy-groups to t-butyl units. The anodic oxidation of α-alkoxy aliphatic acids in anhydrous acetonitrile produces acylals[41] in good yield (Scheme 15).

Peracids.—A number of new peracids have been prepared, and their epoxidizing abilities delineated. O-Benzylmonoperoxycarbonic acid (11) is quite stable in the cold, and shows[42] reactivity intermediate between those of m-chloroperbenzoic acid and perbenzoic acid; the latter has now been

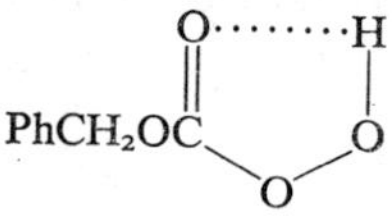

(11) 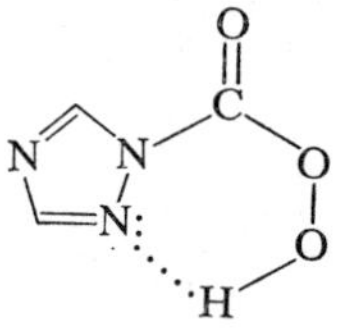   (12)

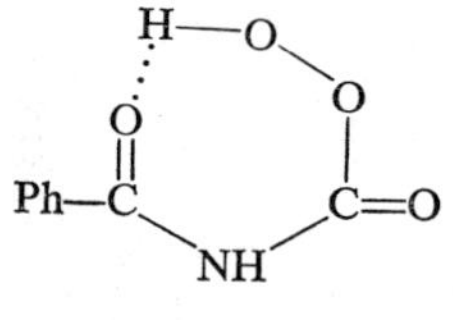

(13) 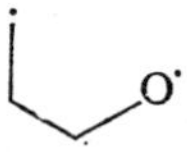   (14)

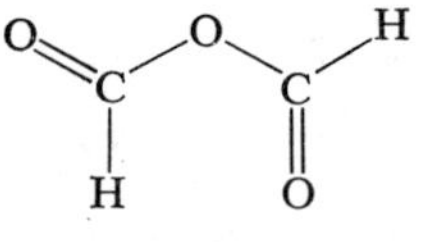

(15)

[40] A. Meisters and T. Mole, *Austral. J. Chem.*, 1974, **27**, 1665; see also refs. 512, 561.
[41] H. G. Thomas and E. Katzer, *Tetrahedron Letters*, 1974, 887.
[42] R. M. Coates and J. W. Williams, *J. Org. Chem.*, 1974, **39**, 3054.

prepared on a polymer support.[43] The peroxycarbamic acids (12) and (13) have been generated *in situ* and (13) has been isolated and crystallized;[44] both efficiently convert olefins into epoxides, with concurrent production of innocuous by-products.

Evidence has been presented[45] for the intervention of the 1,4-biradical 1-oxatetramethylene (14) in the thermodecarboxylation of γ-peroxylactones.

2 Carboxylic Acid Anhydrides

A dynamic n.m.r. spectroscopic investigation[46] of formic anhydride has revealed the (*EZ*)-configuration (15) in solution, in accord with gas-phase electron-diffraction data; the low barrier (18 ± 1 kJ mol^{-1}) for topomerization is taken as further evidence for the diminished importance of resonance in anhydrides in comparison to esters and imides. The conformations of some *gem*-dimethyl-substituted cyclic anhydrides have been determined.[47]

The acylating abilities of some polymer-supported mixed carbonic–carboxylic anhydrides have been investigated.[48]

3 Lactones

Preparation.—The most significant advance in this area has been achieved on the challenging problem of formation of macrocyclic lactones from long-chain hydroxy-acids. Corey[49] has reported that thiopyridone esters (16), on heating in xylene under conditions of high dilution, undergo an electrostatically-driven cyclization to afford the desired lactones in high yield (Scheme 16). This method appears to be generally applicable to a wide range[50] of hydroxy-acid substrates. A related silver-ion-induced process has been described.[51]

Free-radical oxidation[52] of 4-phenylbutyric acid leads in high yield to 4-phenyl-γ-butyrolactone. Full details have been given[53] of the preparation of γ-lactones by free-radical oxidation of olefins with metal carboxylates. The use of thallium(I) salts of $\Delta^{3,4}$-unsaturated acids appears to be the method of choice[54] for formation of iodolactones, especially where the latter are unstable or are the products of kinetic control. Product-selective electrolytic

---

[43] C. R. Harrison and P. Hodge, *J.C.S. Chem. Comm.*, 1974, 1009.
[44] J. Rebek, S. F. Wolf, and A. B. Mossman, *J.C.S. Chem. Comm.*, 1974, 711.
[45] W. Adam and L. M. Szendrey, *J. Amer. Chem. Soc.*, 1974, **96**, 7135; see also W. Adam, *Angew. Chem. Internat. Edn.*, 1974, **13**, 619.
[46] E. A. Noe and M. Raban, *J.C.S. Chem. Comm.*, 1974, 479.
[47] G. Borgen, *Acta Chem. Scand.*, 1974, **B28**, 13.
[48] M. B. Shambu and G. A. Digenis, *J.C.S. Chem. Comm.*, 1974, 619.
[49] E. J. Corey and K. C. Nicolaou, *J. Amer. Chem. Soc.*, 1974, **96**, 5614.
[50] E. J. Corey, K. C. Nicolaou, and L. S. Melvin, *J. Amer. Chem. Soc.*, 1975, **97**, 653, 654.
[51] H. Gerlach and A. Thalmann, *Helv. Chim. Acta*, 1974, **57**, 2661.
[52] A. Clerici, F. Minisci, and O. Porta, *Tetrahedron Letters*, 1974, 4183.
[53] E. I. Heiba, R. M. Dessau, and P. G. Rodewald, *J. Amer. Chem. Soc.*, 1974, **96**, 7977.
[54] R. C. Cambie, R. C. Hayward, J. L. Roberts, and P. S. Rutledge, *J.C.S. Perkin I*, 1974, 1120, 1858, 1864.

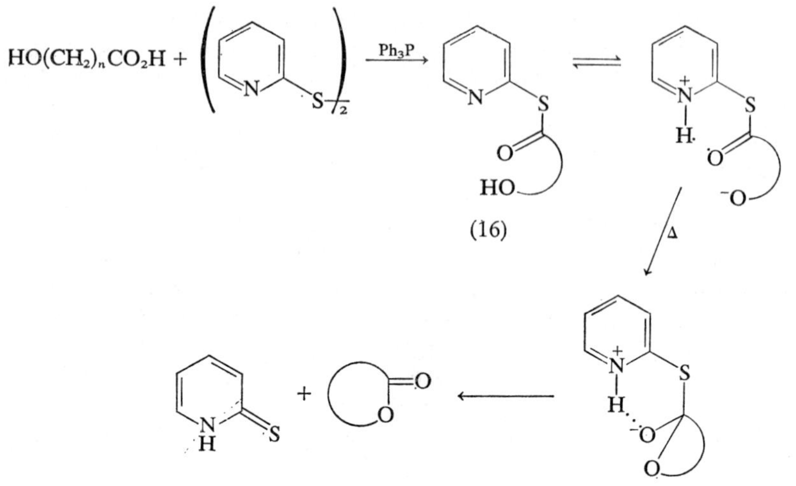

Scheme 16

decarboxylation[55] of γ-substituted paraconic acid salts (17) leads to either γ-substituted γ-butyrolactones or γ-substituted $\Delta^{2,3}$-butenolides (Scheme 17).

Lithio-oxazolines react[56] with epoxides at low temperatures to give, after hydrolysis of the initially formed alkylated intermediates, a variety of substituted γ-lactones (Scheme 18).

Diels–Alder adducts have been used to provide reversible protection for suitable unsaturated substrates; this procedure has been used for the preparation of specifically deuteriated γ-butenolides,[57] and constitutes an improved route[58] to the simplest butenolide, but-2-en-4-olide (Scheme 19).

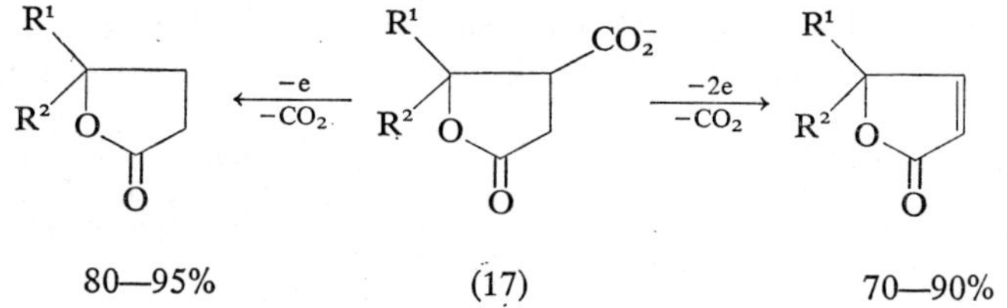

Scheme 17

[55] S. Torii, T. Okamoto, and H. Tanaka, *J. Org. Chem.*, 1974, **39**, 2486.
[56] A. I. Meyers, E. D. Mihelich, and R. L. Nolen, *J. Org. Chem.*, 1974, **39**, 2783.
[57] M. Golfier and T. Prangé, *Bull. Soc. chim. France*, 1974, 1158.
[58] S. Takano and K. Ogasawara, *Synthesis*, 1974, 42.

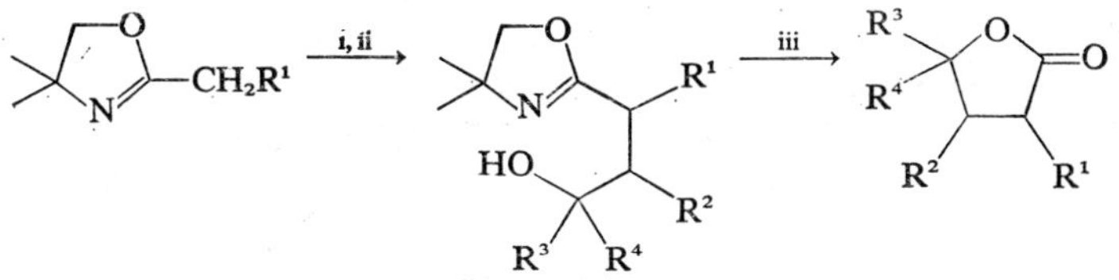

Scheme 18

Reagents: i, Bu^{n}Li; ii, R^{2}CH——CR^{3}R^4 (with O bridging); iii, H$_3$O$^+$

Scheme 19

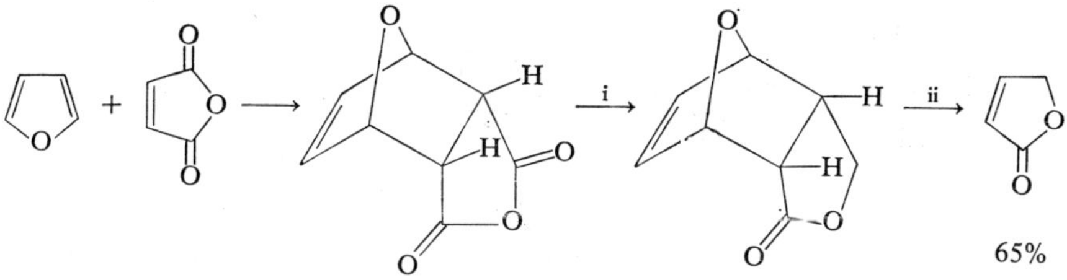

Reagents: i, NaBH$_4$–EtOH, ii, 140—150 °C, reduced pressure

Lewis-acid-catalysed addition of aldehydes to enol γ-lactones leads efficiently[59] to substituted γ-butyrolactones (Scheme 20), and this methodology has been applied in a synthesis[60] of botyrodiplodin.

The readily accessible sulphoxides (18) provide routes[61] to a variety of substituted Δ^{2,3}-butenolides, as shown in Scheme 21.

Cyclopropylcarbinol solvolysis has been used[62] in a route to cycloheptane-fused γ-butyrolactones. Aliphatic α-diazoketones react with dimethylketen to give Δ^{3,4}-butenolides in good yield (Scheme 22); aromatic substrates are less successful.[63]

Various γ-lactones bearing vinylic substituents in the β position can be obtained[64] directly by thermal acid-catalysed condensation of *trans*-alk-2-ene-1,4-diols with orthocarboxylic acid esters (Scheme 23).

---

[59] T. Mukaiyama, J. Hanna, T. Inoue, and T. Sato, *Chem. Letters*, 1974, 381; see also ref. 452.
[60] T. Mukaiyama, M. Wada, and J. Hanna, *Chem. Letters*, 1974, 1181.
[61] K. Iwai, M. Kawai, H. Kosugi, and H. Uda, *Chem. Letters*, 1974, 385; K. Iwai, H. Kosugi, and H. Uda, *ibid.*, p. 1237.
[62] J. A. Marshall, F. N. Tuller, and R. Ellison, *Synthetic Comm.*, 1973, 3, 465; *cf.* P. F. Hudrlik, L. N. Rudnick, and S. H. Korzeniowski, *J. Amer. Chem. Soc.*, 1973, 95, 6848.
[63] W. Ried and R. Kraemer, *Annalen*, 1973, 1952.
[64] K. Kondo and F. Mori, *Chem. Letters*, 1974, 741.

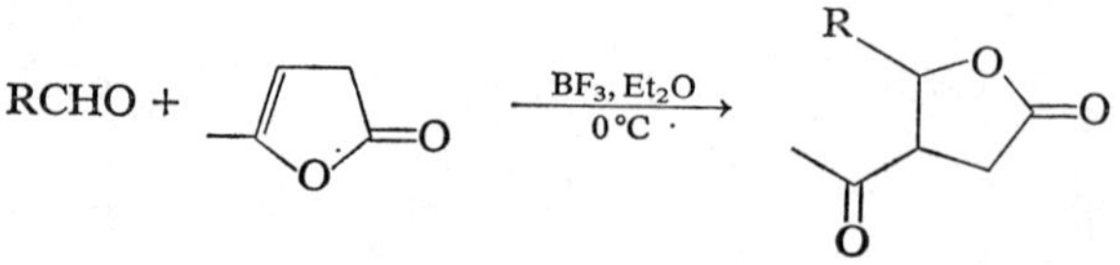

Scheme 20

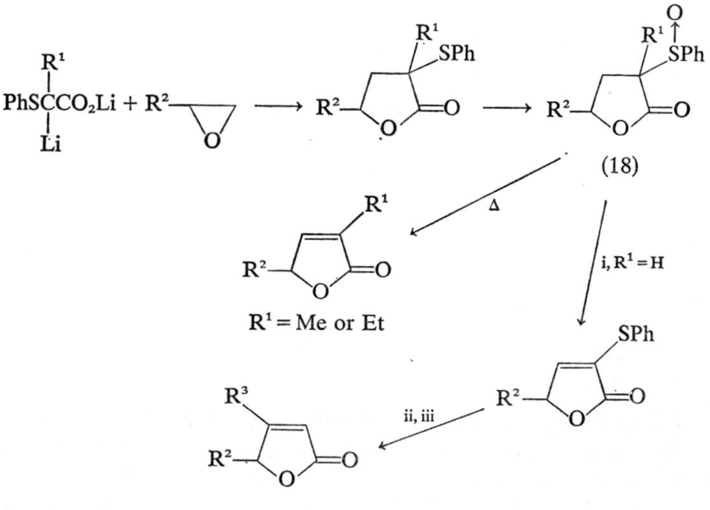

Scheme 21

Reagents: i, Ac$_2$O; ii, LiCuR$^3{}_2$; iii, oxidation, sulphoxide elimination

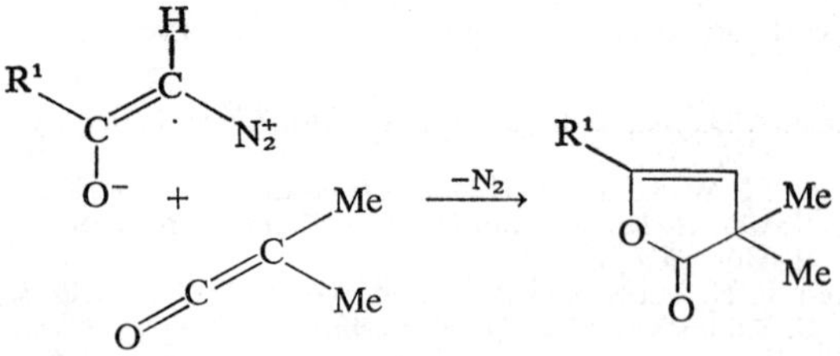

Scheme 22

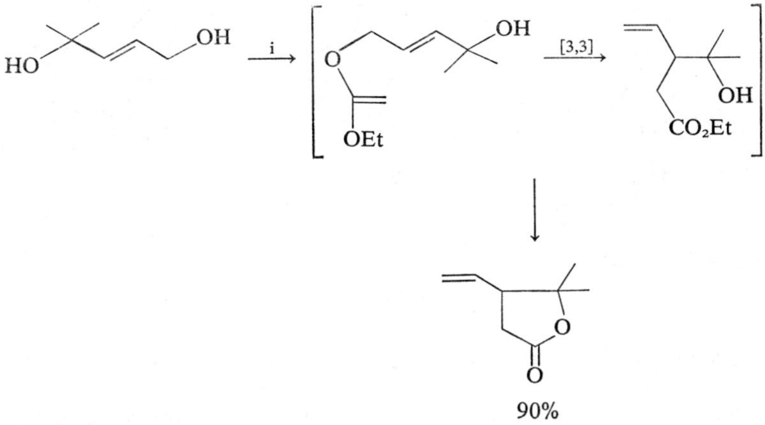

Scheme 23

Reagent: i, MeC(OEt)$_3$

Scheme 23

α-**Methylene-γ- and -δ-lactones.**—The propensity of selenoxides to undergo facile *syn*-elimination at low temperatures provides a new route[65] to *cis*- and *trans*-fused α-methylene-γ-lactones from the parent unsubstituted species; elimination to give pure α-methylene products is achieved by ensuring (Scheme 24) that the α-phenylseleno-group is *anti* to the adjacent methine

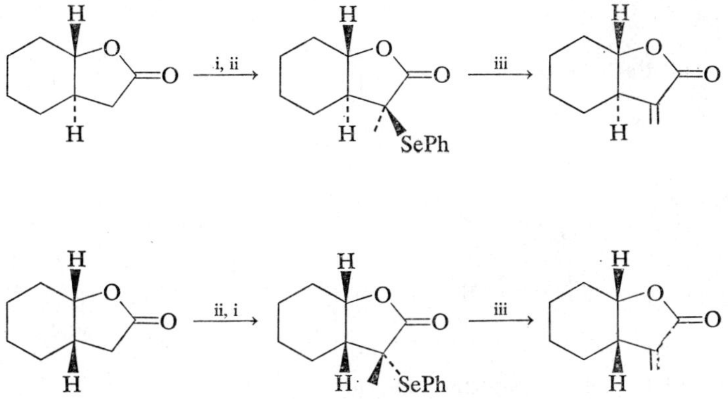

Reagents: i, LiNPr$^i_2$–MeI; ii, LiNPr$^i_2$–Ph$_2$Se$_2$; iii, H$_2$O$_2$–AcOH

Scheme 24

[65] P. A. Grieco and M. Miyashita, *J. Org. Chem.*, 1974, **39**, 120.

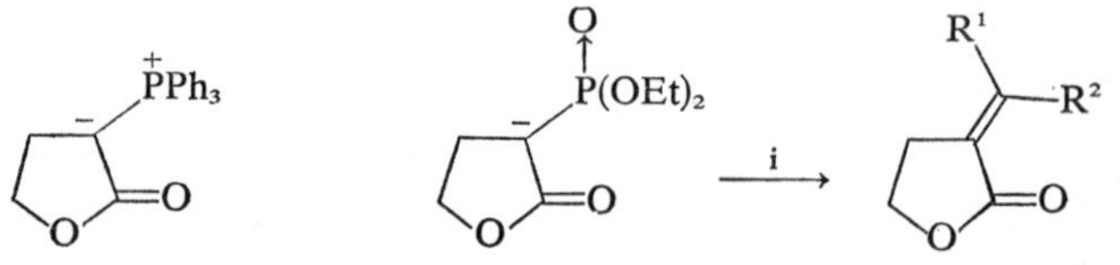

Reagent: i, R^{1}COR^2

Scheme 25

proton in both stereochemical situations. A similar route based on the *syn*-elimination of sulphoxides has been described.[66]

The ylide (19) reacts with paraformaldehyde to give α-methylene-γ-butyrolactone itself; in a probably general reaction, the α-bromo precursor to (19) is obtained[67] in high yield by direct bromination of the lactone enolate generated with lithium di-isopropylamide. The related α-phosphono-γ-butyrolactone (20) reacts[68] with carbonyl compounds to give α-ylidene-γ-butyrolactones (Scheme 25).

The α-ethoxycarbonyl vinylcuprate (21), an ethyl acrylate synthon which reacts only with highly reactive halides, has provided a new route[69] to α-methylene-γ-lactones from allylic halides (Scheme 26).

Ourisson[70] has given full details of eliminative routes to α-methylene-γ-

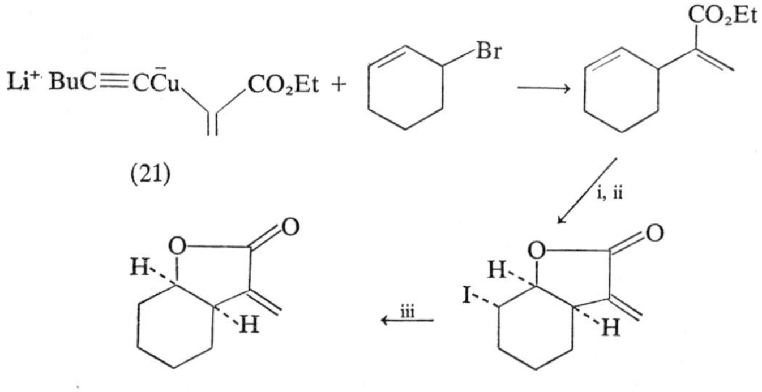

Reagents: i, HO$^-$; ii, KI$_3$–NaHCO$_3$; iii, Bu$_3^n$SnH

Scheme 26

[66] P. A. Grieco and J. J. Reap, *Tetrahedron Letters*, 1974, 1097.
[67] P. A. Grieco and C. S. Pogonowski, *J. Org. Chem.*, 1974, **39**, 1958.
[68] T. Minami, I. Niki, and T. Agawa, *J. Org. Chem.*, 1974, **39**, 3236.
[69] J. P. Marino and D. M. Floyd, *J. Amer. Chem. Soc.*, 1974, **96**, 7139.
[70] A. E. Greene, J.-C. Muller, and G. Ourisson, *J. Org. Chem.*, 1974, **39**, 186.

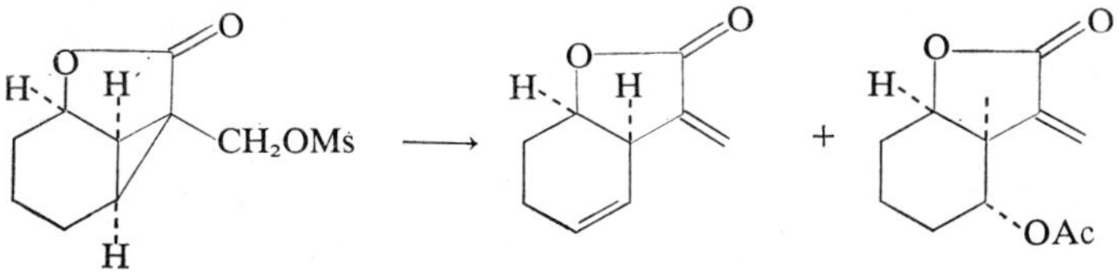

Functional Groups other than Alkanes, Acetylenes, Allenes, and Olefins   95

Scheme 27

lactones, which can also be prepared by oxidative decarboxylation[71] of lactone α-acetic acids. Site-selective solvolysis[72] of the cyclopropylcarbinyl system in (22) yields the lactones (23) and (24) (Scheme 27).

Ketone enamines react with methyl 2-(bromomethyl)acrylate in a general route[73] to α-methylene-δ-lactones (Scheme 28).

The phenylselenide anion has been suggested[74] as a reagent for the protection of α-methylenelactones, although the initial Michael addition proceeds in only modest yield.

Syntheses of Naturally Occurring Lactones.—A number of naturally occurring lactones and tetronic acids have been synthesized, or have had their syntheses improved. These include tetronic acid itself,[75] and some 5-methoxycarbonyl-tetronic acids,[76] including (S)-carlosic acid (25).[77] Pestalotin (26),[78,79] a

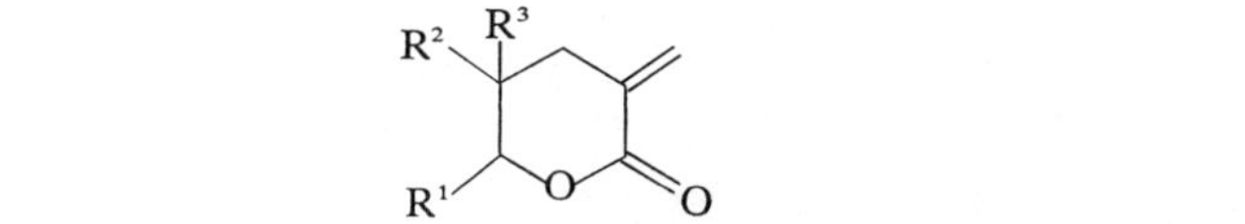

Scheme 28

[71] K. J. Divakar, P. P. Sane, and A. S. Rao, *Tetrahedron Letters*, 1974, 399.
[72] F. E. Ziegler, A. F. Marino, O. A. C. Petroff, and W. L. Studt, *Tetrahedron Letters*, 1974, 2035; cf. E. J. Corey and P. L. Fuchs, *J. Amer. Chem. Soc.*, 1972, **94**, 4014.
[73] H. Marschall, E. Vogel, and P. Weyerstahl, *Chem. Ber.*, 1974, **107**, 2852.
[74] P. A. Grieco and M. Miyashita, *Tetrahedron Letters*, 1974, 1869.
[75] J. V. Greenhill and T. Tomassini, *Tetrahedron Letters*, 1974, 2683.
[76] A. Svendsen and P. M. Boll, *Tetrahedron Letters*, 1974, 2821.
[77] J. L. Bloomer and F. E. Kappler, *J. Org. Chem.*, 1974, **39**, 113.
[78] R. M. Carlson and A. R. Oyler, *Tetrahedron Letters*, 1974, 2615.
[79] D. Seebach and H. Meyer, *Angew. Chem. Internat. Edn.*, 1974, **13**, 77.

(25) (26) (27) (28)

synergist of gibberellins, and strigol (27),[80-82] a germination stimulant of witchweed seeds, have both yielded to elegant total syntheses. Protolichesterinic acid[83] and isotelekin (28)[84] have been prepared; in each case the respective authors have used their previously reported methods for the introduction of the α-methylene moiety.

Properties.—While the $n \rightarrow \pi^*$ Cotton effect of $\Delta^{2,3}$-butenolides is easily influenced by asymmetry external to the lactone ring, the chirality at the γ-carbon of the butenolide is the sole sign-determining factor in the $\pi \rightarrow \pi^*$ c.d. spectrum; the latter absorption is therefore recommended[85] for the determination of absolute configuration of $\Delta^{2,3}$-butenolides. If X is greater than Y in polarizability, then lactone (29) will show a negative effect, and *vice versa*.

Equilibration studies[86] on a number of 2,4-disubstituted γ-butyrolactones

(29)

[80] J. B. Heather, R. S. D. Mittal, and C. J. Sih, *J. Amer. Chem. Soc.*, 1974, **96**, 1976.
[81] G. A. MacAlpine, R. A. Raphael, A. Shaw, A. W. Taylor, and H.-J. Wild, *J.C.S. Chem. Comm.*, 1974, 834.
[82] See also J. M. Cassady and G. A. Howie, *J.C.S. Chem. Comm.*, 1974, 512.
[83] J. Martin, P. C. Watts, and F. Johnson, *J. Org. Chem.*, 1974, **39**, 1676.
[84] R. B. Miller and E. S. Behare, *J. Amer. Chem. Soc.*, 1974, **96**, 8102.
[85] I. Uchida and K. Kuriyama, *Tetrahedron Letters*, 1974, 3761.
[86] S. A. M. T. Hussain, W. D. Ollis, C. Smith, and J. F. Stoddart, *J.C.S. Chem. Comm.*, 1974, 873.

Functional Groups other than Alkanes, Acetylenes, Allenes, and Olefins 97

have indicated that the *cis*-stereoisomer is the thermodynamically more stable in all cases. Detailed kinetic and mechanistic studies[87] of the alkaline hydrolysis of a variety of lactone ring sizes have been reported in detail.

4 Carboxylic Acid Esters

Preparation.—*By Esterification.* A mild and efficient method[51] for the preparation of esters is found in the silver-ion-induced reaction of pyridyl *S*-thioates with alcohols (Scheme 29); this 'push-button' activation has found

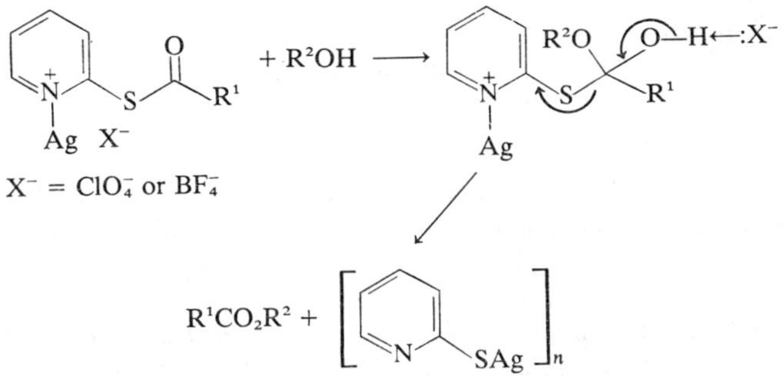

Scheme 29

application in the rapid formation of macrocyclic lactones from hydroxy-acids, at room temperature and under non-basic conditions (see also p. 89).

Sequential treatment of an ester with boron tribromide followed by an alcohol results[88] in clean transesterification; amides can be prepared similarly. Continuing their studies on activating groups, Japanese workers[89] have now shown that *N*-phosphonopyridinium betaines can be used to activate alcohol hydroxy-groups, allowing the alternative sequences shown (Scheme 30).

Electrolytically prepared graphite bisulphate, a blue hygroscopic crystalline substance, functions[90] as an efficient esterification catalyst; reaction seems to occur in the solid phase, which absorbs the carboxylic acid from solution, the graphite gathering as a brown solid. Treatment of this solid with one equivalent of an alcohol gives the corresponding ester in quantitative yield. t-Butyl alcohol is recommended[91] as solvent in the acid-catalysed

---

[87] G. M. Blackburn and H. L. H. Dodds, *J.C.S. Perkin II*, 1974, 377; see also M. Balakrishnan, G. V. Rao, and N. Venkatasubramanian, *ibid.*, p. 1093.
[88] H. Yazawa, H. Nakamura, K. Tanaka, and K. Kariyone, *Tetrahedron Letters*, 1974, 3995.
[89] N. Yamazaki, F. Higashi, and Y. Saito, *Synthesis*, 1974, 495.
[90] J. Bertin, H. B. Kagan, J.-L. Luche, and R. Setton, *J. Amer. Chem. Soc.*, 1974, **96**, 8113.
[91] S. Pavlov, M. Gogovac, and N. V. Arsenijevic, *Bull. Soc. chim. France*, 1974, 2985.

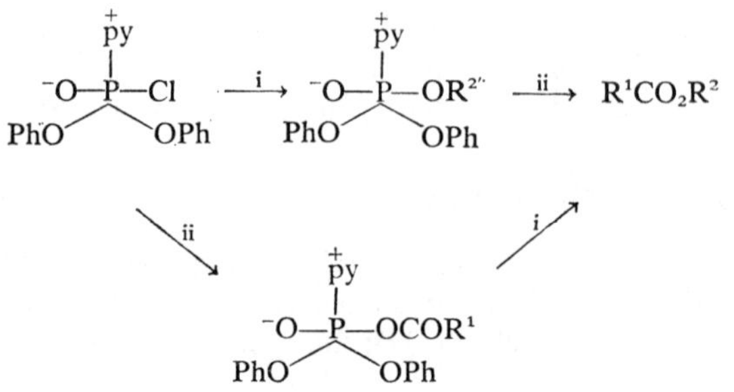

Reagents: i, R^2OH; ii, R^1CO_2H

Scheme 30

esterification of acids with isobutene. It has been shown[92] that, contrary to accepted belief, β-methyl-substituted aliphatic acids undergo acid-catalysed esterification with methanol at greater rates (*ca.* 1.8-fold at 40 °C) than the α-methyl isomers.

Aryl and vinyl bromides and iodides react with CO and an alcohol under mild conditions in the presence of a palladium(II) catalyst to produce esters, appreciable stereospecificity being shown (Scheme 31);[93] benzylic chlorides react similarly, and use of a primary or secondary amine in place of an alcohol produces amides.[94]

Dialkyl oxalates can be obtained[95] in good yield by oxidative carbonylation of alcohols (Scheme 32) in the presence of a copper(I)–copper(II) catalyst system and a dehydrating agent.

Anti-Markovnikoff esterification of terminal alkenes can be achieved[96] in excellent yields using a hydroboration–mercuration–iodination sequence

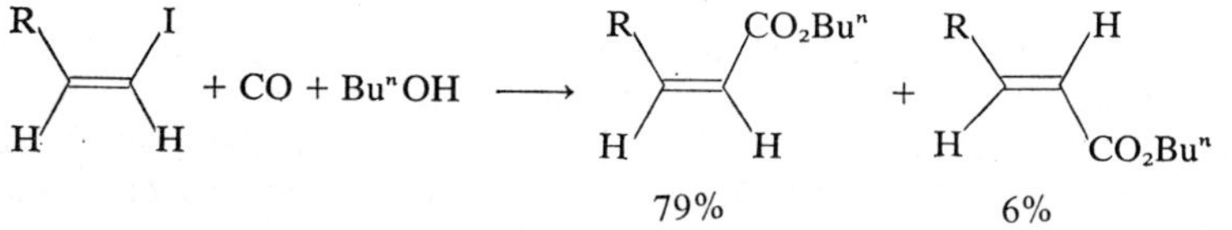

Reagent: i, $(Ph_3P)_2PdI_2$–Et_3N, 80 °C

Scheme 31

[92] P. J. Sniegowski, *J. Org. Chem.*, 1974, **39**, 3141.
[93] A. Schoenberg, I. Bartoletti, and R. F. Heck, *J. Org. Chem.*, 1974, **39**, 3318.
[94] A. Schoenberg and R. F. Heck, *J. Org. Chem.*, 1974, **39**, 3327.
[95] D. M. Fenton and P. J. Steinwand, *J. Org. Chem.*, 1974, **39**, 701.
[96] R. C. Larock, *J. Org. Chem.*, 1974, **39**, 834.

$$2CO + 2ROH + \tfrac{1}{2}O_2 \rightarrow RO_2CCO_2R + H_2O$$

Scheme 32

(Scheme 33); this method fails with more highly substituted olefins. Commercial mixtures of vinyl esters of isomeric acids have been used[97] to synthesize lower vinyl esters by mercury(II) catalysed exchange.

A wide variety of organic halides are converted quantitatively into acetate esters by reaction with the highly nucleophilic 'naked' acetate ion[98] obtained by solubilizing potassium acetate in acetonitrile with crown ethers; phase-transfer catalysis has been utilized[99] to the same ends. Crown ethers have

$$R^1CH{=}CH_2 \xrightarrow{\text{i-iii}} R^1CH_2CH_2O_2CR^2$$

Reagents: i, $\tfrac{1}{3}BH_3$; ii, $Hg(OCOR^2)_2$; iii, I_2

Scheme 33

been used to solubilize and activate potassium carboxylates in an efficient route[100] to carboxylic acid phenacyl esters. The scope of the quantitative esterification of acids by reaction of their salts with alkyl halides in HMPA has been extended.[101] Base-induced transesterification has been used for the preparation[102] of benzoate esters of tertiary alcohols. Thermal decomposition of benzyldimethylanilinium carboxylates produces benzyl esters (Scheme 34) in good yields,[103] even in sterically hindered cases.

$$RCO_2^- \; PhCH_2\overset{+}{N}(Me)_2Ph \xrightarrow{\Delta} RCO_2CH_2Ph$$

Scheme 34

Of β-Hydroxy- and αβ-Unsaturated Esters. Side-reactions in the Reformatsky synthesis of β-hydroxy-esters can be minimized[104] by using a continuous flow system with a heated column of granulated zinc. The scope and limitations of the production of β-hydroxy-esters by reaction between a carbonyl compound, keten, and a titanium(IV) alkoxide have been delineated.[105] In a related reaction, titanium(IV) chloride induces[106] diketen to react with

---

[97] M. A. S. Mondal, R. van der Meer, A. L. German, and D. Heikens, *Tetrahedron*, 1974, **30**, 4205.
[98] C. L. Liotta, H. P. Harris, M. McDermott, T. Gonzalez, and K. Smith, *Tetrahedron Letters*, 1974, 2250; see also refs. 598—600.
[99] T. Toru, S. Kurozumi, T. Tanaka, S. Miura, M. Kobayashi, and S. Ishimoto, *Synthesis*, 1974, 867.
[100] H. D. Durst, *Tetrahedron Letters*, 1974, 2421.
[101] J. E. Shaw and D. C. Kunerth, *J. Org. Chem.*, 1974, **39**, 1968.
[102] R. A. Rossi and R. H. de Rossi, *J. Org. Chem.*, 1974, **39**, 855.
[103] K. Williams and B. Halpern, *Synthesis*, 1974, 727.
[104] J. F. Ruppert and J. D. White, *J. Org. Chem.*, 1974, **39**, 269.
[105] L. Vuitel and A. Jacot-Guillarmod, *Helv. Chim. Acta*, 1974, **57**, 1703; L. Vuitel, R. Tabacchi, and A. Jacot-Guillarmod, *ibid.*, p. 1713.
[106] T. Izawa and T. Mukaiyama, *Chem. Letters*, 1974, 1189.

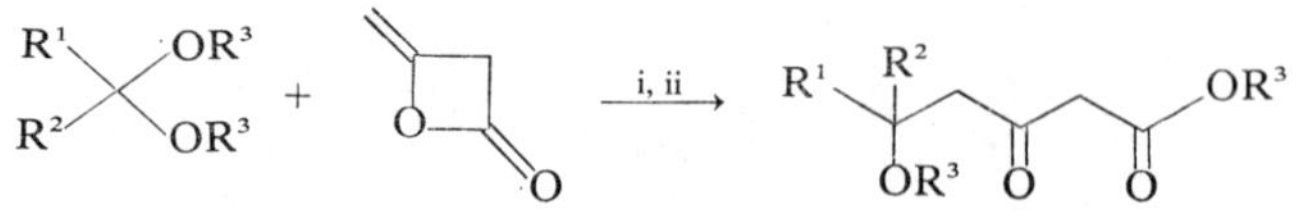

Scheme 35

Reagents: i, TiCl$_4$; ii, R^{3}OH

acetals at low temperature to yield δ-alkoxy-β-keto-esters (Scheme 35) in good yield.

The stereoselective production of (E)-αβ-unsaturated esters by dehydration of β-hydroxy-esters with aluminium reagents has been studied.[107] A number of reports[108,109] have described the direct conversion of an aldehyde or ketone into a homologated αβ-unsaturated ester (Scheme 36) in a method based on the well documented affinity of suitably substituted silyl groups for alkoxide anions.

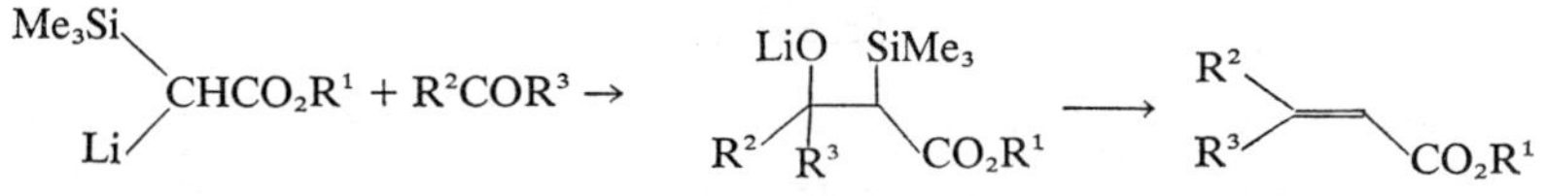

Scheme 36

Stereoselective syntheses of (Z)- and (E)-αβ-unsaturated esters can be achieved *via* hydroboration[110] of ethyl propiolate, as shown in Scheme 37; the migrating group R retains its stereochemistry.

The generation and reactivity of the α-ethoxycarbonyl vinylcuprate shown in Scheme 38 has been described;[69] it reacts only with allylic or propargylic halides: an application to the synthesis of α-methylene-γ-lactones was illustrated in Scheme 26.

General. Ester homologation can be effected[111] *via* the methyl methylthiomethyl sulphoxide anion, as shown in Scheme 39.

2,3-Dibromoethyl carbonate and 1,3-dibromo-2-propyl ethyl carbonate equilibrate[112] thermally with concomitant formation of 3-bromopropylene carbonate and ethyl bromide. In a continuing study, Japanese authors[113] have reported that selenium oxidizes formates in the presence of alkoxides to give dialkyl carbonates in excellent yield at room temperature. The anodic oxidation of several acyclic esters in acetonitrile results in clean conversion[114] into monoacetamido-esters, with considerable (ω − 1) selectivity.

---

[107] J. A. Katzenellenbogen and T. Utawanit, *J. Amer. Chem. Soc.*, 1974, **96**, 6153.
[108] K. Shimoji, H. Taguchi, K. Oshima, H. Yamamoto, and H. Nozaki, *J. Amer. Chem. Soc.*, 1974, **96**, 1620; H. Taguchi, K. Shimoji, H. Yamamoto, and H. Nozaki, *Bull. Chem. Soc. Japan*, 1974, **47**, 2529.
[109] S. L. Hartzell, D. F. Sullivan, and M. W. Rathke, *Tetrahedron Letters*, 1974, 1403.
[110] E. Negishi, G. Lew, and T. Yoshida, *J. Org. Chem.*, 1974, **39**, 2321.
[111] K. Ogura, S. Furukawa, and G. Tsuchihashi, *Chem. Letters*, 1974, 659.
[112] R. G. Pews, *J.C.S. Chem. Comm.*, 1974, 659.
[113] K. Kondo, N. Sonoda, and H. Sakurai, *Tetrahedron Letters*, 1974, 803.
[114] L. L. Miller and V. Ramachandran, *J. Org. Chem.*, 1974, **39**, 369.

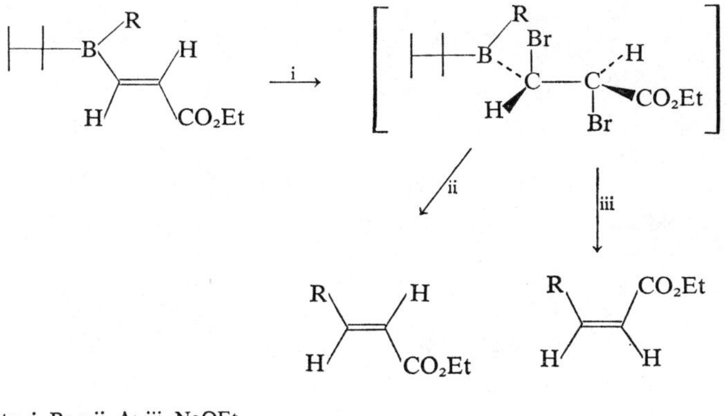

Functional Groups other than Alkanes, Acetylenes, Allenes, and Olefins 101

Reagents: i, Br$_2$; ii, Δ; iii, NaOEt

Scheme 37

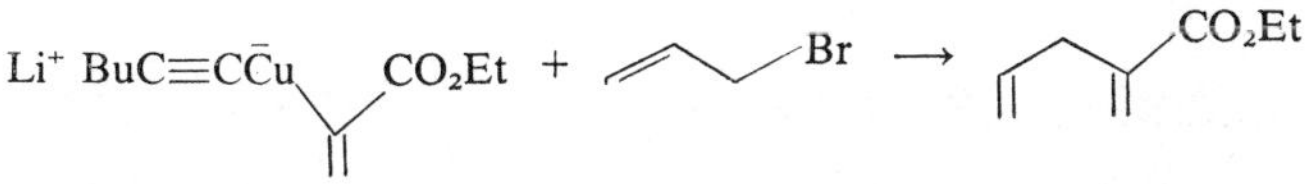

Scheme 38

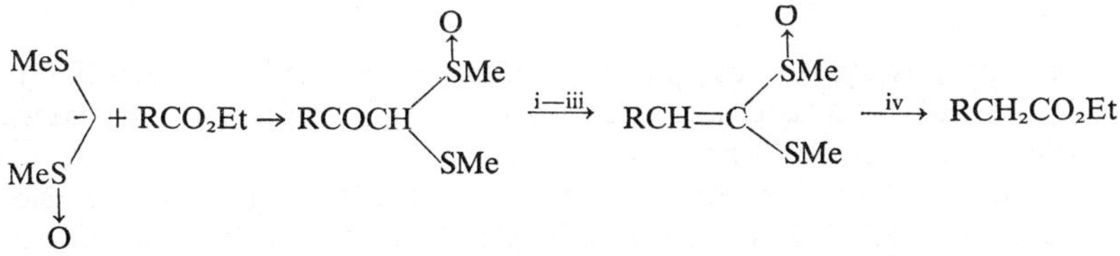

Reagents: i, NaBH$_4$; ii, Ac$_2$O–py; iii, Et$_3$N; iv, EtOH–H$^+$

Scheme 39

In a mechanistically interesting process,[115] the diesters shown rearrange thermally to substituted glutarates (Scheme 40).

Chiral rhodium complexes catalyse[116] the asymmetric hydrosilylation of α-keto-esters, leading to optically active α-hydroxy-esters. The scope of the addition reaction of the zinc organometallic derived from diethyl methylbromomalonate with propargylic compounds has been extended,[117] as exemplified in Scheme 41.

[115] P. Dowd and K. Kang, *J.C.S. Chem. Comm.*, 1974, 258.
[116] I. Ojima, T. Kogure, and Y. Nagai, *Tetrahedron Letters*, 1974, 1889.
[117] M. T. Bertrand, G. Courtois, and L. Miginiac, *Tetrahedron Letters*, 1974, 1945.

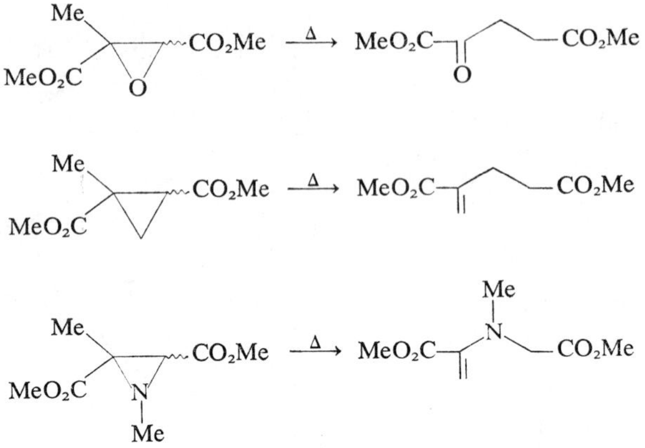

Scheme 40

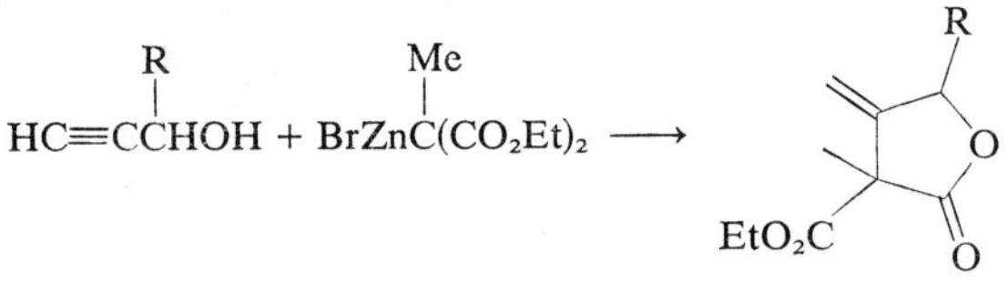

Scheme 41

A copper isocyanide complex[118] induces the cyclization of 1,3-di-iodopropane with αβ-unsaturated esters to give a new route to cyclopentanecarboxylic acid esters (Scheme 42).

Further studies[119] on the conjugate reduction of αβ-unsaturated esters have been reported. In the presence of aluminium trichloride, trimethylsilyl ethers of alkanethiols react[120] with esters to give the corresponding thiol esters in good yield. Two other preparations of thiol esters[121,122] have been described; one[122] is induced direct reaction of an acid with a thiol (Scheme 43). Synthetic routes to orthoesters have been reviewed.[123]

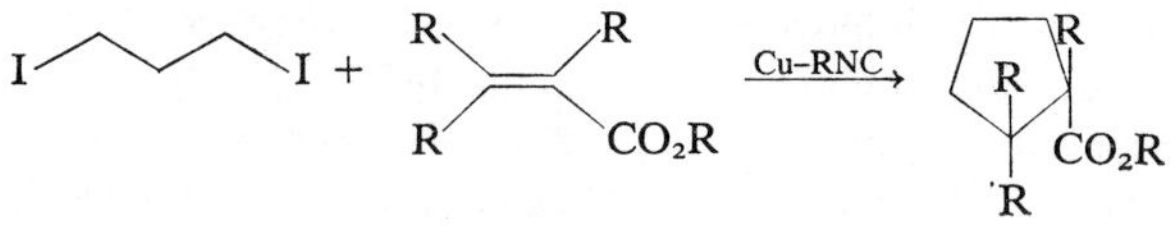

Scheme 42

[118] Y. Ito, K. Nakayama, K. Yonezawa, and T. Saegusa, *J. Org. Chem.*, 1974, **39**, 3273.
[119] J. H. Schauble, G. J. Walter, and J. G. Morin, *J. Org. Chem.*, 1974, **39**, 755.
[120] T. Mukaiyama, T. Takeda, and K. Atsumi, *Chem. Letters*, 1974, 187.
[121] R. A. Gorski, D. J. Dagli, V. A. Patronik, and J. Wemple, *Synthesis*, 1974, 811.
[122] S. Yamada, T. Yokayama, and T. Shiori, *J. Org. Chem.*, 1974, **39**, 3302.
[123] R. H. DeWolfe, *Synthesis*, 1974, 153.

Functional Groups other than Alkanes, Acetylenes, Allenes, and Olefins 103

$$R^1CO_2H + R^2SH \xrightarrow{i} R^1COSR^2$$

Reagent: i, $(EtO)_2\overset{O}{\overset{\uparrow}{P}}CN$ or $(PhO)_2\overset{O}{\overset{\uparrow}{P}}N_3-Et_3N-DMF$

Scheme 43

Alkylation.—The utility of ester α-anions as precursors to malonates, phosphonacetates, and α-selenyl and α-sulphinyl esters has been described[124] once again. Varying regioselectivity is seen[125] in the alkylation of lithium and copper enolates of butenoic acid esters with allyl halides; the lithium enolate gives mainly the product of α-alkylation, whereas the copper enolate gives an equal mixture of α- and γ-alkylated products, perhaps by alternative decomposition paths of an intermediate such as (30) (Scheme 44).

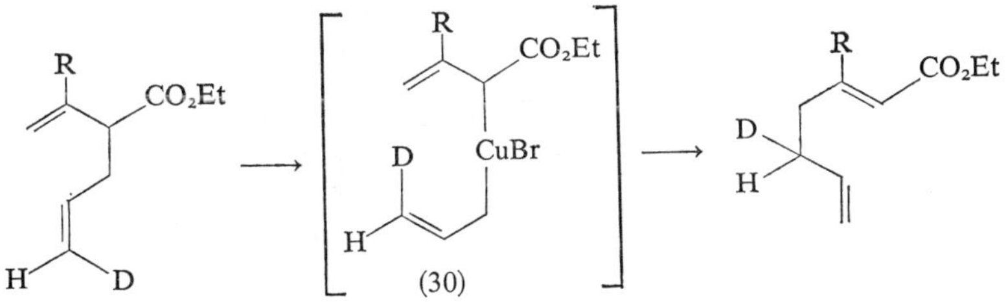

Scheme 44

Further studies on the alkylation[126] and acylation[127] of acetoacetic ester enolate anions have been described. The use of phase-transfer catalysis with benzene as organic phase has been reported[128] to give exclusive C-alkylation of such esters. The acid-catalysed reaction of β-keto-esters with isopropenyl acetate leads almost entirely to (Z) enol acetates, whereas the (E) isomers are formed in the triethylamine promoted reaction with acetyl chloride in HMPA; this is attributed to the respective reactive forms being the (Z) enol and the (E) solvent-separated enolate anion. Subsequent reaction with a lithium dialkylcuprate provides highly stereoselective syntheses[129] of αβ-unsaturated esters (Scheme 45).

In a related process,[130] ethyl (E)- or (Z)-3-methylhept-2-enoate can be prepared in good yield from the corresponding ethyl 3-phenylthio compound (Scheme 46).

---

[124] T. J. Brocksom, N. Petragnani, and R. Rodrigues, *J. Org. Chem.*, 1974, **39**, 2114.
[125] J. A. Katzenellenbogen and A. L. Crumrine, *J. Amer. Chem. Soc.*, 1974, **96**, 5662.
[126] F. Guibé, P. Sarthou, and G. Bram, *Tetrahedron*, 1974, **30**, 3139.
[127] R. Gelin, S. Gelin, and A. Galliaud, *Bull. Soc. chim. France*, 1973, 3416.
[128] H. D. Durst and L. Liebeskind, *J. Org. Chem.*, 1974, **39**, 3271.
[129] C. P. Casey and D. F. Marten, *Tetrahedron Letters*, 1974, 925.
[130] S. Kobayashi, H. Takei, and T. Mukaiyama, *Chem. Letters*, 1973, 1097.

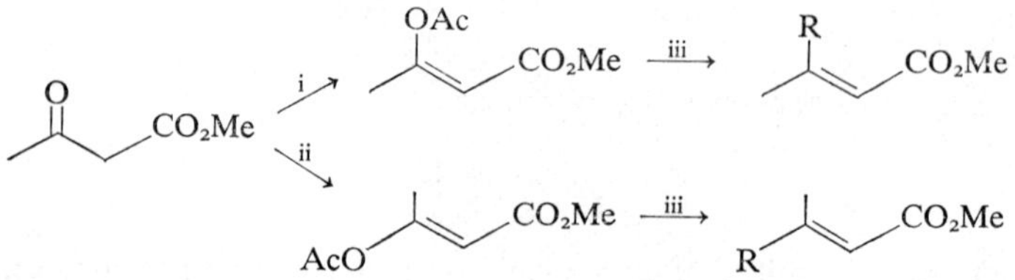

Reagents: i, 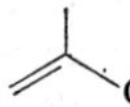 ⤳OAc–H$^+$; ii, AcCl–Et$_3$N–HMPA; iii, LiCuR$_2$

Scheme 45

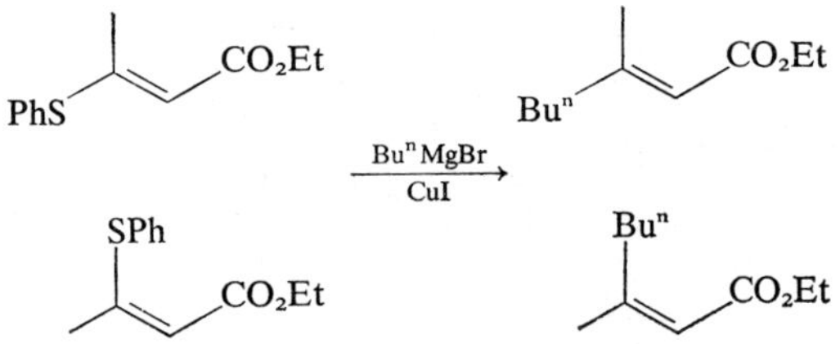

Scheme 46

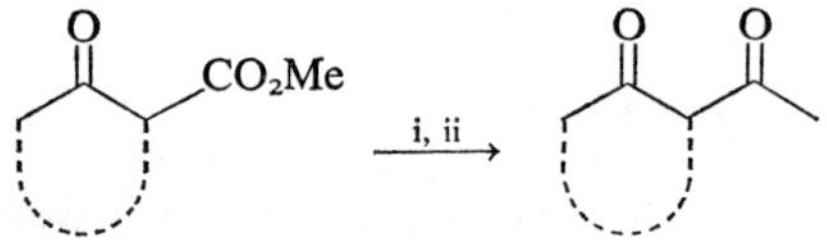

Reagents: i, NaH; ii, 2 MeLi

Scheme 47

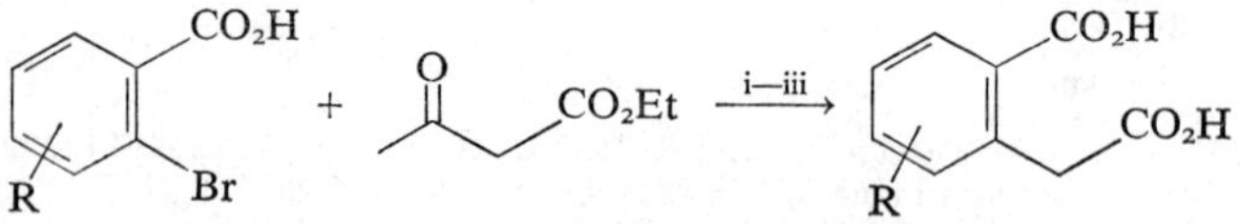

Reagents: i, NaH–CuBr; ii, NaOH; iii, H$_3$O$^+$

Scheme 48

Functional Groups other than Alkanes, Acetylenes, Allenes, and Olefins 105

The action of two equivalents of methyl-lithium on simple β-keto-ester anions produces β-diketones in good yield (Scheme 47); this represents a facile method[131] for C-acylation of a ketone.

β-Keto-ester anions undergo a copper-catalysed direct reaction with 2-bromobenzoic acids; hydrolysis of the intermediates gives homophthalic acids (Scheme 48).[132]

Full descriptive details of the specific γ-alkylation[133] of β-keto-ester dianions have been published, as have reports of γ-acylation[134] and γ-aldolization;[135] the last reaction provides a route (Scheme 49) to γδ-unsaturated β-keto-esters, though not to the simplest member (31; $R^1 = R^2 =$ H) of this series of annelating agents. A convenient alternative[136] to the high-temperature pyrolysis system required in the best current route[137] to (31; $R^1 = R^2 =$ H) is provided by sulphoxide *syn* elimination (Scheme 49).

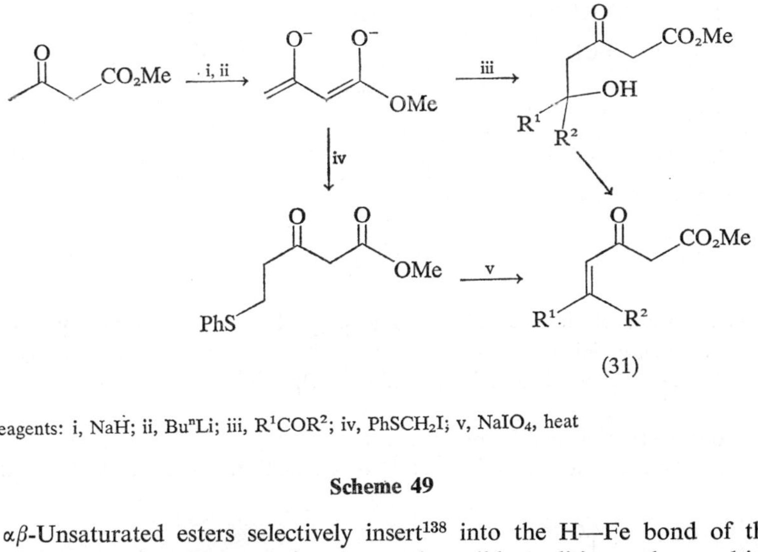

Reagents: i, NaH; ii, Bu^{n}Li; iii, R^1COR^2; iv, PhSCH$_2$I; v, NaIO$_4$, heat

Scheme 49

αβ-Unsaturated esters selectively insert[138] into the H—Fe bond of the hydridotetracarbonylferrate(-II) anion under mild conditions; the resulting complexes yield β-keto-esters, as shown in Scheme 50.

Ester α-anions react with nitro-enamines[139] such as (32) to afford vinylogous β-keto-esters (Scheme 51).

---

[131] S. N. Huckin and L. Weiler, *Canad. J. Chem.*, 1974, **52**, 1379.
[132] A. Bruggink and A. McKillop, *Angew. Chem. Internat. Edn.*, 1974, **13**, 340.
[133] S. N. Huckin and L. Weiler, *J. Amer. Chem. Soc.*, 1974, **96**, 1082.
[134] S. N. Huckin and L. Weiler, *Canad. J. Chem.*, 1974, **52**, 1343.
[135] S. N. Huckin and L. Weiler, *Canad. J. Chem.*, 1974, **52**, 2157.
[136] B. A. Trost and R. A. Kunz, *J. Org. Chem.*, 1974, **39**, 2648; *cf.* ref. 404.
[137] G. Stork and R. N. Guthikonda, *Tetrahedron Letters*, 1972, 2755.
[138] T. Mitsudo, Y. Watanabe, M. Yamashita, and Y. Takegami, *Chem. Letters*, 1974, 1385.
[139] H. Lerche, D. König, and T. Severin, *Chem. Ber.*, 1974, **107**, 1509; *cf.* ref. 334.

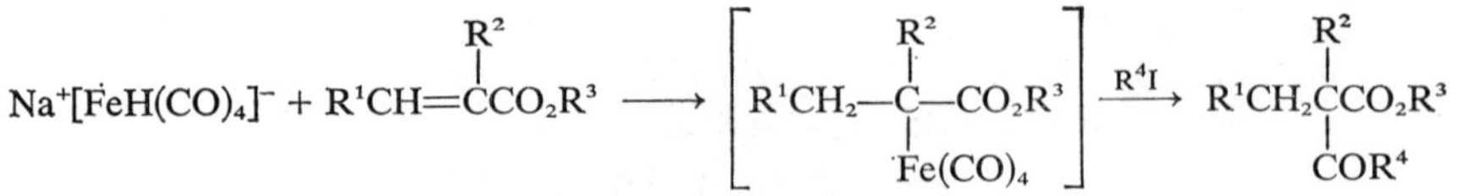

$$Na^+[FeH(CO)_4]^- + R^1CH=\overset{R^2}{\underset{|}{C}}CO_2R^3 \longrightarrow \left[R^1CH_2-\overset{R^2}{\underset{\underset{|}{Fe(CO)_4}}{\overset{|}{C}}}-CO_2R^3 \right] \xrightarrow{R^4I} R^1CH_2\overset{R^2}{\underset{\underset{|}{COR^4}}{\overset{|}{C}}}CO_2R^3$$

Scheme 50

Cleavage.—Sodium cyanide in wet HMPA cleaves[140] methyl esters in the presence of ethyl esters, even in apparently unfavourable cases (Scheme 52); this selective *O*-alkyl fission by cyanide ion is general for esters of both aliphatic and aromatic acids. 1,1-Dimethylhydrazine has been observed[141] to mono-demethylate a range of dimethyl esters at room temperature to afford, on acidification, diacid half esters.

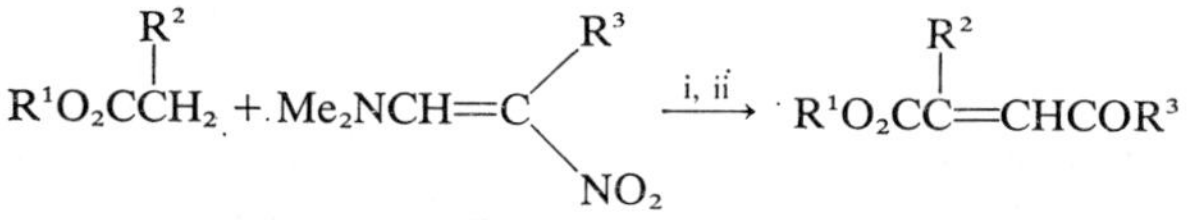

$$R^1O_2C\overset{R^2}{\underset{|}{C}}H_2 + Me_2NCH=C\overset{R^3}{\underset{NO_2}{\diagup}} \xrightarrow{i, ii} R^1O_2C\overset{R^2}{\underset{|}{C}}=CHCOR^3$$

(32) R^3 = H or Me

Reagents: i, LiNPr$^i_2$; ii, SiO$_2$ chromatography

Scheme 51

The trithiocarbonate dianion[142] smoothly cleaves 2-halogenoethyl esters (Scheme 53), which suggests that such esters can be used as acid protecting groups; in a related reaction, sodium sulphide[143] cleaves ω-chloroalkyl esters.

Sodium chloride accelerates the rate of dealkoxycarbonylation of *gem*-diesters and β-keto-esters in wet DMSO, but in most cases it is quite

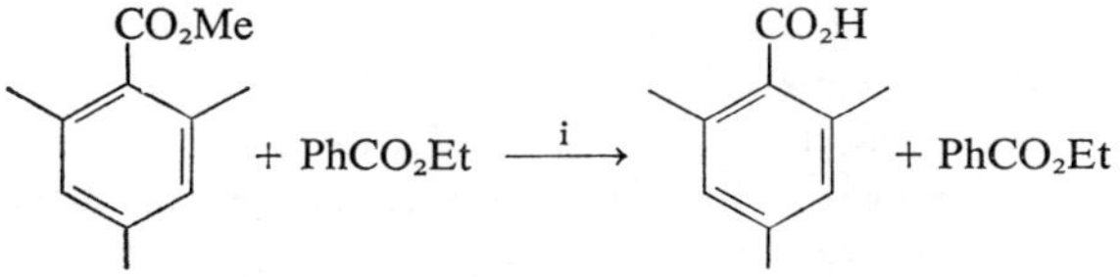

[Ar-CO$_2$Me] + PhCO$_2$Et $\xrightarrow{i}$ [Ar-CO$_2$H] + PhCO$_2$Et

85% 93%

Reagent: i, NaCN–HMPA, heat

Scheme 52

[140] P. Müller and B. Siegfried, *Helv. Chim. Acta*, 1974, **57**, 987.
[141] J. Nematollahi and S. Kasina, *J.C.S. Chem. Comm.*, 1974, 775.
[142] T.-L. Ho, *Synthesis*, 1974, 715.
[143] T.-L. Ho and C. M. Wong, *Synthetic Comm.*, 1974, **4**, 307.

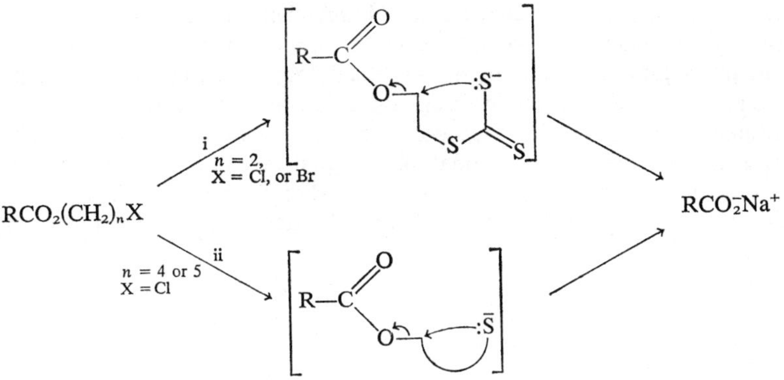

Scheme 53

Reagents: i, Na_2CS_3; ii, Na_2S

unnecessary as wet DMSO itself constitutes an excellent system.[144,145] 1,4-Diazabicyclo[2,2,2]octane cleaves β-keto-esters[146a] with at least one α-hydrogen atom, and vinylogous β-keto-esters[146b] (Scheme 54) in a process which does not seem to involve initial *O*-alkyl fission, as even n-butyl esters are cleaved. β-Keto-esters react with phosphorus pentachloride to give β-chloro-αβ-unsaturated acids directly.[147]

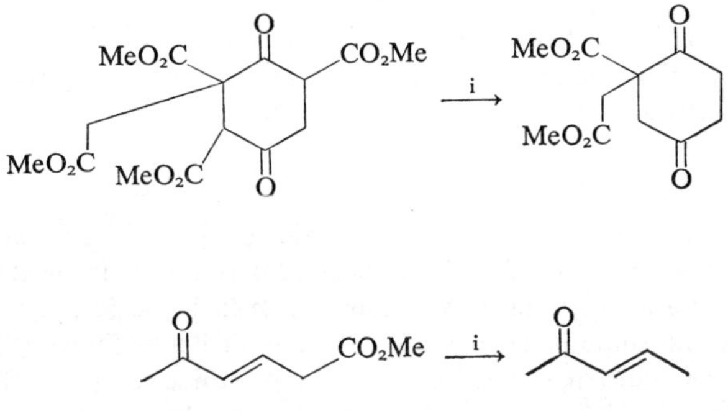

Reagent: i, DABCO–*o*-xylene, heat

Scheme 54

[144] A. P. Krapcho, E. G. E. Jahngen, A. J. Lovey, and F. W. Short, *Tetrahedron Letters*, 1974, 1091.
[145] C. L. Liotta and F. L. Cook, *Tetrahedron Letters*, 1974, 1095.
[146] (a) B.-S. Huang, E. J. Parish, and D. H. Miles, *J. Org. Chem.*, 1974, 39, 2647; (b) E. J. Parish, N. V. Mody, P. A. Hedin, and D. H. Miles, *ibid.*, p. 1592; see also D. H. Miles and E. J. Parish, *A.C.S. Sept. 1974, Orgn. 133* for use of 3-quinuclidinol.
[147] A.-H. Youssef and H. M. Abdel-Maksoud, *J.C.S. Chem. Comm.*, 1974, 288.

108   *Aliphatic Chemistry*

General Properties and Reactions.—N.m.r. chemical shift data have been used to evaluate[148] an acidity function applicable to aliphatic esters for aqueous solutions containing 60—98% sulphuric acid; the function appears adequately to define the protonation behaviour of several simple unsubstituted esters, but unique responses are seen in those examples with substituents near the site of protonation. The first case of epoxide formation by action of a diazoalkane upon an ester carbonyl group has been noted.[149] The preparation of ^{14}C-labelled aldehydes by the route outlined (Scheme 55)

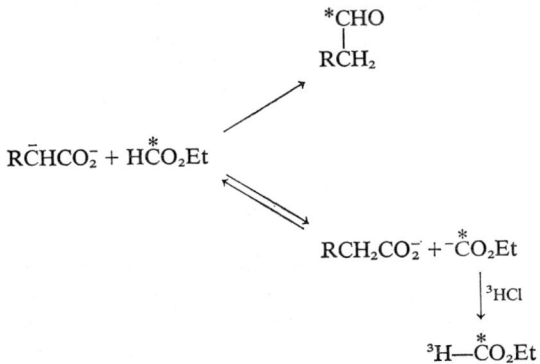

Scheme 55

shows a remarkable barrier to yield improvement; this is rationalized by proposing[150] the competing alternative pathway shown, with formation of the hitherto unknown ethyl formate anion.

5 α-Amino-acids

Preparation.—Constant potential cathodic electrolysis of certain alkyl halides in the presence of a Schiff base (33) results[151] in electroreductive coupling and a new synthesis of α-amino-acids (Scheme 56).

The enamino-sulphoxide (34), derived from nitriles as shown (Scheme 57), undergoes an unusual rearrangement[152] to the thioester (35) on treatment with acetic anhydride; subsequent reductive desulphurization leads to *N*-acetylamino-acids.

A partial asymmetric synthesis of amino-acids has been described which uses chiral isocyanides[153] and the derived chiral lithio-aldimines (Scheme 58);

[148] D. G. Lee and M. H. Sadar, *J. Amer. Chem. Soc.*, 1974, **96**, 2862.
[149] F. M. Deans and B. K. Park, *J.C.S. Chem. Comm.*, 1974, 162.
[150] G. K. Kochi and J. M. M. Kop, *Tetrahedron Letters*, 1974, 603.
[151] T. Iwasaki and K. Harada, *J.C.S. Chem. Comm.*, 1974, 338.
[152] K. Ogura and G. Tsuchihashi, *J. Amer. Chem. Soc.*, 1974, **96**, 1960.
[153] N. Hirowatari and H. M. Walborsky, *J. Org. Chem.*, 1974, **39**, 604.

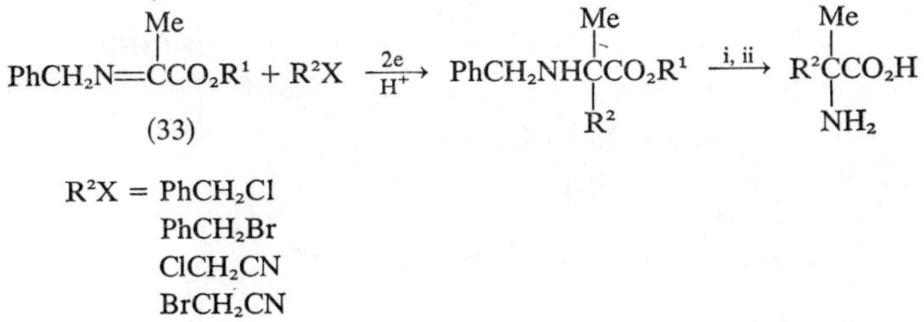

Scheme 56

PhCH₂N=C(Me)CO₂R¹ + R²X $\xrightarrow{2e/H^+}$ PhCH₂NHC(Me)(R²)CO₂R¹ $\xrightarrow{i, ii}$ R²C(Me)(NH₂)CO₂H

(33)

R²X = PhCH₂Cl
PhCH₂Br
ClCH₂CN
BrCH₂CN

Reagents: i, H_3O^+; ii, H_2–Pd–C

Scheme 56

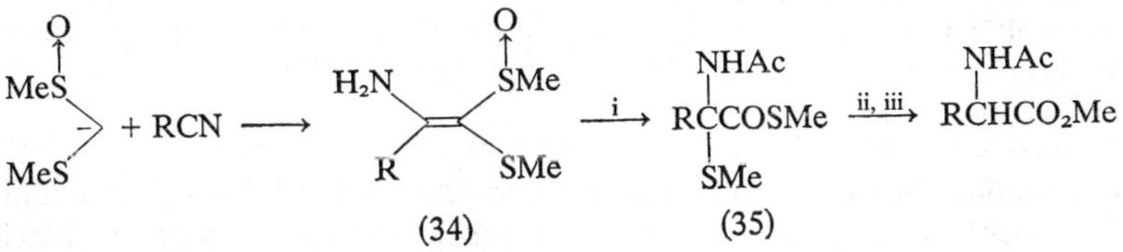

MeS(C=O)CH(MeS) + RCN ⟶ H₂N–C(R)=C(SMe)(SMe) $\xrightarrow{i}$ RC(NHAc)(COSMe)(SMe) $\xrightarrow{ii, iii}$ RCH(NHAc)CO₂Me

(34)   (35)

Reagents: i, Ac_2O; ii, Et_3N–MeOH; iii, Ni

Scheme 57

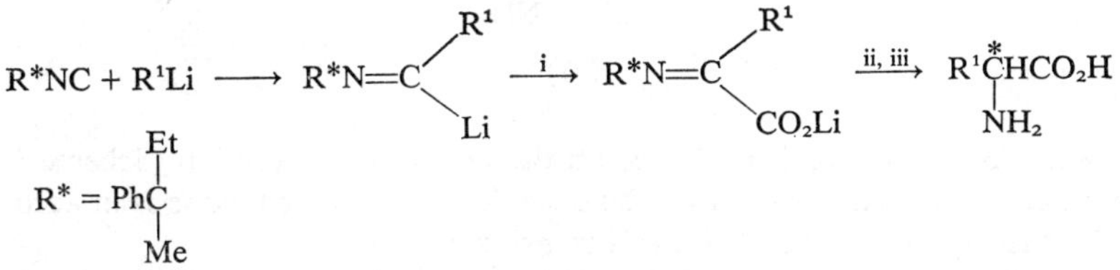

R*NC + R¹Li ⟶ R*N=C(R¹)(Li) $\xrightarrow{i}$ R*N=C(R¹)(CO₂Li) $\xrightarrow{ii, iii}$ R¹C*H(NH₂)CO₂H

R* = PhC(Et)(Me)

Reagents: i, CO_2; ii, B_2H_6 or equivalent; iii, H_2–Pd–C

Scheme 58

optical and chemical yields are good. Optically pure vinyl glycine[154] has been prepared.

Protection and Deprotection.—The nitro enol ether (36) serves[155] as an amine-protecting group as shown (Scheme 59). The scope and limitations of boron tribromide[156] and trifluoromethanesulphonic acid[157] for the cleavage of

---

[154] P. Friis, P. Helboe, and P. O. Larsen, *Acta Chem. Scand.*, 1974, **B28**, 317.
[155] P. L. Southwick, R. F. Dufresne, and J. J. Lindsey, *J. Org. Chem.*, 1974, **39**, 3351.
[156] A. M. Felix, *J. Org. Chem.*, 1974, **39**, 1427.
[157] H. Yajima, N. Fujii, H. Ogawa, and H. Kawatani, *J.C.S. Chem. Comm.*, 1974, 107.

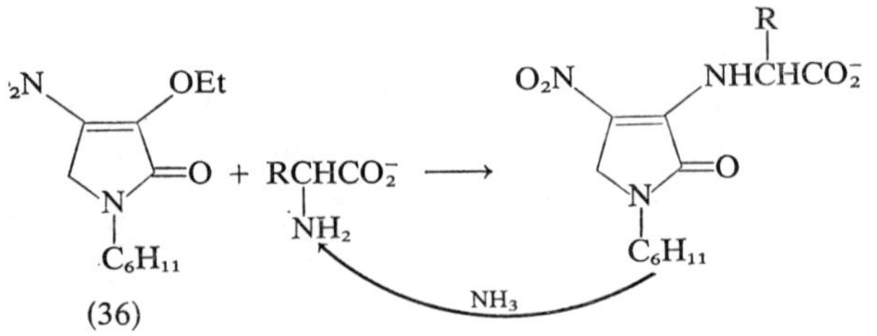

Scheme 59

protecting groups have been delineated. Formic acid (85%) selectively cleaves[158] N-t-butoxycarbonyl groups in the presence of t-butyl esters.

The o-nitrocinnamoyl group has been used as an amine blocking function; removal[159] is by hydrogenolysis. A simple method for the preparation of N-t-butoxycarbonyl amino-acids has been described.[160]

Properties and Reactions.—Information regarding the rotamer populations of amino-acids can be obtained by considering the ^{13}C–C–C–H coupling constant[161] between the α-CO_2^- and the β-protons. A second chlorine-containing natural (L)-amino-acid (37) has been isolated.[162]

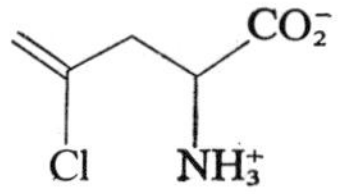

Kinetic and mechanistic studies on the Dakin–West reaction (Scheme 60) have convinced Allinger[163] that the oxazolone mechanism alone is in accord with the experimental facts on this transformation.

A chiral host molecule capable of achieving the total optical resolution of amino-acid ester salts by specific molecular complexation has been described.[164] Sulphoxide elimination has been employed in a new route[165]

---

[158] H. Kinoshita and H. Kotake, *Chemistry Letters*, 1974, 631.
[159] G. Just and G. Rosebery, *Synthetic Comm.*, 1973, **3**, 447.
[160] Z. Grzonka and B. Lammek, *Synthesis*, 1974, 661.
[161] J. Feeney, P. E. Hansen, and G. C. K. Roberts, *J.C.S. Chem. Comm.*, 1974, 465.
[162] S.-I. Hatanaka, S. Kaneko, Y. Niimura, F. Kinoshita, and G. Soma, *Tetrahedron Letters*, 1974, 3931.
[163] N. L. Allinger, G. L. Wang, and B. B. Dewhurst, *J. Org. Chem.*, 1974, **39**, 1730; see also J. Lepschy, G. Hofle, L. Wilschowitz, and W. Steglich, *Annalen*, 1974, 1753.
[164] L. R. Sousa, D. H. Hoffman, L. Kaplan, and D. J. Cram, *J. Amer. Chem. Soc.* 1974, **96**, 7100; see also J. M. Timko, R. C. Helgeson, M. Newcomb, G. W. Gokel, and D. J. Cram, *ibid.*, p. 7097; M. Newcomb, R. C. Helgeson, and D. J. Cram, *ibid.*, p. 7367.
[165] D. H. Rich, J. Tam, P. Mathiaparanam, J. A. Grant, and C. Mabuni, *J.C.S. Chem. Comm.*, 1974, 897.

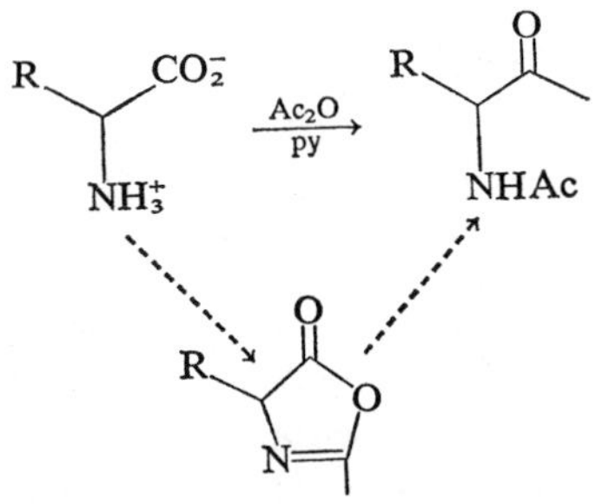

Functional Groups other than Alkanes, Acetylenes, Allenes, and Olefins 111

Scheme 60

to αβ-didehydroamino-acids and peptides. Oxidative[166] and reductive[167] routes to *N*-hydroxyamino-acids and derivatives have been described.

6 Carboxylic Acid Amides

Amide (Peptide) Bond Formation.—New developments in peptide bond formation have been discussed.[168,169] Extensive studies[170] of the use of *N*-phosphonium salts of pyridines as coupling agents in peptide synthesis have been described. The reactive species are generated from phosphorus acid esters and pyridine, with or without the presence of an oxidizing agent, which can be iodine,[171] as exemplified in Scheme 61; sulphurous acid esters[172] can also be used. Ureas or thioureas can be prepared[173] by this method from carbon dioxide or carbon disulphide and amines under ambient conditions.

Sulphonates of strongly acidic *N*-hydroxy-compounds are recommended[174] as coupling agents for amide bond formation, as are some oxadiazoline

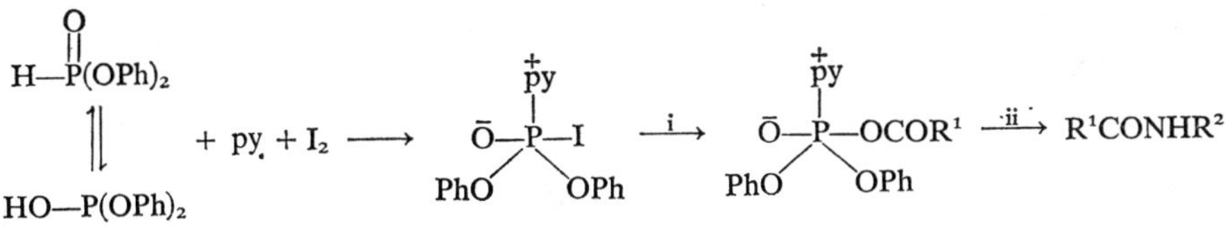

Reagents: i, R^1CO_2H; ii, R^2NH_2

Scheme 61

[166] T. Polonski and A. Chimiak, *Tetrahedron Letters*, 1974, 2453.
[167] A. Ahmad, *Bull. Chem. Soc. Japan*, 1974, **47**, 1819.
[168] J. H. Jones, *Chem. and Ind.*, 1974, 723.
[169] Y. S. Klausner and M. Bodanszky, *Synthesis*, 1974, 549.
[170] N. Yamazaki and F. Higashi, *Bull. Chem. Soc. Japan*, 1973, **46**, 3821; 1974, **47**, 170, and references therein.
[171] N. Yamazaki, F. Higashi, and S. A. Kazaryan, *Synthesis*, 1974, 436.
[172] N. Yamazaki, F. Higashi, and M. Niwano, *Tetrahedron*, 1974, **30**, 1319.
[173] N. Yamazaki, F. Higashi, and T. Iguchi, *Tetrahedron Letters*, 1974, 1191.
[174] M. Itoh, H. Nojima, and J. Notani, *Tetrahedron Letters*, 1974, 3089.

$$(Me_2N)_3\overset{+}{P}-N_3 \; PF_6^-$$
$$(38)$$

derivatives.[175] Further studies of the reactivity of pentafluorophenyl esters in peptide synthesis have been described.[176] The oxidation–reduction sequence of peptide coupling with triphenylphosphine and 2,2′-bipyridyl disulphide has now been applied to solid-phase peptide synthesis.[177] Full descriptive details[178] have been given on the use of diphenylphosphoryl azide as a peptide coupling agent; the intermediate reactive acyl azide can also be obtained by use of the phosphonium azide salt (38).[179] Tetrabutylammonium azide is recommended[180] as a stable source of azide ions soluble in organic solvents.

While it has been reported that the initial acylating agent in dicyclohexyl-carbodi-imide-mediated peptide synthesis is an *O*-acylisourea, the reaction seems[181] to follow the alternative path in the solid phase, where the acylating agent is a symmetrical acid anhydride. Thallium(I) ethoxide catalyses[182] the cleavage of sterically hindered peptides from Merrifield resins by trans-esterification with 2-dimethylaminoethanol.

General Preparation.—Adsorption of aldoximes on to chromatographic silica gel, followed by heating, results in their clean conversion[183] into amides; at no stage is nitrile detected, which suggests the mechanism shown (Scheme 62) for this preparatively useful reaction.

Nitriles are converted smoothly into amides by hydrolysis with acetic acid in the presence of titanium(IV) chloride;[184] other mild reagent systems[185,186] for this conversion have been described. The synthesis of allenic

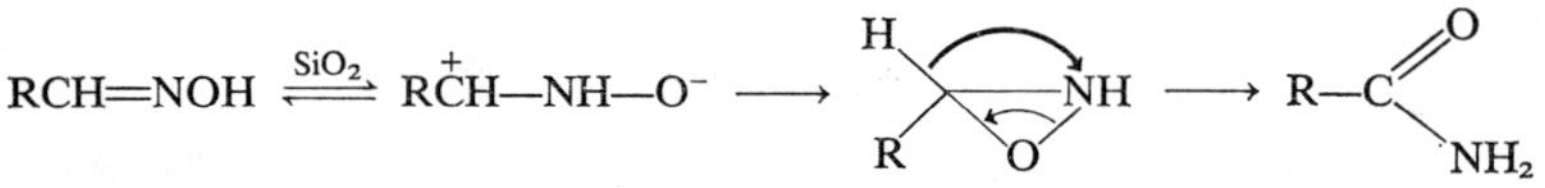

$$RCH=NOH \xrightleftharpoons{SiO_2} RC\overset{+}{H}-NH-O^- \longrightarrow \overset{H}{\underset{R}{\bigvee}}\overset{\frown}{\underset{O}{\bigvee}}NH \longrightarrow R-C\overset{O}{\underset{NH_2}{\diagdown}}$$

Scheme 62

[175] H. Fukuda, T. Endo, and M. Okawara, *Chem. Letters*, 1973, 1181.
[176] L. Kisafaludy, I. Schön, T. Szirtes, O. Nyéki, and M. Löw, *Tetrahedron Letters*, 1974, 1785.
[177] R. Matsueda, H. Maruyama, E. Kitazawa, H. Takahagi, and T. Mukaiyama, *Bull. Chem. Soc. Japan*, 1973, **46**, 3240.
[178] K. Ninomiya, T. Shioiri, and S. Yamada, *Tetrahedron*, 1974, **30**, 2151.
[179] B. Castro and J. R. Dormoy, *Bull. Soc. chim. France*, 1973, 3359.
[180] A. Brändström, B. Lamm, and I. Palmertz, *Acta Chem. Scand.*, 1974, **B28**, 699.
[181] J. Rebek and D. Feitler, *J. Amer. Chem. Soc.*, 1974, **96**, 1606; *cf.* 1973, **95**, 4052.
[182] J. Y. Savoie and M. A. Barton, *Canad. J. Chem.*, 1974, **52**, 2832; *cf.* M. A. Barton, R. U. Lemieux, and J. Y. Savoie, *J. Amer. Chem. Soc.*, 1973, **95**, 4501.
[183] J. B. Chattopadhyaya and A. V. Rama Rao, *Tetrahedron*, 1974, **30**, 2899.
[184] T. Mukaiyama, K. Kamio, S. Kobayashi, and H. Takei, *Chem. Letters*, 1973, 357.
[185] S. E. Diamond, B. Grant, G. M. Tom, and H. Taube, *Tetrahedron Letters*, 1974, 4025.
[186] S. Paraskewas, *Synthesis*, 1974, 574.

Functional Groups other than Alkanes, Acetylenes, Allenes, and Olefins 113

amides from allenic nitriles has been reported.[187] Barton[188] has given full details of the generalized Ritter reaction for the transformation of alcohols into amides under conditions of thermodynamic control.

The action of light enhances the rate of reaction of alkyl isocyanates with alcohols to give carbamates; the rate can be further enhanced by transition-metal catalysts such as ferrocene.[189] A non-aqueous low-temperature modification of the Hofmann rearrangement has been successfully applied[190] even to the demanding case shown (Scheme 63).

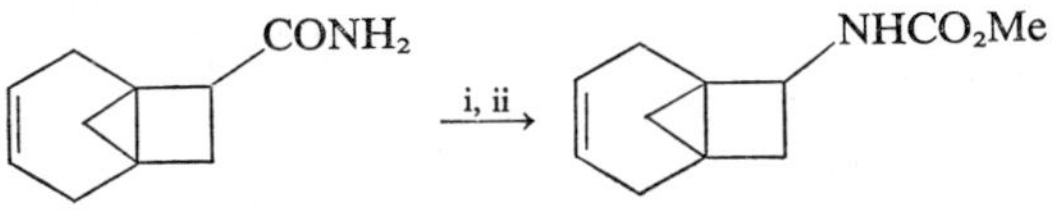

Reagents: i, NaOMe–MeOH–Br$_2$, −40 to −15 °C; ii, 50 °C

Scheme 63

Various allylic alcohols are transformed into homologous $\beta\gamma$-unsaturated NN-dimethylamides by thermolysis of their NN-dimethylformamide mixed acetals (Scheme 64), presumably via [2,3]sigmatropic rearrangement of the carbene (39); preparatively useful yields are realized[191] with all but $\gamma\gamma$-disubstituted allylic alcohols. This rearrangement complements the well-known [3,3]sigmatropic rearrangement observed when the allylic alcohol NN-dimethylacetamide mixed acetals are heated to yield $\gamma\delta$-unsaturated amides.

A stereospecific synthesis of allyl amides and amines is illustrated in Scheme 65; vinyl groups migrate readily with retention of configuration, allowing obtention of sterically defined products.[192]

The cycloaddition of ynamines to the carbonyl group of saturated lactones and esters leads (Scheme 66) to β-alkoxyacrylamides,[193] precursors of β-oxoamides. In a minor variant of their previous work, French authors[194] have reported on the Reformatsky reaction of α-bromoamides with Schiff bases.

---

[187] P. M. Greaves, P. D. Landor, S. R. Landor, and O. Odyek, *Tetrahedron*, 1974, **30**, 1427.
[188] D. H. R. Barton, P. D. Magnus, J. A. Garbarino, and R. N. Young, *J.C.S. Perkin I*. 1974, 2101.
[189] S. P. McManus, H. S. Bruner, H. D. Coble, and G. Choudhary, *J.C.S. Chem. Comm.* 1974, 253.
[190] P. Radlick and L. R. Brown, *Synthesis*, 1974, 290.
[191] G. Büchi, M. Cushman, and H. Wuest, *J. Amer. Chem. Soc.*, 1974, **96**, 5563.
[192] A. Pelter, A. Arase, and M. G. Hutchings, *J.C.S. Chem. Comm.*, 1974, 346.
[193] J. Ficini, J.-P. Genêt, and J.-C. Depezay, *Bull. Soc. chim. France*, 1973, 3367, 3369.
[194] F. Dardioze and M. Gaudemar, *Bull. Soc. chim. France*, 1974, 939.

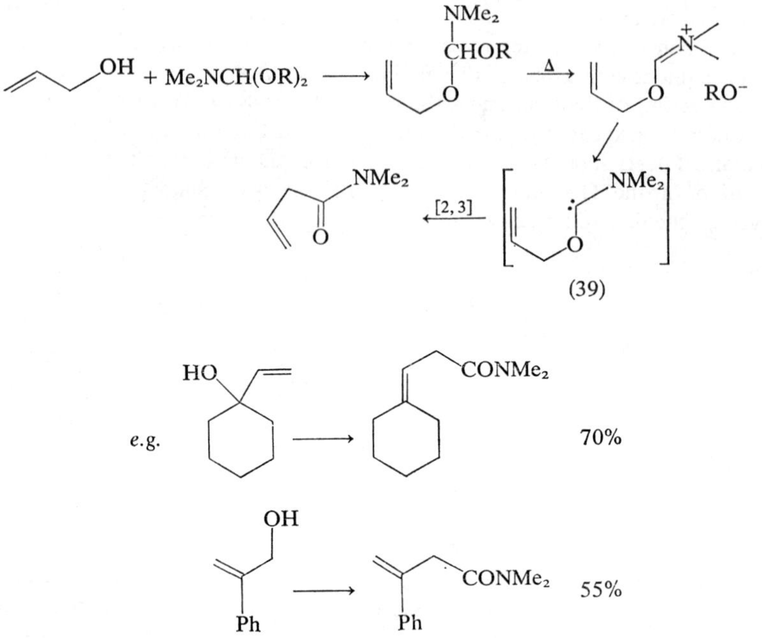

Scheme 64

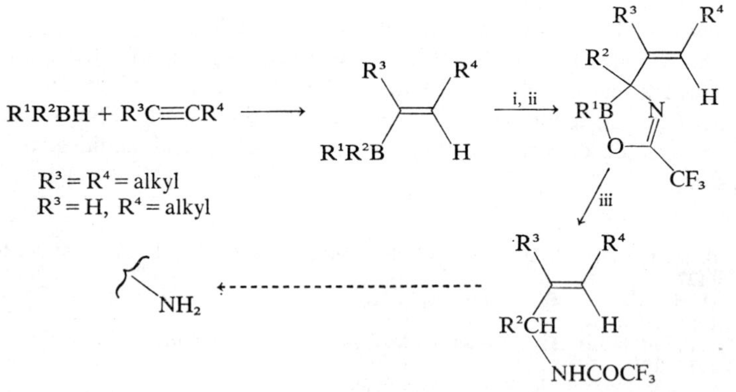

Scheme 65

Reagents: i, KCN; ii, (CF$_3$CO)$_2$O; iii, 2N–NaOH

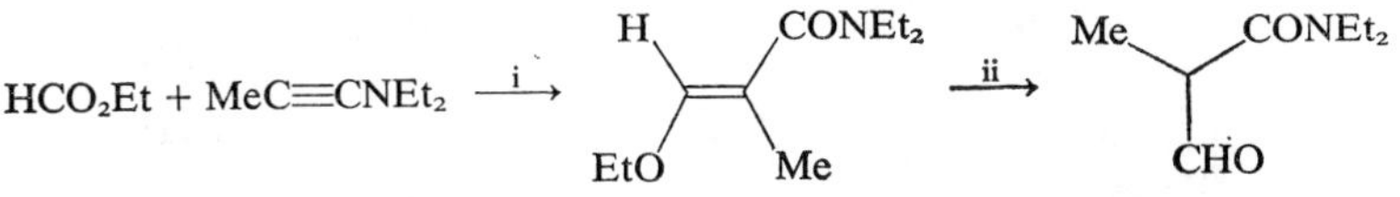

Functional Groups other than Alkanes, Acetylenes, Allenes, and Olefins 115

$$HCO_2Et + MeC{\equiv}CNEt_2 \xrightarrow{i} \underset{EtO}{\overset{H}{>}}{=}\underset{Me}{\overset{CONEt_2}{<}} \xrightarrow{ii} \underset{CHO}{\overset{Me}{>}}{<}\overset{CONEt_2}{}$$

Reagents: i, MgBr$_2$; ii, H$_3$O$^+$

Scheme 66

Metallated N-phenylketenimines react[195] with benzaldehyde and $\alpha\beta$-unsaturated aldehydes *via* carbonyl olefination to give α-phenylacrylanilides (Scheme 67).

αα-Dicyanoethyl acetate is reported[196] to function as a reagent for selective N-acetylation in the presence of hydroxy-groups, although pertinent examples are lacking. The antibiotic hydroxamic acid actinonin (40) has been synthesized.[197]

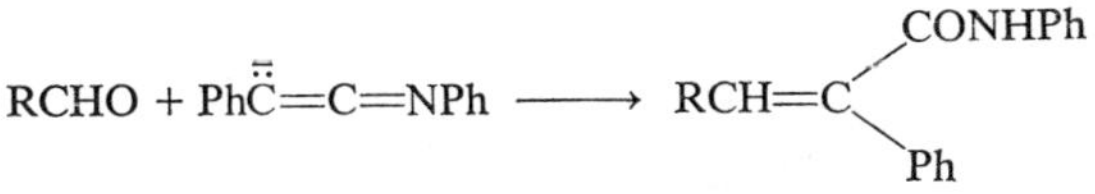

$$RCHO + Ph\ddot{C}{=}C{=}NPh \longrightarrow RCH{=}C{<}\overset{CONHPh}{\underset{Ph}{}}$$

Scheme 67

Hydrolysis and Protonation.—Although there is now considerable evidence for the formation of a tetrahedral intermediate in carbonyl substitution reactions, the instability of this intermediate and the complicated kinetics of its simultaneous formation and breakdown have precluded much detailed analysis. It has now been reported[198] that the reaction of the imidate cation (41) with nucleophiles proceeds *via* the *stable* tetrahedral species (42), and a thorough mechanistic study has been carried out.

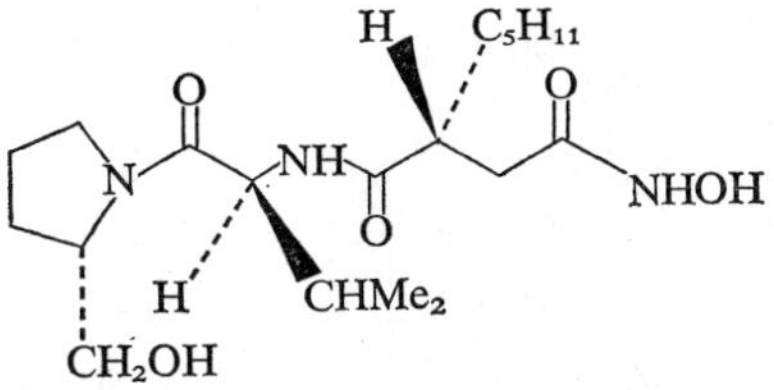

(40)

[195] U. Schöllkopf and I. Hoppe, *Annalen*, 1974, 1655.
[196] T.-L. Ho, *Synthetic Comm.*, 1974, 4, 351.
[197] N. H. Anderson, W. D. Ollis, J. E. Thorpe, and A. D. Ward, *J.C.S. Chem. Comm.* 1974, 420; J. P. Devlin, W. D. Ollis, J. E. Thorpe, R. J. Wood, B. J. Broughton, P. J. Warren, K. R. H. Wooldridge, and D. E. Wright, *ibid.*, p. 421.
[198] N. Gravitz and W. P. Jencks, *J. Amer. Chem. Soc.*, 1974, 96, 489, 499, 507.

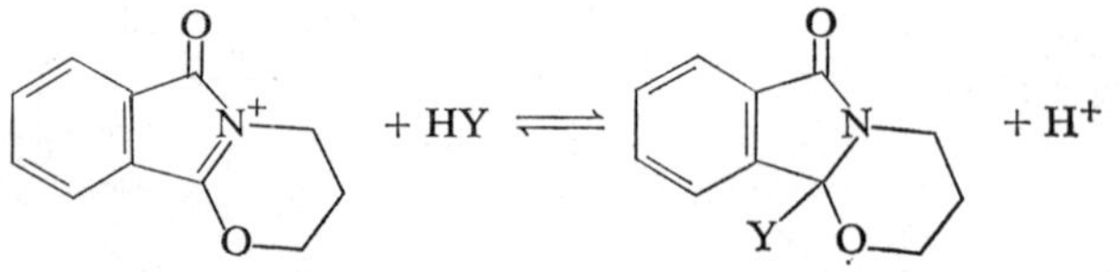

(41) + HY ⇌ (42) + H⁺

An *ab initio* study[199] of the reactivity of aminodihydroxymethane has provided theoretical evidence for the operation of marked stereoelectronic effects, which substantiates Deslongchamps' postulate[200] that the direction of cleavage of the tetrahedral intermediate is controlled by lone-pair orbitals being orientated antiperiplanar to the bond being cleaved. The comparison of imidate and amide hydrolysis in concentrated acids may be unwarranted; evidence has been presented that O-alkyl cleavage occurs[201] in the imidate case and that neutral amide is the leaving group (Scheme 68). Other studies

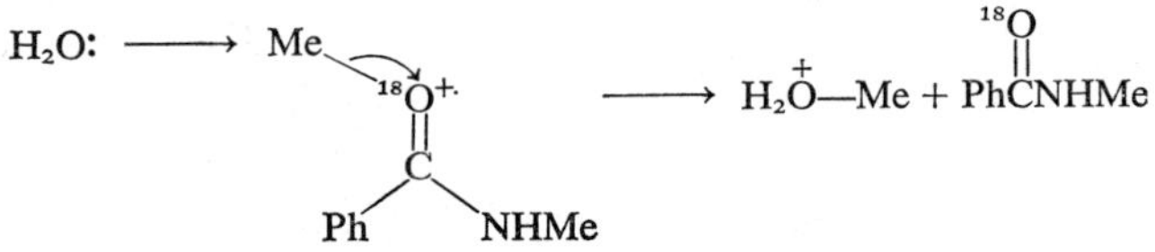

$$H_2O: \longrightarrow Me-{}^{18}O^+-C(Ph)(NHMe) \longrightarrow H_2\overset{+}{O}-Me + PhC({}^{18}O)NHMe$$

Scheme 68

of amide[202] and thioamide[203] hydrolysis have been reported. The stereochemistry[204] of open chain O-methyl imidates and large-membered cyclic lactim ethers has been revised to (E).

From a careful study of the two sets of amides (43) and (44), using criteria such as ring size, electronic effects, change in hybridization during reaction, and conjugation, Kresge[205] has discounted N-protonated conjugate acids,

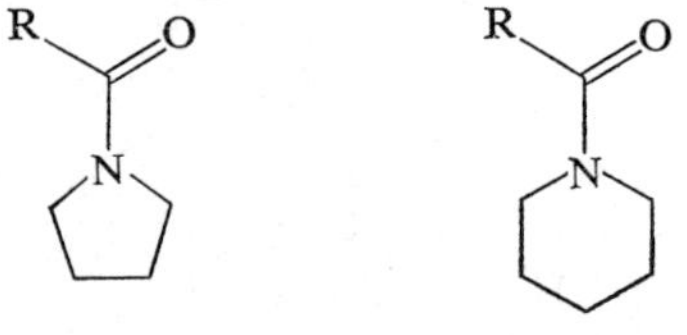

(43)   (44)

[199] J. M. Lehn and G. Wipff, *J. Amer. Chem. Soc.*, 1974, **96**, 4048.
[200] P. Deslongchamps, P. Atlani, D. Fréhel, and A. Malaval, *Canad. J. Chem.*, 1972, **50**, 3405; P. Deslongchamps, C. Lebreux, and R. Taillefer, *ibid.*, 1973, **51**, 1665.
[201] R. A. McLelland, *J. Amer. Chem. Soc.*, 1974, **96**, 3690.
[202] B. C. Challis and S. P. Jones, *J.C.S. Chem. Comm.*, 1974, 748.
[203] A. J. Hall and D. P. N. Satchell, *Chem. and Ind.*, 1974, 527.
[204] C. O. Messe, W. Walter, and M. Berger, *J. Amer. Chem. Soc.*, 1974, **96**, 2259.
[205] A. J. Kresge, P. H. Fitzgerald, and Y. Chiang, *J. Amer. Chem. Soc.*, 1974, **96**, 4698,

Functional Groups other than Alkanes, Acetylenes, Allenes, and Olefins 117

not only as the principal products of equilibrium protonation of amides in dilute and moderately concentrated aqueous acids, but also as essential intermediates in acid-catalysed amide hydrolysis. The predominant protonated form of DMF in aqueous acid is the O-protonated amide, and a doublet methyl signal is observed[206] in ^{13}C n.m.r. spectroscopy at all acidities. Further evidence has been presented[207] for S-protonation of primary thioamides. Liler[208] has published full details of her work on the site of protonation of benzamide, and has promised a detailed response to the many critics of N-protonation.

Properties and Reactions.—A spectroscopic investigation[209] has shown that whereas bistrimethylsilylformamide has the amide structure (45), all other bistrimethylsilylamides studied are in the imidate form (46).

(45)   (46)

An *ab initio* study of hydrogen bonding involving the amide linkage has been reported,[210] as has an n.m.r. spectroscopic study of hydrogen exchange[211] in amidium ions. Hartree–Fock SCF calculations[212] on amido radicals[213] have predicted the π ground state (47), in accord with e.s.r. spectroscopic assignments.

Some spectral consequences[214] of non-planarity of amide bonds have been noted. The activation parameters for C—N bond rotation in some amides, thioamides, and amidinium ions have been determined[215] by n.m.r. spectroscopy.

N-Substituted maleimides are obtained[216] in good yield by direct coupling of alkyl or aralkyl halides with silver(I) maleimide (Scheme 69). Interestingly, adamantyl-1-bromide gave the O-alkyl maleimide (48) at room temperature; use of higher temperatures resulted in formation of the expected N-alkyl

---

[206] R. A. McLelland and W. F. Reynolds, *J.C.S. Chem. Comm.*, 1974, 824.
[207] W. Walter, M. F. Sieveking, and E. Schaumann, *Tetrahedron Letters*, 1974, 839.
[208] M. Liler, *J.C.S. Perkin II*, 1974, 71.
[209] C. H. Yoder, W. C. Copenhafer, and B. DuBeshter, *J. Amer. Chem. Soc.*, 1974, 96, 4283.
[210] A. Johansson, P. Kollman, S. Rothenberg, and J. McKelvey, *J. Amer. Chem. Soc.*, 1974, 96, 3794.
[211] C. L. Perrin, *J. Amer. Chem. Soc.*, 1974, 96, 5629, 5631.
[212] T. Koenig, J. A. Hoobler, C. E. Klopfenstein, G. Hedden, F. Sunderman, and B. R. Russell, *J. Amer. Chem. Soc.*, 1974, 96, 4573.
[213] J. N. S. Tam, R. W. Yip, and Y. L. Chow, *J. Amer. Chem. Soc.*, 1974, 96, 4543, 4573.
[214] M. Tichý, E. Dušková, and K. Bláha, *Tetrahedron Letters*, 1974, 237.
[215] R. C. Neuman and V. Jonas, *J. Org. Chem.*, 1974, 39, 925, 929.
[216] A. L. Schwartz and L. M. Lerner, *J. Org. Chem.*, 1974, 39, 21.

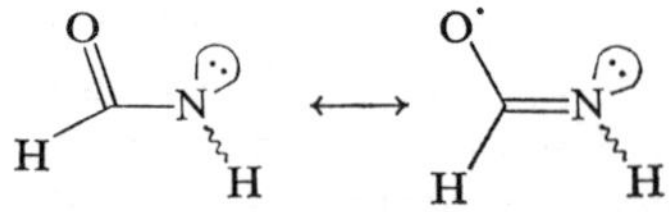

(47)

product. Some results[217] on the regioselectivity of lactam enolate alkylation have been described.

Formamides are oxidized by selenium[218] in the presence of alkoxides to give carbamates. Substituted α-lactams react[219] with t-butyl-lithium as shown in Scheme 70.

NN-Dimethyl-N'-alkyl formamidines can be obtained[220] by reaction of aliphatic iminophosphoranes with DMF. NN-Disubstituted β-ketoamides undergo photocyclization to give pyrrolidinones (Scheme 71) in good yield.[221]

The properties and preparation of some N-substituted di-iodoacetamides have been described.[222] N-Monochloro or N-monobromo derivatives of primary amides (Hofmann rearrangement intermediates) and carbamates can be obtained under carefully controlled[223] conditions. Hydrogen cyanide

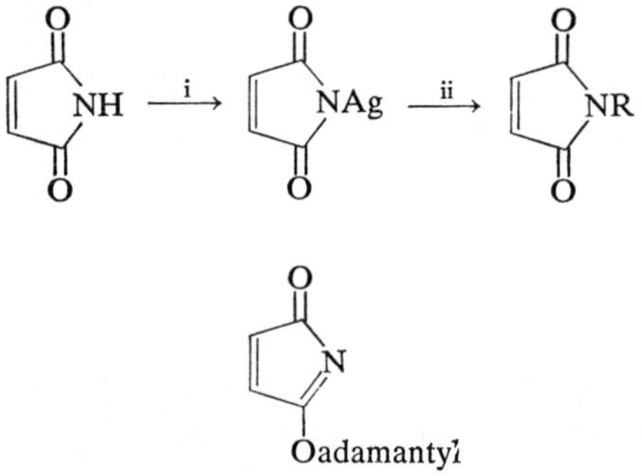

Oadamantyl
(48)

Reagents: i, AgNO$_3$–NaOH; ii, RX–PhCH$_3$, heat

Scheme 69

[217] B. M. Trost and R. A. Kunz, *J. Org. Chem.*, 1974, **39**, 2475.
[218] K. Kondo, N. Sonoda, and H. Sakurai, *J.C.S. Chem. Comm.*, 1974, 160; cf. ref. 113.
[219] E. R. Talaty and C. M. Utermoehlen, *J.C.S. Chem. Comm.*, 1974, 204.
[220] J. M. Muchowski, *Canad. J. Chem.*, 1974, **52**, 2255.
[221] T. Hasegawa and H. Aoyama, *J.C.S. Chem. Comm.*, 1974, 743.
[222] A. Stephen, G. Schultz, and H. Reinshagen, *Annalen*, 1974, 363.
[223] C. Bachand, H. Driguez, J. M. Paton, D. Touchard, and J. Lessard, *J. Org. Chem.*, 1974, **39**, 3136; cf. S. C. Czapf, H. Gottlieb, G. F. Whitfield, and D. Swern, *J. Org. Chem.*, 1973, **38**, 2555.

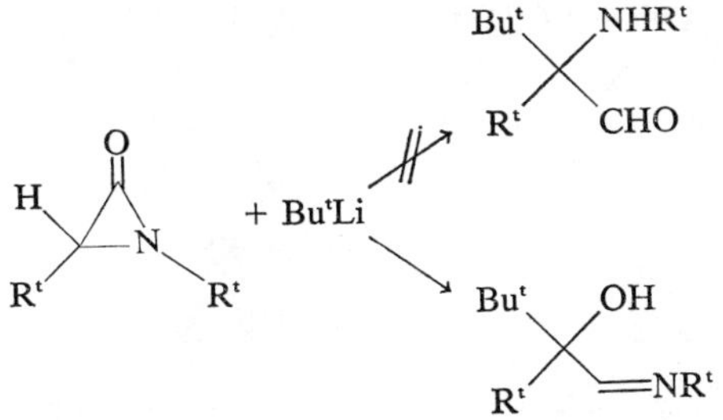

Scheme 70

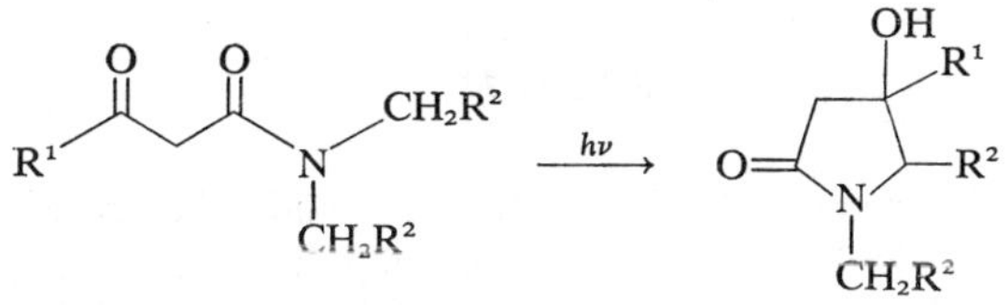

Scheme 71

is present in concentrations of 10^{-5} to 10^{-3} mol l^{-1} in crude DMF, which explains some hitherto peculiar results.[224]

7 Nitriles and Isocyanides

Preparation.—Mild dehydrative routes to nitriles continue to be of interest. Aldoximes give nitriles when treated with alkyl orthoformates under acid catalysis; the primary products are oxime dialkyl orthoesters (Scheme 72), which can be isolated[225] prior to undergoing Beckmann fragmentation. This system is also well-suited for the fragmentation of α-oximino-ketone acetals, as shown. Phosgene immonium chlorides[226] and carbodi-imides[227] both smoothly dehydrate aldoximes; copper(II) catalysis is required in the latter case. The generality of use of hydroxylamine O-sulphonic acid for oxidative conversion of aldehydes into nitriles has been determined.[228]

Trialkylboranes react with 2-bromo-6-lithiopyridine; stereospecific alkylative cleavage[229] occurs and 5-alkyl-(2Z, 4E)-pentadiene nitriles are formed in high yield (Scheme 73).

---

[224] J. C. Trisler, B. F. Freasier, and S.-M. Wu, *Tetrahedron Letters*, 1974, 687.
[225] M. M. Rogić, J. F. Van Peppen, K. P. Klein, and T. R. Demmin, *J. Org. Chem.*, 1974, **39**, 3424.
[226] V. P. Kukhar and V. I. Pasternak, *Synthesis*, 1974, 563.
[227] E. Vowinkel and J. Bartel, *Chem. Ber.*, 1974, **107**, 1221.
[228] C. Fizet and J. Streith, *Tetrahedron Letters*, 1974, 3187.
[229] K. Utimoto, N. Sakai, and H. Nozaki, *J. Amer. Chem. Soc.*, 1974, **96**, 5602.

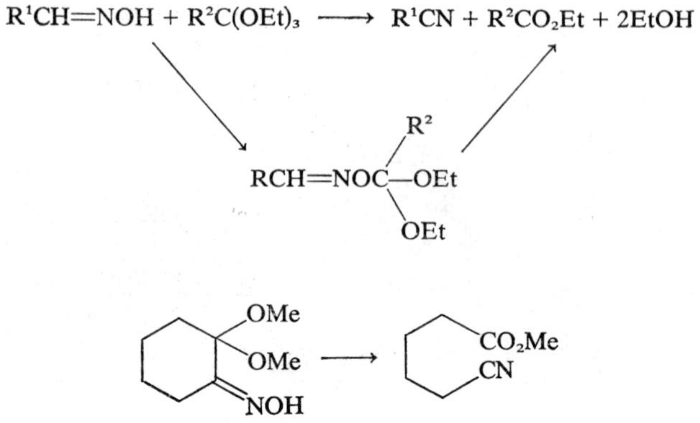
R¹CH=NOH + R²C(OEt)₃
R¹CN + R²CO₂Et + 2EtOH
R²
RCH=NOC—OEt
OEt
OMe
OMe
NOH
CO₂Me
CN

Scheme 72

Stereoselective routes to $\alpha\beta$-unsaturated[230] and $\alpha\beta$-epoxy[231] nitriles have been explored, *via* Wadsworth–Horner–Emmons and Darzens condensations, respectively. Phase transfer catalysis[232] has been used for the preparation of aroyl nitriles in excellent yields.

α-Anions.—Cyclizations involving carbanion attack on an electrophile usually result in formation of a five- rather than a six-membered ring, and seldom a four-membered ring. Stork[233] has reported a process of 'epoxy-nitrile cyclization' in which these tendencies are reversed (Scheme 74); the reversals are ascribed to the geometric constraints imposed by the oxiran

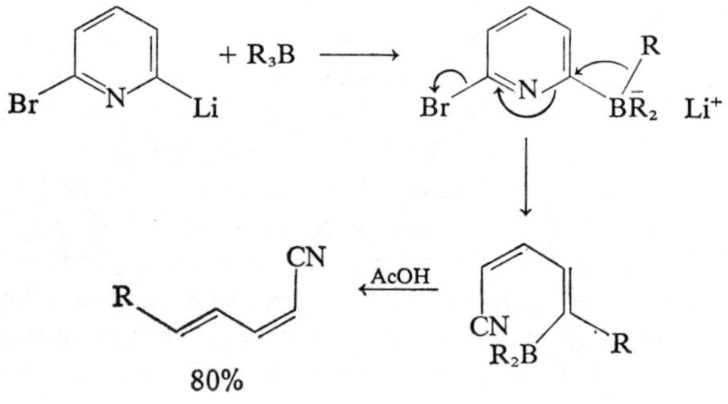
+ R₃B
Br
N
Li
R
BR₂
Li⁺
CN
AcOH
R
80%
R₂B

Scheme 73

[230] A. Redjal and J. Seyden-Penne, *Tetrahedron Letters*, 1974, 1733.
[231] G. Kyriakakou and J. Seyden-Penne, *Tetrahedron Letters*, 1974, 1737.
[232] K. E. Koenig and W. P. Weber, *Tetrahedron Letters*, 1974, 2275.
[233] G. Stork, L. D. Cama, and D. R. Coulson, *J. Amer. Chem. Soc.*, 1974, **96**, 5268; G. Stork and J. F. Cohen, *ibid.*, p. 5270.

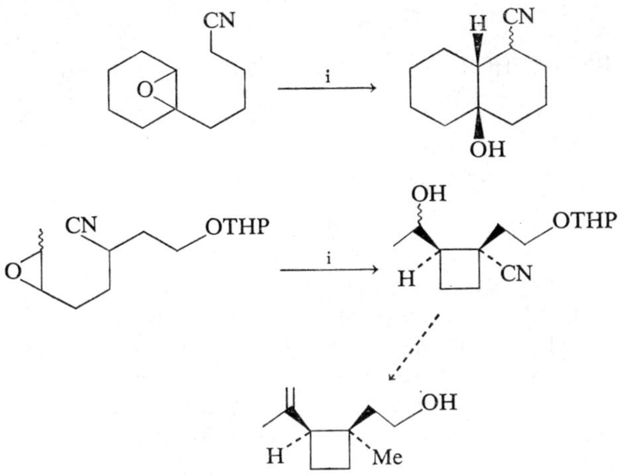

Functional Groups other than Alkanes, Acetylenes, Allenes, and Olefins 121

Reagent: i, NaN(SiMe₃)₂

Scheme 74

ring in each case, which make it difficult for the nitrile anion and the epoxide C—O bond to come into line for five-membered ring formation. The second reaction shown is highly stereoselective, and has been employed in a synthesis of (±)-grandisol (49).

α-Sulphenylation or α-selenylation introduces[234] αβ-unsaturation into nitriles (Scheme 75); two equivalents of base are required, to preclude complications arising from proton transfer between the product and the unchanged starting anion.

A general synthesis[235] of α-substituted acrylonitriles has been outlined. α-Silyl nitrile α-anions react as expected with carbonyl compounds to afford[236] αβ-unsaturated nitriles, as shown in Scheme 76.

The monoalkylation[237] of primary nitriles using lithium dialkylamide bases has been described. ω-Halogenonitriles can be prepared by alkylation[238]

$$RCH_2CH_2CN \xrightarrow{i-iii} RCH=CHCN$$

Reagents: i, 2 LiNR₂; ii, (PhSe)₂, PhSeBr, or (MeS)₂; iii, H₂O₂, heat

Scheme 75

[234] D. N. Brattesani and C. H. Heathcock, *Tetrahedron Letters*, 1974, 2279.
[235] R. B. Miller and B. F. Smith, *Synthetic Comm.*, 1973, 3, 413.
[236] I. Ojima, M. Kumagai, and Y. Nagai, *Tetrahedron Letters*, 1974, 4005.
[237] D. S. Watt, *Tetrahedron Letters*, 1974, 707.
[238] M. Larchevêque, A. Debal, and T. Cuvigny, *Bull. Soc. chim. France*, 1974, 1710.

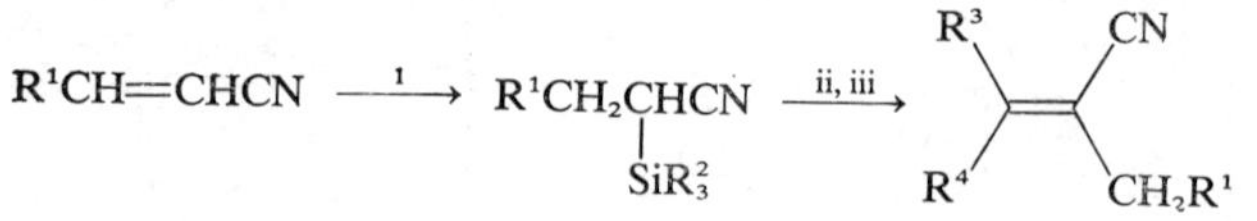

$R^1CH=CHCN \xrightarrow{i} R^1CH_2\underset{SiR_3^2}{CHCN} \xrightarrow{ii, iii} \underset{R^4}{\overset{R^3}{>}}=\underset{CH_2R^1}{\overset{CN}{<}}$

Reagents: i, $R_3^2SiH-(Ph_3P)_3RhCl$; ii, $LiNPr_2^i$; iii, R^3COR^4

Scheme 76

of nitrile α-anions with αω-dihalogenoalkanes. Nitrile α-anions react with trimethylsilyl chloride as expected, to give α-silyl nitriles. If, however, t-butyldimethylsilyl chloride is employed, the anions are trapped in their ketenimine form. This provides an efficient route[239] for the oxidative decyanation of secondary aralkyl and diaryl nitriles to ketones (Scheme 77).

The utility of α-metallated isocyanides[240] for the introduction of, *inter alia*, masked aminoalkyl groups has been reviewed.

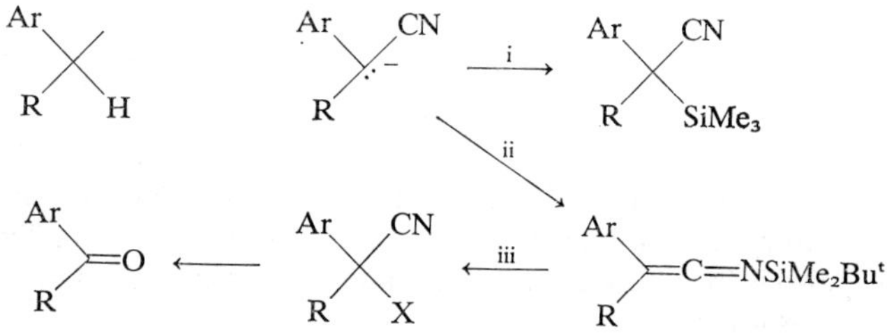

Reagents: i, Me_3SiCl; ii, Bu^tMe_2SiCl; iii, I_2, Br_2, or $PhSCl$

Scheme 77

Properties and Reactions.—The microwave spectrum[241] of methoxyacetonitrile has been determined. Silane reduction of nitrilium ions produces aldimines, and thence aldehydes[242] (Scheme 78); this complements the known borohydride reduction of such ions to amines.

Syntheses of α-amino-acids from nitriles[152] and new mild methods for the hydrolysis of nitriles[184-186] to amides are described above.

Walborsky[243] has given full details of the utility of metalloaldimines as masked acyl carbanions; the required species are obtained by the α-addition

---

[239] D. S. Watt, *J. Org. Chem.*, 1974, **39**, 2799; S. J. Selikson and D. S. Watt, *Tetrahedron Letters*, 1974, 3029.
[240] D. Hoppe, *Angew. Chem. Internat. Edn.*, 1974, **13**, 789.
[241] R. Kewley, *Canad. J. Chem.*, 1974, **52**, 509.
[242] J. L. Fry, *J.C.S. Chem. Comm.*, 1974, 45.
[243] G. E. Niznik, W. H. Morrison, and H. M. Walborsky, *J. Org. Chem.*, 1974, **39**, 600.

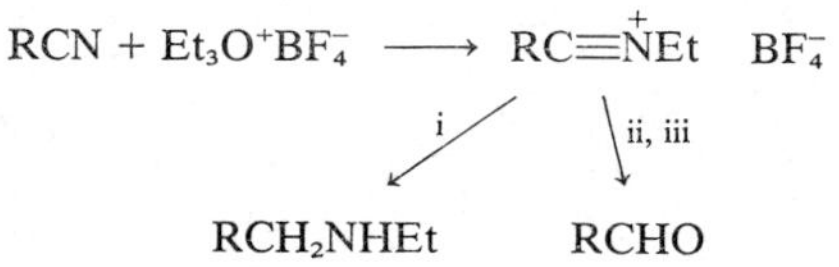

$$RCN + Et_3O^+BF_4^- \longrightarrow RC\equiv\overset{+}{N}Et \quad BF_4^-$$

with i giving RCH_2NHEt and ii, iii giving $RCHO$

Reagents: i, $NaBH_4$; ii, Et_3SiH; iii, H_3O^+

Scheme 78

of organometallic reagents to isocyanides which do not possess α-hydrogen atoms. Applications include a partial asymmetric synthesis[153] of α-amino-acids, and a new route[244] to secondary and tertiary nitriles (Scheme 79).

Toluene-*p*-sulphonylmethyl isocyanide reacts with ketones to give a variety of products, depending on the conditions used[245] (Scheme 80); this results in, *inter alia*, a new synthesis of α-hydroxyaldehydes from ketones, where the reagent effectively acts as the anion of formaldehyde.

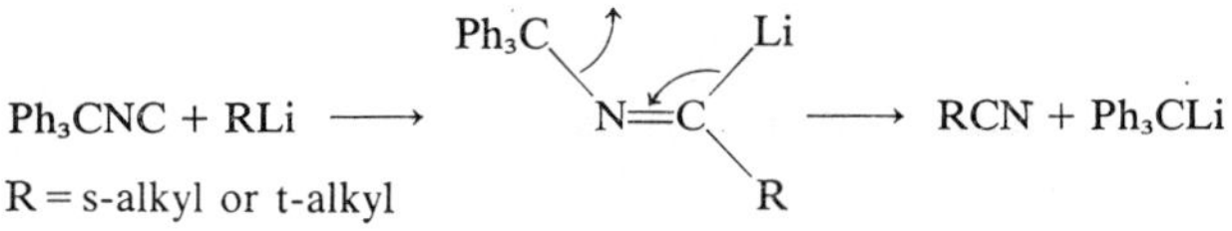

$$Ph_3CNC + RLi \longrightarrow \underset{R}{Ph_3C\diagdown N\overset{Li}{=}C\diagup} \longrightarrow RCN + Ph_3CLi$$

R = s-alkyl or t-alkyl

Scheme 79

The bis-π-allylnickel complex (50), formed from butadiene and allene, undergoes an insertion reaction[246] with t-butyl isocyanide to give the cyclic imine (51), hydrolysis and hydrogenation of which provides a high yield synthesis of (±)-muscone (Scheme 81).

The isocyanides (52), isomers of cyanohydrin esters, have been prepared[247] for the first time (Scheme 82).

8 Aldehydes and Ketones

Preparation.—*By Oxidation.* A detailed mechanistic study[248] of the dimethyl sulphide–*N*-chlorosuccinimide oxidation of alcohols has confirmed the intramolecular nature of proton abstraction in the transition state (53). While secondary and tertiary 1,2-diols normally undergo cleavage on oxidation, the use of sulphonium or sulphoxonium salts allows[249] their clean conversion into α-hydroxyketones (Scheme 83); such selectivity is ascribed

---

[244] M. P. Periasamy and H. M. Walborsky, *J. Org. Chem.*, 1974, **39**, 611.
[245] O. H. Oldenziel and A. M. van Leusen, *Tetrahedron Letters*, 1974, 163, 167.
[246] R. Baker, R. C. Cookson, and J. R. Vinson, *J.C.S. Chem. Comm.*, 1974, 515.
[247] G. Höfle, *Angew. Chem. Internat. Edn.*, 1974, **13**, 677.
[248] J. P. McCormick, *Tetrahedron Letters*, 1974, 1701.
[249] E. J. Corey and C. U. Kim, *Tetrahedron Letters*, 1974, 287.

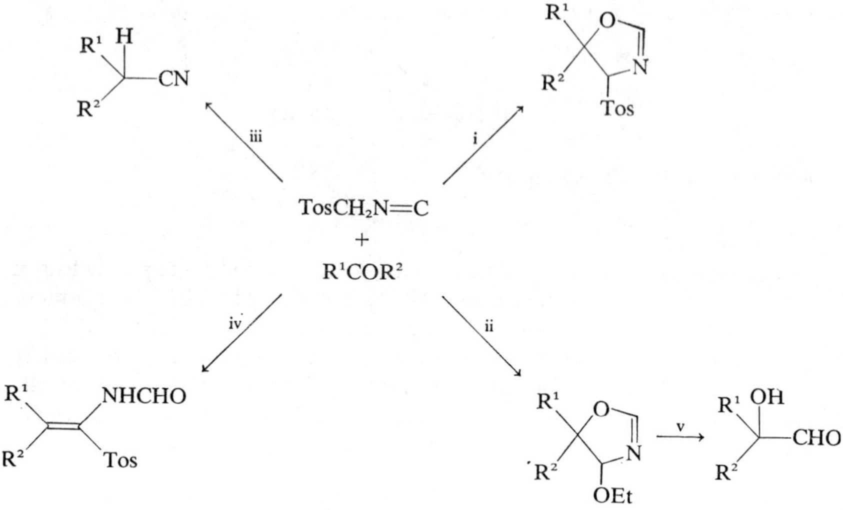

Reagents: i, K_2CO_3–MeOH; ii, TlOEt–EtOH–DME; iii, $KOBu^t$–DME; iv, $KOBu^t$–THF, −5 °C; v, H_3O^+

Scheme 80

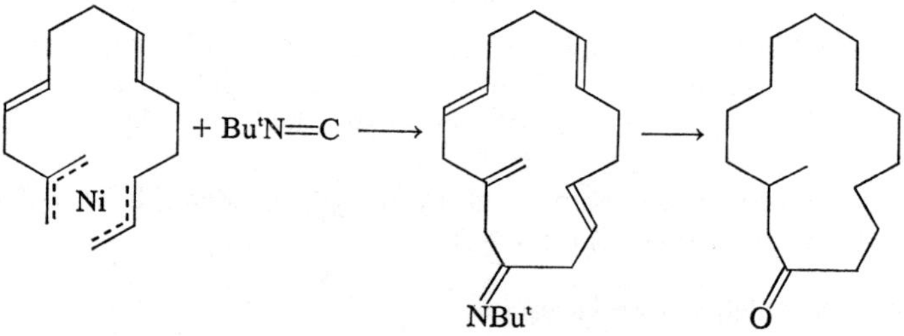

(50) (51)

Scheme 81

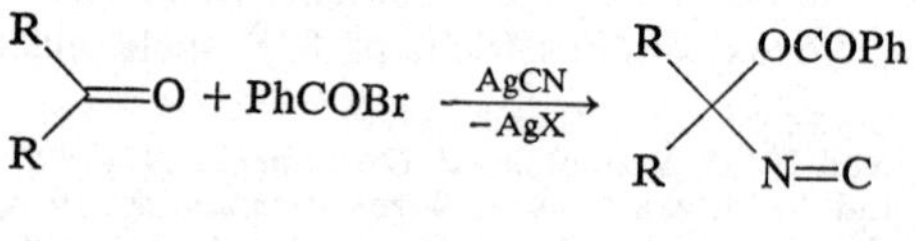

(52)

Scheme 82

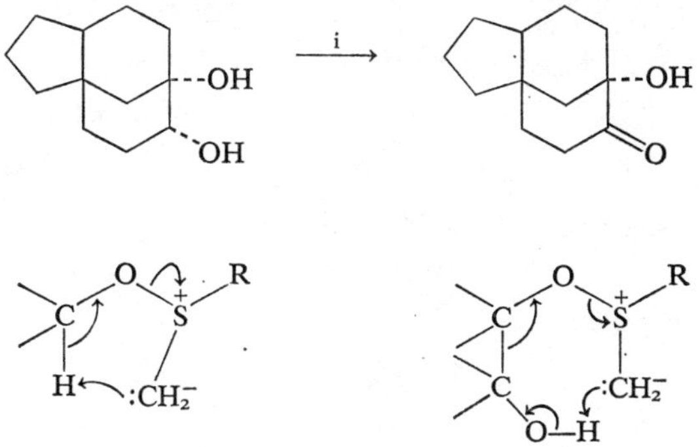

Functional Groups other than Alkanes, Acetylenes, Allenes, and Olefins 125

(53)   (54)

Reagent: i, PhSMe–Cl$_2$ or Me$_2$SO–Cl$_2$

Scheme 83

to entropy favouring the five-membered oxidation transition state (53) rather than the seven-membered cleavage transition state (54).

The range of electrophiles suitable for the activation of dimethyl sulphoxide as an alcohol oxidizing agent has been extended[250] to include silver(I) ions.[251] A solution of sodium dichromate and sulphuric acid in dimethyl sulphoxide efficiently oxidizes primary and secondary alcohols to aldehydes and ketones respectively; dimethyl sulphoxide is not consumed.[252] A pyridine–chlorine complex cleanly converts[253] alcohols into carbonyl compounds, and secondary alcohols are oxidized preferentially (Scheme 84).

Few methods are available for the direct selective oxidation of a primary hydroxy-group in the presence of a secondary one. Triphenylphosphine dibromide in DMF cleanly converts primary alcohols into bromides (which can be thought of as masked aldehydes) and secondary alcohols into the corresponding formates (Scheme 85); this selectivity is applicable intramolecularly, and increases when the alcohols are adjacent.[254] Full details[255]

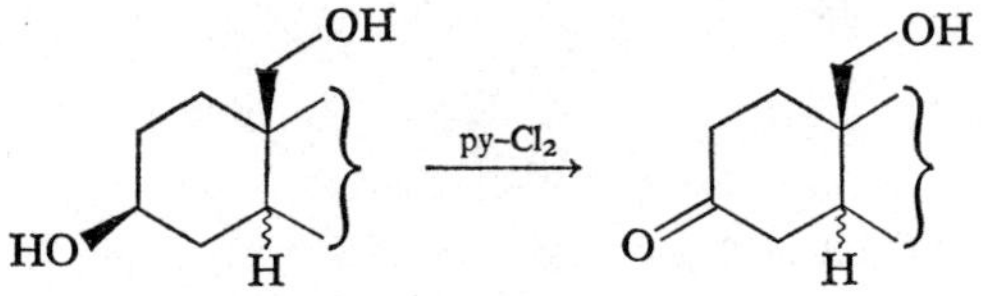

Scheme 84

[250] J. D. Albright, *J. Org. Chem.*, 1974, **39**, 1977.
[251] B. Ganem and R. K. Boeckman, *Tetrahedron Letters*, 1974, 917.
[252] Y. S. Rao and R. Filler, *J. Org. Chem.*, 1974, **39**, 3304.
[253] J. Wicha and A. Zarecki, *Tetrahedron Letters*, 1974, 3059.
[254] R. K. Boeckman and B. Ganem, *Tetrahedron Letters*, 1974, 913.
[255] F. J. Kakis, M. Fetizon, N. Douchkine, M. Golfier, P. Mourges, and T. Prange, *J. Org. Chem.*, 1974, **39**, 523.

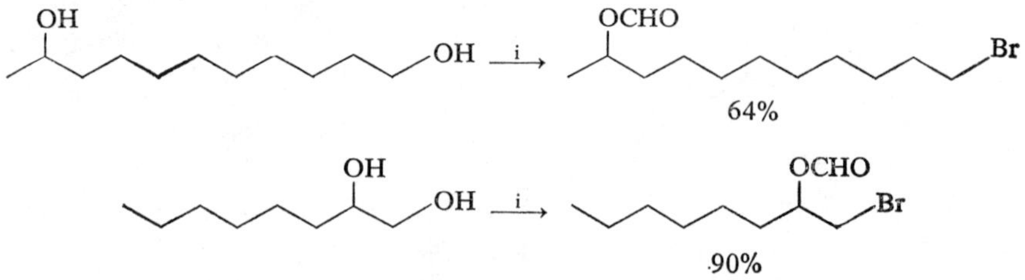

Scheme 85

Reagent: i, Ph₃PBr₂–DMF

have been given of the mechanism of oxidation of alcohols with silver(I) carbonate on Celite; a concerted process is implicated.

A safer and more convenient procedure has been described[256] for the direct oxidation of olefins to α-diketones with potassium permanganate in acetic anhydride. The palladium(II) chloride-induced oxidation of terminal olefins to methyl ketones has been studied further,[257] and aqueous sulpholane is recommended as solvent. Cyclic olefins can be oxidatively cleaved to di-aldehydes,[258] often in acceptable yield, by amino-group assisted cleavage of the derived 2-aminocycloalkyl nitrate esters (Scheme 86).

The double bond of a silyl enol ether is so electron rich that selective ozonolysis[259] in the presence of other double bonds is possible; the particular example shown (Scheme 87) is part of a synthesis of (±)-vernolepin. The selective ozonolysis of the double bond of an enyne has been described;[260] the amount of ozone used does not appear to be very critical.

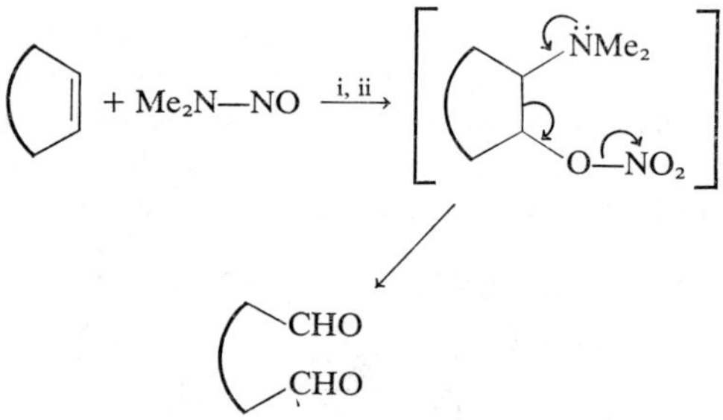

Reagents: i, hν–HCl–O₂; ii, pH 9

Scheme 86

[256] H. P. Jensen and K. B. Sharpless, *J. Org. Chem.*, 1974, **39**, 2314.
[257] D. R. Fahey and E. A. Zuech, *J. Org. Chem.*, 1974, **39**, 3276.
[258] K. S. Pillay, R. N. Lockhart, T. Tezuka, and Y. L. Chow, *J.C.S. Chem. Comm.*, 1974, 80.
[259] R. D. Clark and C. H. Heathcock, *Tetrahedron Letters*, 1974, 1713, 2027.
[260] P. McCurry and K. Abe, *Tetrahedron Letters*, 1974, 1387.

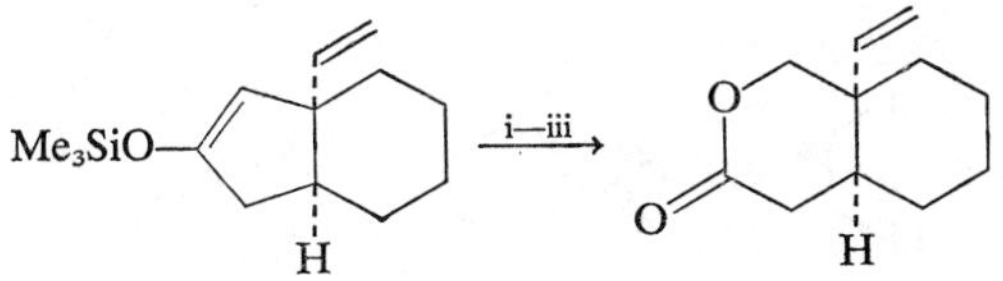

Functional Groups other than Alkanes, Acetylenes, Allenes, and Olefins 127

Reagents: O_3–MeOH; ii, $NaBH_4$; iii, H_3O^+

Scheme 87

Improved procedures for the allylic oxidation[261] of cyclohexene to cyclohexenone, and for the conversion of benzyl halides into benzaldehydes,[262] have been described. Alkoxides are oxidized by singlet oxygen to carbonyl compounds in good yield; the intermolecular mechanism shown (Scheme 88) is favoured,[263] as the reaction is sluggish in aprotic solvents.

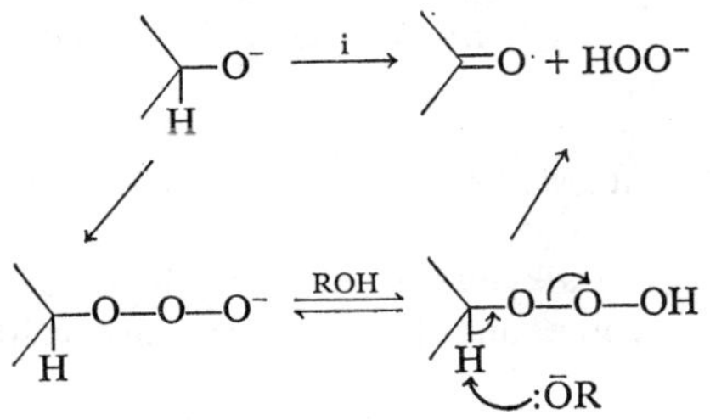

Reagent: i, O_2–$h\nu$–sensitizer

Scheme 88

The presence of sodium acetate greatly enhances the yields in the Pummerer rearrangement[264] of functionalized sulphoxides to protected aldehydes. Methyl stearate and homologous esters are converted into mono-keto derivatives by chromium trioxide–acetic anhydride–acetic acid; modest selectivity[265] for attack at biologically interesting positions was observed.

By Addition, Insertion, and Rearrangement Routes. The formation of cyclopentanone from ethylene and CO has been reported.[266] The titanium(IV) metallocycle (55) reacts with CO to give cyclopentanone in 80% yield; reaction of (56) with ethylene gives an intermediate which behaves in an identical manner to the metallocycle (55), although yields are lower; this suggests the titanium(II)–titanium(IV) equilibrium shown (Scheme 89).

[261] C. S. Sharma, S. C. Sethi, and S. Dev, *Synthesis*, 1974, 45.
[262] A. McKillop and M. E. Ford, *Synthetic Comm.*, 1974, **4**, 45.
[263] H. H. Wasserman and J. E. Van Verth, *J. Amer. Chem. Soc.*, 1974, **96**, 585.
[264] S. Iriuchijima, K. Maniwa, and G. Tsuchihashi, *J. Amer. Chem. Soc.*, 1974, **96**, 4280; see also ref. 641.
[265] C. R. Eck, D. J. Hunter, and T. Money, *J.C.S. Chem. Comm.*, 1974, 865.
[266] J. X. McDermott and G. M. Whitesides, *J. Amer. Chem. Soc.*, 1974, **96**, 947.

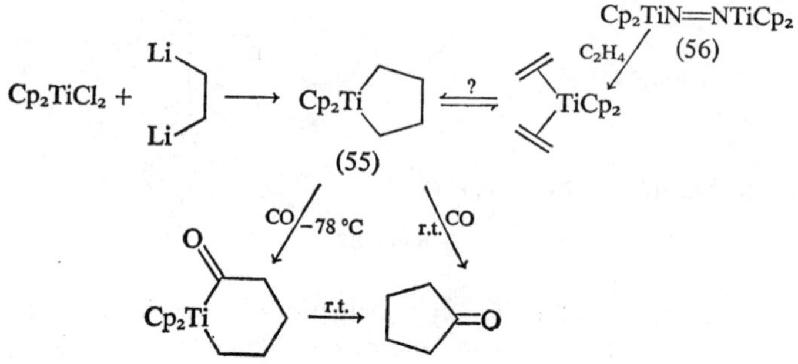

Scheme 89

Olefins, particularly if strained, react[267] with iron pentacarbonyl as shown (Scheme 90); a related reaction[268] with butadienes leads to cyclopentenones, albeit in modest yield.

In a process mechanistically analogous to hydroformylation, the metal hydride mediated synthesis of ketones from terminal olefins[269] has been described (Scheme 91).

Rearrangement of 1-alkylidene-2-alkoxycyclopropanes leads[270] to $\beta\gamma$-unsaturated carbonyl compounds (Scheme 92), free from their $\alpha\beta$-unsaturated isomers.

Acid-catalysed rearrangement of dichlorocyclopropyl carbinols produces[271] cyclopent-2-enones (Scheme 93), possibly *via* thermal conrotatory closure of the pentadienyl cation (57); phase transfer catalysis is used in the preparation of the starting compounds, and the choice of surfactant affects the regioselectivity of carbene addition in polyolefin cases.

Seebach has described a variant of his method of conversion of cyclic

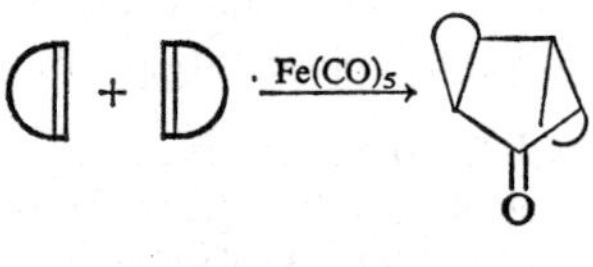

Scheme 90

$R^1CH{=}CH_2 + HCo(N_2)L_3 + R^2COCl \rightarrow R^1CH_2CH_2COR^2 + L_3CoCl$

Scheme 91

---

[267] J. Mantzaris and E. Weissberger, *J. Amer. Chem. Soc.*, 1974, **96**, 1873, 1880.
[268] B. F. G. Johnson, J. Lewis, and D. J. Thompson, *Tetrahedron Letters*, 1974, 3789.
[269] J. Schwartz and J. B. Cannon, *J. Amer. Chem. Soc.*, 1974, **96**, 4721.
[270] M. S. Newman and G. M. Fraunfelder, *J. Org. Chem.*, 1974, **39**, 251; M. S. Newman, and M. C. Vander Zwan, *ibid.*, p. 1186.
[271] T. Hiyama, M. Tsukanaka, and H. Nozaki, *J. Amer. Chem. Soc.*, 1974, **96**, 3713.

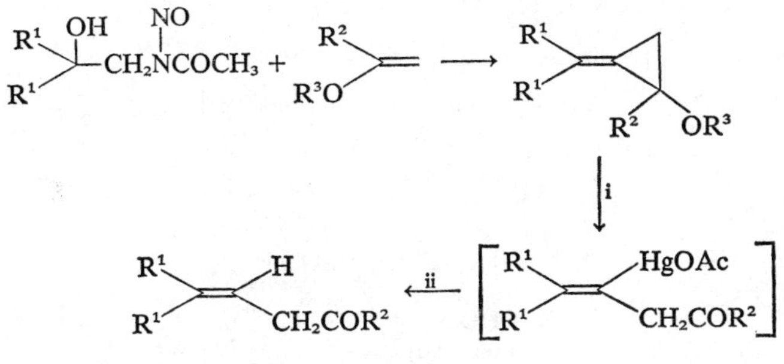

Functional Groups other than Alkanes, Acetylenes, Allenes, and Olefins   129

Scheme 92

Reagents: i, Hg(OAc)$_2$; ii, H$_2$S

olefins into homologated ketones.[272] Intramolecular addition[273] of the furanoid keto-carbene (58) affords a novel ring-opened product (Scheme 94) in good yield.

A number of [2,3]-sigmatropic rearrangement routes to carbonyl compounds have been described, including rearrangement of the sulphonium ylide (59) to Artemesia ketone,[274] of the ylide (60) in a synthesis of α-sinensal,[275] of allyl cyanohydrin ethers (61) to βγ-unsaturated neopentyl ketones,[276] and of γ-chloroallyl sulphoxides (62) and amine oxides to αβ-unsaturated ketones[277] (Scheme 95); the last conversion is related to the process studied extensively by Evans.[278]

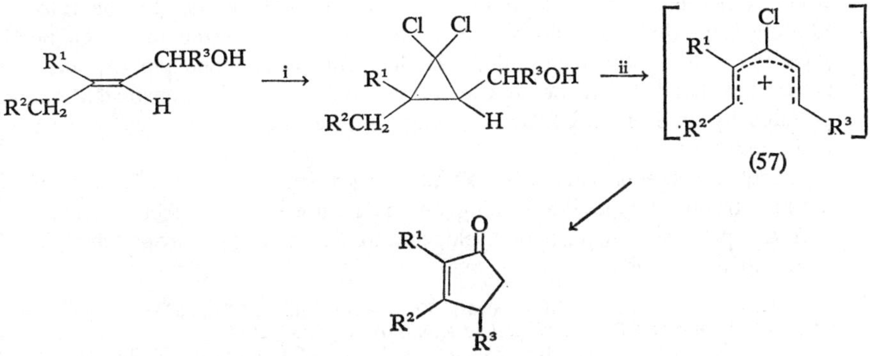

Reagents: i, :CCl$_2$; ii, HBr, 100 °C

Scheme 93

[272] D. Seebach and H. Neumann, *Chem. Ber.*, 1974, **107**, 847.
[273] M. N. Nwaji and O. S. Onyiriuka, *Tetrahedron Letters*, 1974, 2255.
[274] D. Michelot, G. Linstrumelle, and S. Julia, *J.C.S. Chem. Comm.*, 1974, 10.
[275] G. Büchi and H. Wuest, *J. Amer. Chem. Soc.*, 1974, **96**, 7573.
[276] B. Cazes and S. Julia, *Tetrahedron Letters*, 1974, 2077.
[277] P. T. Lansbury and J. E. Rhodes, *J.C.S. Chem. Comm.*, 1974, 21.
[278] D. A. Evans and G. C. Andrews, *Accounts Chem. Res.*, 1974, **7**, 147.

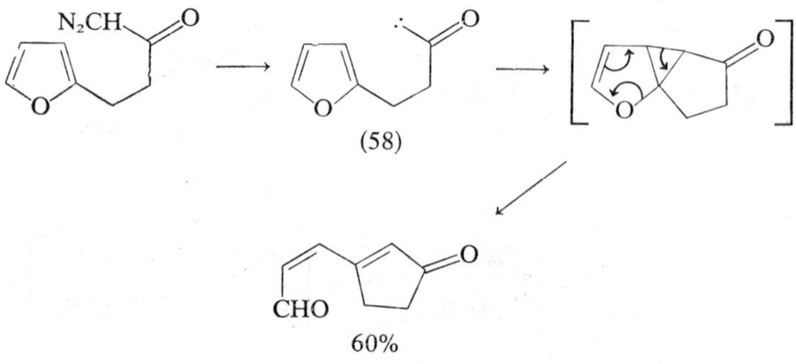

Scheme 94

The [3,3]sigmatropic rearrangement of S-allyl dithiocarbamates has been explored[279] as a novel route to αβ-unsaturated aldehydes (Scheme 96). Certain acetonyl allyl ethers undergo a base-induced [2,3]sigmatropic rearrangement to afford α-hydroxy-ketones in moderate yield.[280] α-Alkoxy-ketones are available[281] by hydroboration of acetylenic acetals (Scheme 97).

Acyl tetrafluoroborates react with alkynes in non-nucleophilic solvents to give substituted cyclopentenones[282] in good yield (Scheme 98).

The cationic dienyl iron complex (63) reacts regiospecifically with trialkylalkynylborate salts, leading[283] to a range of carbonyl compounds (Scheme 99).

The copper-catalysed vapour phase rearrangement of allylic alcohols to saturated ketones appears[284] to proceed by initial dehydrogenation, which generates a small amount of αβ-unsaturated ketone. In the rate-limiting step, the C-3 hydrogen of the alcohol is transferred to the enone β-carbon atom (Scheme 100). Hydridochlorotris(triphenylphosphine)-ruthenium catalyses a homogeneous version[285] of this transformation, perhaps by an intramolecular equivalent of the known hydrogen transfer reaction.[286]

Grignard reagents react with CO at high pressure[287] to give a variety of ketone structural types, depending on the nature of the reagent. Acetate buffered photochlorination of cyclopentanol leads to 5-chloropentanal in quantitative yield.[288]

[279] T. Nakai, H. Shiono, and M. Okawara, *Tetrahedron Letters*, 1974, 3625.
[280] A. F. Thomas and R. Dubini, *Helv. Chim. Acta*, 1974, **57**, 2084.
[281] G. Zweifel, A. Horng, and J. E. Plamondon, *J. Amer. Chem. Soc.*, 1974, **96**, 316.
[282] A. A. Schegolev, W. A. Smit, G. V. Roitburd, and V. F. Kucherov, *Tetrahedron Letters*, 1974, 3373.
[283] A. Pelter, K. J. Gould, and L. A. P. Kane-Maguire, *J.C.S. Chem. Comm.*, 1974, 1029.
[284] G. Eadon and M. Y. Shiekh, *J. Amer. Chem. Soc.*, 1974, **96**, 2288.
[285] Y. Sasson and G. L. Rempel, *Tetrahedron Letters*, 1974, 4133; *Canad. J. Chem.*, 1974, **52**, 3824.
[286] G. Brieger and T. J. Nestrick, *Chem. Rev.*, 1974, **74**, 567.
[287] W. J. J. M. Sprangers, A. P. van Sweeten, and R. Louw, *Tetrahedron Letters*, 1974, 3377.
[288] N. C. Deno, K. A. Eisenhardt, D. G. Pohl, H. H. Spinelli, and R. C. White, *J. Org. Chem.*, 1974, **39**, 520.

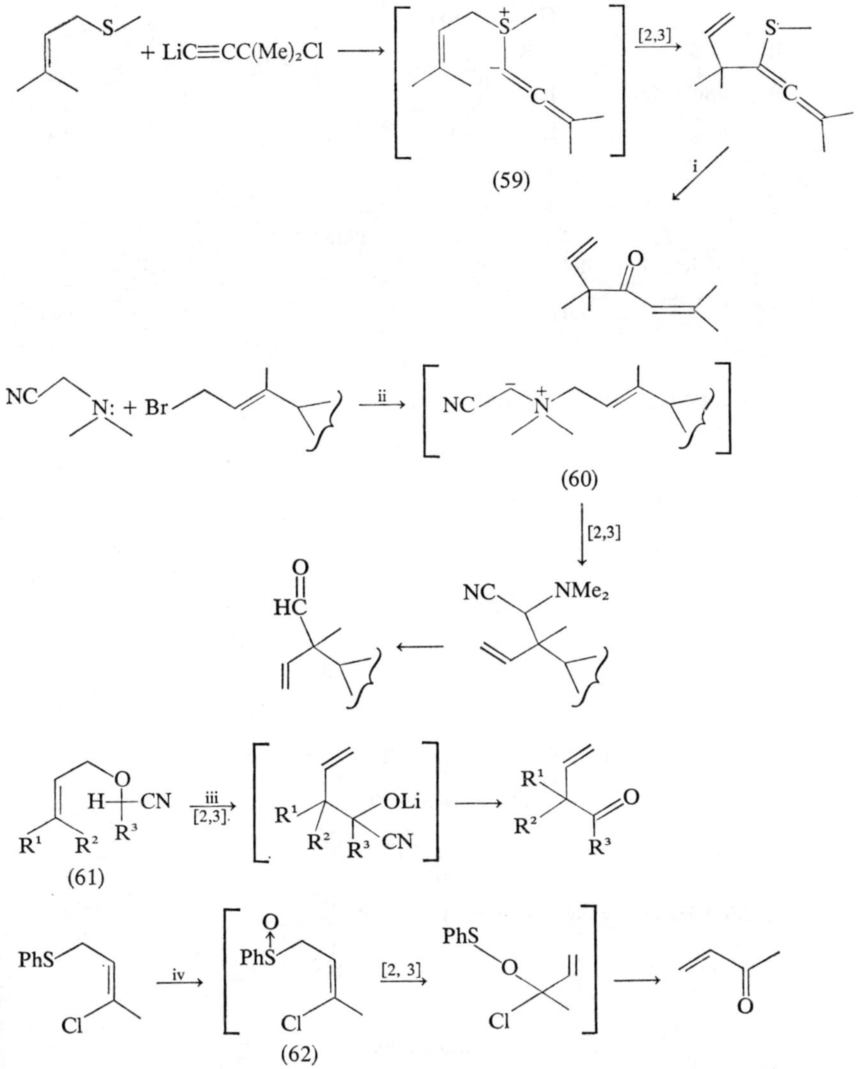

Functional Groups other than Alkanes, Acetylenes, Allenes, and Olefins 131

Reagents: i, HgCl$_2$–H$_2$O; ii, KOBu^t; iii, LiNPr$^i_2$; iv, *m*-chloroperbenzoic acid

Scheme 95

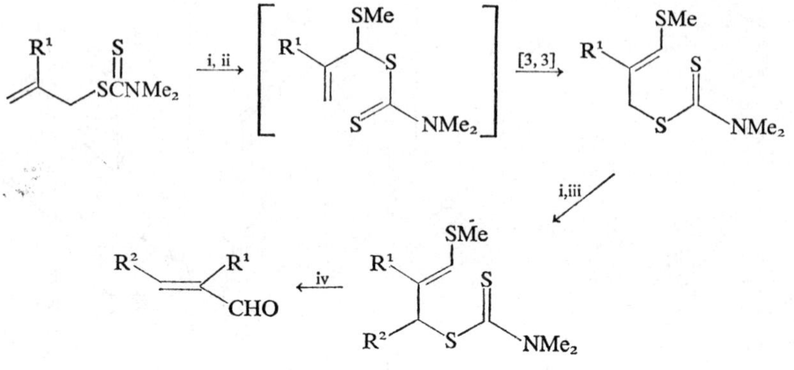

Reagents: i, LiNPr$^i_2$; ii, (MeS)$_2$; iii, R^{2}X; iv, Hg^{++}

Scheme 96

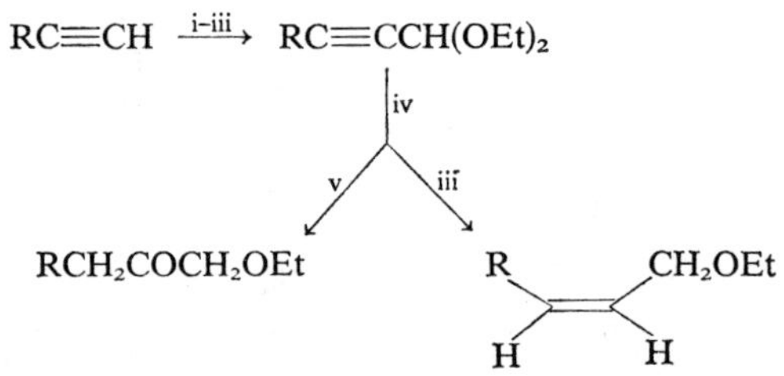

$$RC{\equiv}CH \xrightarrow{i-iii} RC{\equiv}CCH(OEt)_2$$

Reagents: i, EtMgBr; ii, HCO$_2$Et; iii, AcOH; iv, R$^1_2$BH; v, NaOH–H$_2$O$_2$

Scheme 97

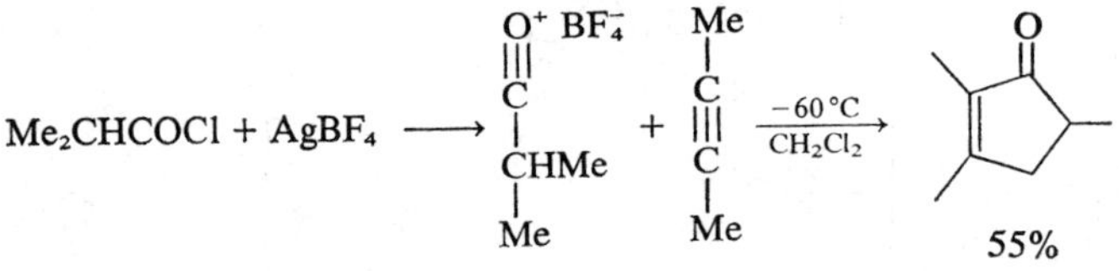

Scheme 98

By Regeneration. Modifications of the Nef transformation of nitro-compounds into carbonyl species include ozonolysis[289] of the derived nitronate salts and the use of methanol[290] as solvent; the latter case produces the product as its dimethylacetal. A palladium–oxygen complex smoothly liberates[291]

[289] J. E. McMurry, J. Melton, and H. Padgett, *J. Org. Chem.*, 1974, **39**, 259.
[290] R. M. Jacobson, *Tetrahedron Letters*, 1974, 3215.
[291] K. Maeda, I. Moritani, T. Hosokawa, and S.-I. Murahashi, *Tetrahedron Letters* 1974, 797.

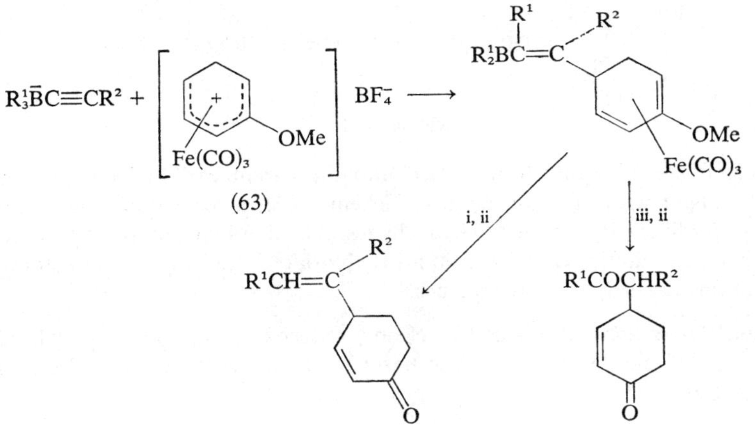

Functional Groups other than Alkanes, Acetylenes, Allenes, and Olefins   133

Reagents: i, H⁺; ii, Ce^{iv}; iii, Me₃NO

Scheme 99

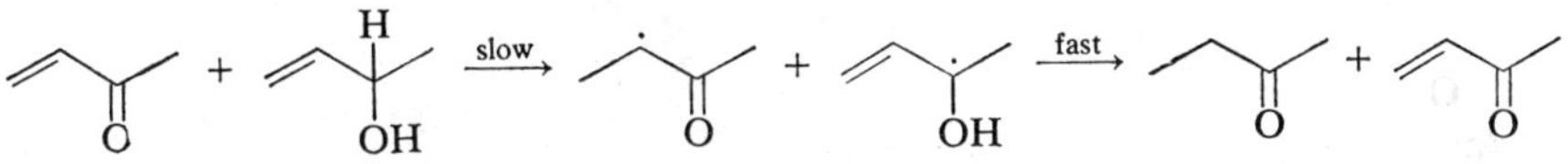

Scheme 100

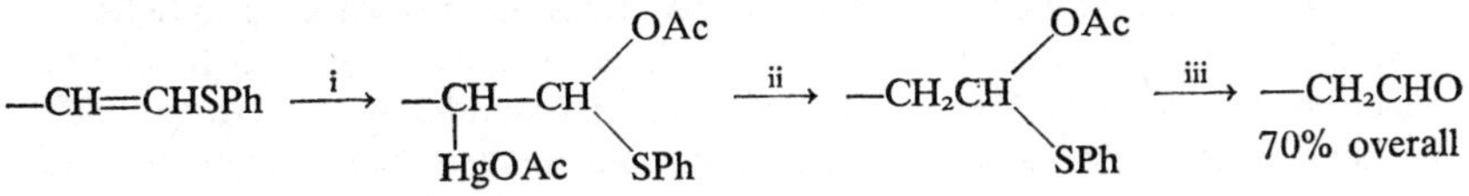

Reagents: i, Hg(OAc)₂–MeCN; ii, Al/Hg–THF; iii, K₂CO₃–H₂O

Scheme 101

ketones from ketoximes; the mild conditions are well suited to unsaturated and/or labile substrates. Chromium trioxide and periodic acid based reagents also effect[292] such an oxidative deoximation under more vigorous conditions. The titanium(IV)-mediated hydrolysis of vinyl chlorides[293] to ketones has been reported, as has a mildly basic method for the liberation of carbonyl compounds from vinyl sulphides[294] (Scheme 101).

[292] H. Araújo, G. A. L. Ferreira, and J. R. Mahajan, *J.C.S. Perkin I*, 1974, 2257.
[293] T. Mukaiyama, T. Imamoto, and S. Kobayashi, *Chem. Letters*, 1973, 261, 715.
[294] I. Vlattas and A. O. Lee, *Tetrahedron Letters*, 1974, 4451.

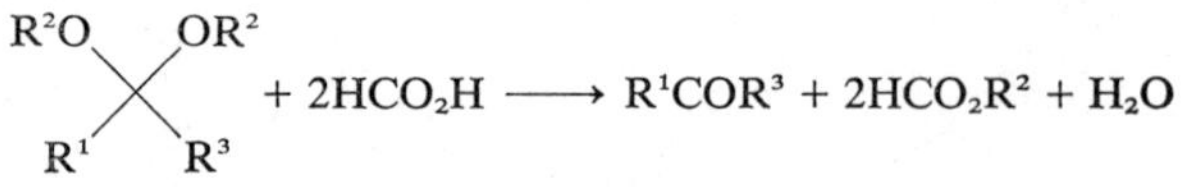

$$\underset{R^1}{\overset{R^2O}{\times}}\underset{R^3}{\overset{OR^2}{}} + 2HCO_2H \longrightarrow R^1COR^3 + 2HCO_2R^2 + H_2O$$

Scheme 102

Acetals are smoothly transformed into the parent carbonyl compounds by treatment with pure formic acid (Scheme 102), a technique[295] suited to the *in situ* liberation of sensitive aldehydes. Alkaline hypochlorite,[296] within its obvious limitations, and titanium(III) chloride[297] both regenerate ketones from toluene-*p*-sulphonylhydrazones.

General Preparation. Acids can be cleanly reduced to aldehydes by hydride reduction of the derived *N*-acylsaccharins[298] or of the triazolium salts (64) (Scheme 103).[299]

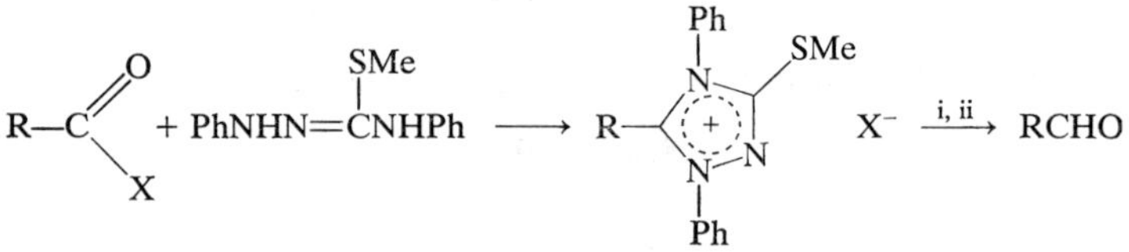

$$R-\overset{O}{\underset{X}{C}} + PhNHN=\overset{SMe}{\underset{}{C}}NHPh \longrightarrow R-\underset{Ph}{\overset{Ph}{\underset{N-N}{\overset{N^+}{\diamond}}}}SMe \quad X^- \xrightarrow{i, ii} RCHO$$

(64)

Reagents: i, NaBH$_4$–MeOH–H$_2$O; ii, H$_3$O$^+$

Scheme 103

Another new method[300] for such a transformation is illustrated in Scheme 104; preliminary results have indicated the feasibility of the alkyl-lithium reaction also shown. The reduction of nitriles to aldehydes was noted[242] earlier; nitriles react with trimethylaluminium under nickel acetylacetonate catalysis[301] to give methyl ketones.

Organocuprates react with *S*-alkyl and *S*-aryl thioesters in a new general route[302] to ketones (Scheme 105); acid anhydrides may also be suitable substrates. The reaction of bis-acid chlorides with t-alkyl Grignard reagents in the presence of copper(I) salts yields the expected diketones.[303] Vinyl

---

[295] A. Gorgues, *Bull. Soc. chim. France*, 1974, 529.
[296] T.-L. Ho and C. M. Wong, *J. Org. Chem.*, 1974, **39**, 3453.
[297] B. P. Chandrasekhar, S. V. Sunthankar, and S. G. Telang, *Chem. and Ind.*, 1975, 87.
[298] N. S. Ramegowda, M. N. Modi, A. K. Koul, J. M. Bora, C. K. Narang, and N. K Mathur, *Tetrahedron*, 1973, **29**, 3985.
[299] G. Doleschall, *Tetrahedron Letters*, 1974, 2649.
[300] D. J. Raber and W. C. Guida, *Synthesis*, 1974, 808.
[301] L. Bagnell, E. A. Jeffrey, A. Meisters, and T. Mole, *Austral. J. Chem.*, 1974, **27**, 2577.
[302] R. J. Anderson, C. A. Henrick, and L. D. Rosenblum, *J. Amer. Chem. Soc.*, 1974, **96**, 3654.
[303] P. W. Ford, *Austral. J. Chem.*, 1974, **27**, 2525.

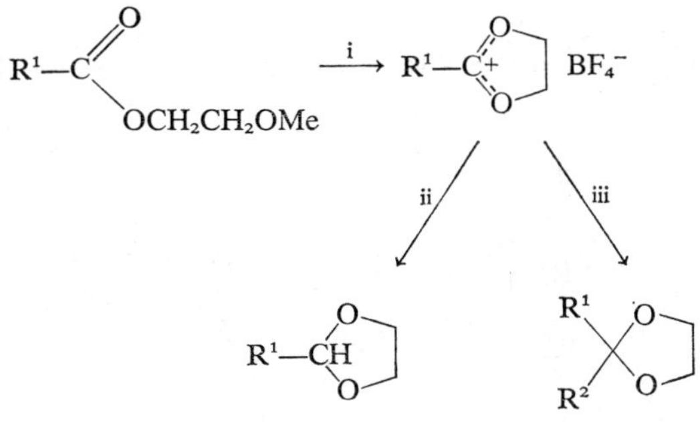

Functional Groups other than Alkanes, Acetylenes, Allenes, and Olefins

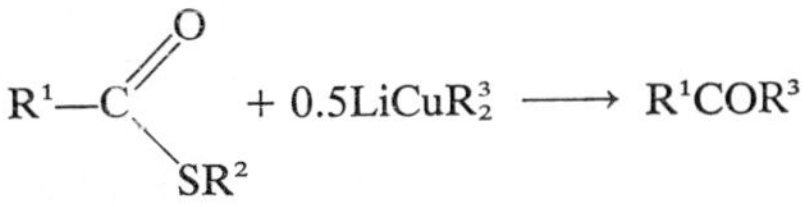

Scheme 104

Reagents: i, $Et_3O^+BF_4^-$; ii, $Bu_4N^+BH_4^-$; iii, R^2Li

Scheme 105

$$R^1-C(=O)SR^2 + 0.5LiCuR^3_2 \longrightarrow R^1COR^3$$

ketones are obtained when carboxylic acids are treated with vinyl-lithium.[304] Acid chlorides, on treatment with excess vinylmagnesium chloride, ultimately yield 2-allyl-1,3-diketones[305] *via* conjugate addition of the organometallic reagent to the intermediate vinyl ketone.

Ketones can be prepared in good yield by low-temperature reaction of Grignard reagents with mixed carboxylic acid anhydrides derived[306] from *o*-substituted benzoic and pivalic acids and simple carboxylic acids; the reaction is regioselective and γ- and ε-oxo-acids can be used. Full details have been given on the use of carboxylic acids as their esters with pyridine-2-thiol[307] and 8-hydroxyquinoline[308] in such ketone producing Grignard reactions. The direct synthesis of ketones by a modified Reformatsky reaction on nitriles has been described.[309] Treatment of N-substituted α-lactams with organolithium reagents at low temperatures results[310] in regiospecific cleavage of the acyl-nitrogen bond and production of N-substituted α-aminoketones.

[304] J. C. Floyd, *Tetrahedron Letters*, 1974, 2877.
[305] K. Suga, T. Fujita, S. Watanabe, and Y. Takahashi, *Synthesis*, 1974, 133.
[306] M. Araki and T. Mukaiyama, *Chem. Letters*, 1974, 663; M. Araki, S. Sakata, H. Takei, and T. Mukaiyama, *ibid.*, p. 687.
[307] M. Araki, S. Sakata, T. Takei, and T. Mukaiyama, *Bull. Soc. Chem. Japan*, 1974, **47**, 1777.
[308] T. Sakan, Y. Mori, and T. Yamazaki, *Chem. Letters*, 1973, 713.
[309] M. Bellassoued and M. Gaudemar, *J. Organometallic Chem.*, 1974, **80**, 139.
[310] E. R. Talaty, L. M. Pankow, D. D. Delling, and C. M. Utermoehlen, *Synthetic Comm.*, 1974, **4**, 143.

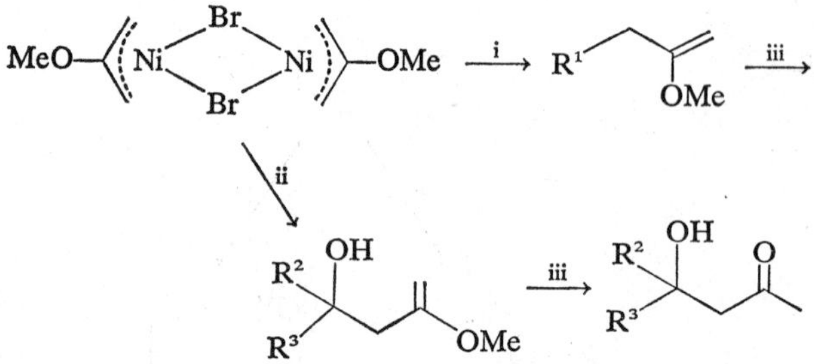

Reagents: i, R¹X; ii, R²COR³; iii, H₃O⁺

Scheme 106

The palladium(II)-catalysed formylation[311] of a number of aryl, heterocyclic, and vinyl halides has been described. The scope and reactivity of π-(2-methoxyallyl)nickel(I) bromide, a reagent for the introduction of the acetonyl functional group (Scheme 106), have been outlined;[312] the reagent reacts with a variety of halide types, and, to a lesser extent, with carbonyl compounds.

The isolation and powerful acylating ability[313] of the halogenoacetylium salts (65) have been described.

$$XCH_2\overset{+}{C}=O \quad SbF_6^-$$

(65)

Further examples[314] of the ring expansion reaction summarized in Scheme 107 have been reported; a related rearrangement involving 1,2-migration of an ethynyl group[315] is also illustrated.

Following the discovery of a method[316] for the *in situ* generation of dibromomethyl-lithium, and earlier work on the rearrangement of α-halogeno-α-lithiomethyl alkoxides, the efficient ring-expansion reaction shown in Scheme 108 has been devised.[317]

Cyclic α-chloroketones react with excess vinyl magnesium chloride to give the 1,2-divinyl species[318] shown (Scheme 109), which in turn undergo a facile ring-expanding [3,3]sigmatropic rearrangement.

[311] A. Schoenberg and R. F. Heck, *J. Amer. Chem. Soc.*, 1974, **96**, 7761.
[312] L. Hegedus and R. K. Stiverson, *J. Amer. Chem. Soc.*, 1974, **96**, 3250.
[313] G. A. Olah, H. C. Lin, and A. Germain, *Synthesis*, 1974, 895.
[314] A. J. Sisti and G. M. Rusch, *J. Org. Chem.*, 1974, **39**, 1182.
[315] J. J. Riehl, A. Smolikiewicz, and L. Thil, *Tetrahedron Letters*, 1974, 1451.
[316] H. Taguchi, H. Yamamoto, and H. Nozaki, *J. Amer. Chem. Soc.*, 1974, **96**, 3010.
[317] H. Taguchi, H. Yamamoto, and H. Nozaki, *J. Amer. Chem. Soc.*, 1974, **96**, 6510.
[318] M. Nishino, H. Kondo, and A. Miyake, *Chem. Letters*, 1973, 667.

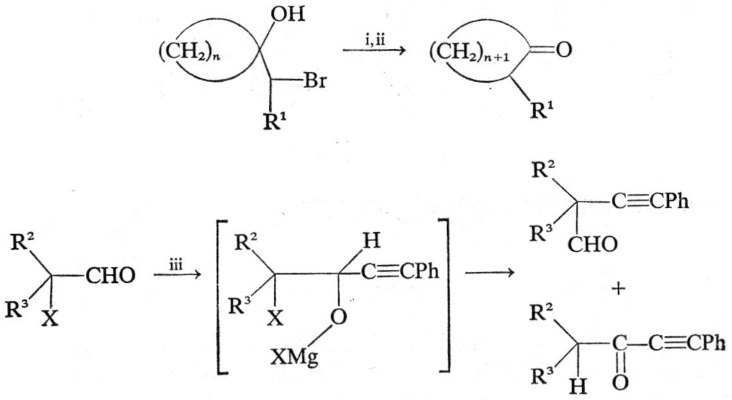

Reagents: i, Pr^{i}MgBr; ii, heat; iii, PhC≡CMgBr

Scheme 107

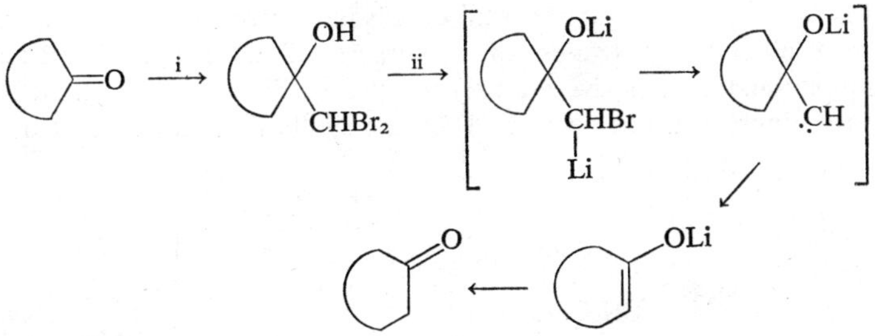

Reagents: i, CH$_2$Br$_2$–LiNR$_2$; ii, Bu^{n}Li, −78 °C

Scheme 108

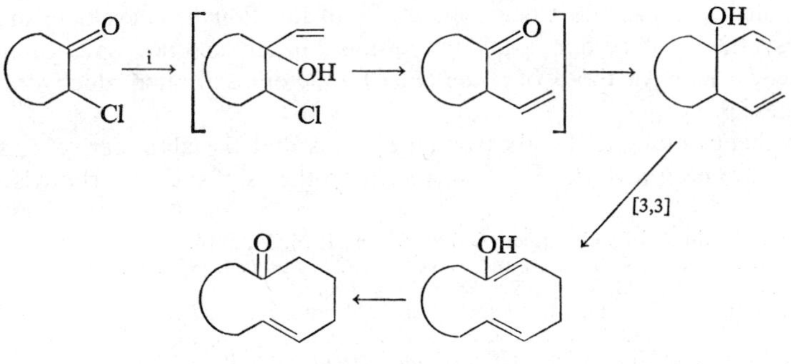

Reagent: i, 2 CH$_2$=CHCH$_2$MgCl

Scheme 109

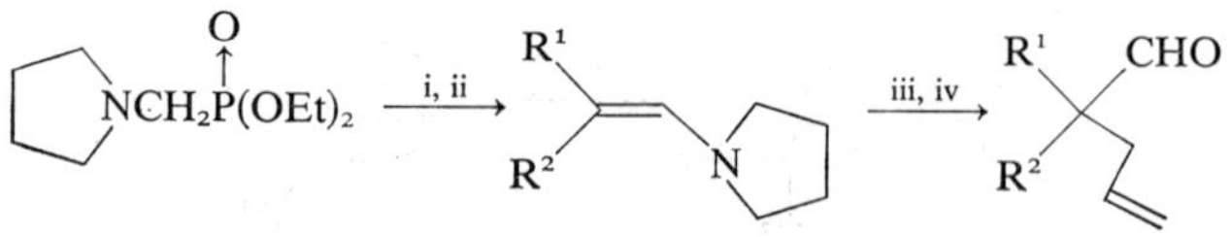

Reagents: i, Bu^{n}Li; ii, R^{1}COR^2; iii, ⟋⟍Br; iv, H$_3$O$^+$

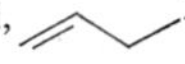

Scheme 110

Ketones can be transformed into homologous α-substituted aldehydes[319] via the enamine (66) (Scheme 110). The conversion of ketones into α-hydroxyaldehydes by reaction with toluene-*p*-sulphonylmethyl isocyanide is illustrated above (Scheme 80).

The chloromethyloxazine (3) yields Wittig reagents suitable for the homologation[7] of carbonyl compounds to αβ-unsaturated homologues; *in situ* generation of the acetal ylide (67) is recommended[320] for a similar purpose, as is the use of nitromethane.[321] Full details have been given[322] of the use of lithium trialkylalkynylborates as precursors of αβ-unsaturated ketones. The simple butadiene derivative (68) has proved[323] to be of value in Diels–Alder reactions, and has been used for the synthesis of a range of usefully

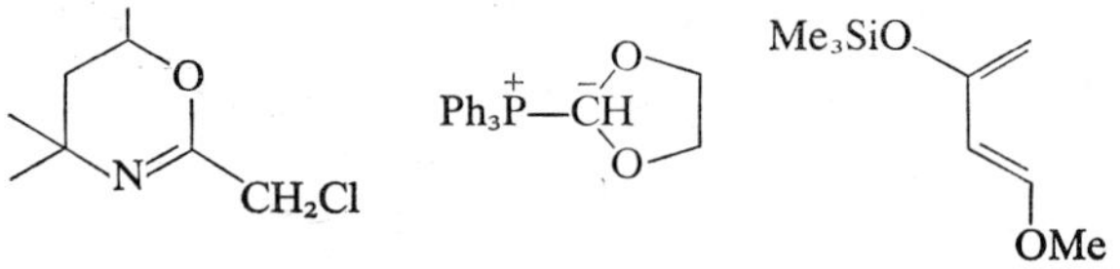

(3)   (67)   (68)

functionalized cyclohexenones. β-Dicarbonyl dianions condense with nitro-olefins to provide a new synthesis[324] of functionalized cyclohexanones. The syntheses of certain γ-alkenyl-αβ-unsaturated ketones have been described,[325] as have those of some (*E*, *Z*)-αβγδ-diunsaturated aldehydes and ketones.[326]

Synthetic routes to 1,4-dicarbonyl systems and the aldol-derived jasmonoids have been reviewed.[327] Some recent syntheses of such dicarbonyls have

---

[319] S. F. Martin and R. Gompper, *J. Org. Chem.*, 1974, **39**, 2814.
[320] T. M. Cresp, M. V. Sargent, and P. Vogel, *J.C.S. Perkin I*, 1974, 37.
[321] T.-L. Ho and C. M. Wong, *Synthesis*, 1974, 196.
[322] M. Naruse, T. Tomita, K. Utimoto, and H. Nozaki, *Tetrahedron*, 1974, **30**, 835.
[323] S. Danishefsky and T. Kitahara, *J. Amer. Chem. Soc.*, 1974, **96**, 7807.
[324] D. Seebach and V. Ehrig. *Angew. Chem. Internat. Edn.*, 1974, **13**, 400.
[325] H. O. House, W. C. Liang, and P. D. Weeks, *J. Org. Chem.*, 1974, **39**, 3102.
[326] F. Näf and R. Decorzant, *Helv. Chim. Acta*, 1974, **57**, 1309.
[327] T.-L. Ho, *Synthetic Comm.*, 1974, **4**, 265.

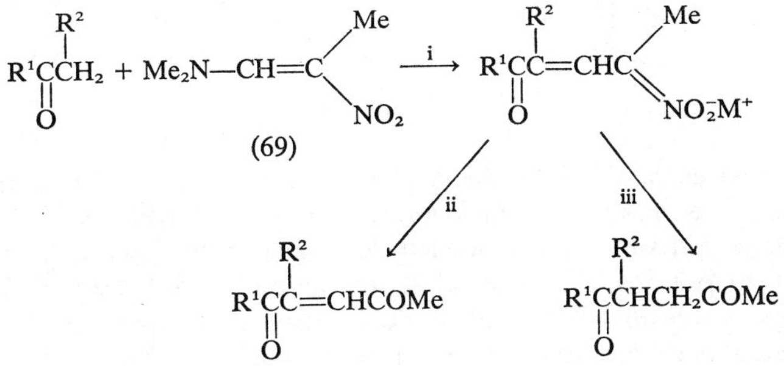

Functional Groups other than Alkanes, Acetylenes, Allenes, and Olefins 139

$$R^1\overset{O}{\underset{\|}{C}}CH_2 + Me_2N-CH=C\overset{R^2}{\underset{NO_2}{\diagdown}}\overset{Me}{\xrightarrow{i}} R^1\overset{O}{\underset{\|}{C}}C=CH\overset{R^2}{\underset{NO_2^-M^+}{\diagdown}}\overset{Me}{}$$
(69)

$$\overset{ii}{\swarrow} \qquad \overset{iii}{\searrow}$$

$$R^1\overset{O}{\underset{\|}{C}}C=CHCOMe \qquad R^1\overset{O}{\underset{\|}{C}}CHCH_2COMe$$
$$\overset{R^2}{} \qquad \overset{R^2}{}$$

Reagents: i, base; ii, SiO$_2$ or ascorbic acid; iii, ascorbic acid + Cu powder.

Scheme 111

involved oxidative addition of ketones to enol acetates,[328] alkylation of enamines,[329] of metallated ketimines,[330] of acyloin dienolates,[331] and of β-ketosulphoxide dianions;[332] Kolbe electrolysis[333] of β-keto-acid salts leads to symmetrical 1,4-diketones. A general route[334] to saturated and unsaturated 1,4-dicarbonyls utilizes the nitro enamines (69) and active methylene compounds such as ketones and esters (Scheme 111).

The syntheses of some 1,5-diketones[335,336] and of linear 1,5-polyketones[337] have been described. The preparation and properties of a number of 1,2,3-tricarbonyl compounds[338] have been given. Harris[339] has published a welcome review of his work on the synthesis of β-polycarbonyl compounds[340] and their cyclization to polyketide metabolites. The δ-tetraketone (70) has been prepared and identified with 'Harries' tetraketone',[341] isolated in 1914 by degradation of caoutchouc.

The utility of ynamines in the synthesis of polycarbonyl compounds has

[328] E. I. Heiba and R. M. Dessau, *J. Org. Chem.*, 1974, **39**, 3456, 3457.
[329] K. U. Acholonu and D. K. Wedegaertner, *Tetrahedron Letters*, 1974, 3253.
[330] Th. Cuvigny, M. Larchevêque, and H. Normant, *Tetrahedron Letters*, 1974, 1237.
[331] T. Wakamatsu, K. Akasaka, and Y. Ban, *Tetrahedron Letters*, 1974, 3879, 3883.
[332] P. A. Grieco and C. S. Pogonowski, *J. Org. Chem.*, 1974, **39**, 732.
[333] D. Lelandais and M. Chkir, *Tetrahedron Letters*, 1974, 3113.
[334] T. Severin and D. König, *Chem. Ber.*, 1974, **107**, 1499; H. Lerche, D. König, and T. Severin, ibid., p. 1509.
[335] A. Hercouet and M. Le Corre, *Tetrahedron Letters*, 1974, 2491.
[336] K. Narasaka, K. Soai, and T. Mukaiyama, *Chemistry Letters*, 1974, 1223.
[337] J. Ferard and P.-F. Casals, *Tetrahedron Letters*, 1974, 2483.
[338] F. Dayer, H. L. Dao, H. Gold, H. Rodé-Gowal, and H. Dahn, *Helv. Chim. Acta*, 1974, **57**, 2201; H. Rodé-Gowal, H. L. Dao, and H. Dahn, ibid., p. 2209; H. L. Dao, F. Dayer, L. Duc, H. Rodé-Gowal, and H. Dahn, ibid., p. 2215.
[339] T. M. Harris, C. M. Harris, and K. B. Hindley, *Fortschr. Chem. org. Naturstoffe*, 1974, **31**, 217.
[340] T. M. Harris, T. P. Murray, C. M. Harris, and M. Gumulka, *J.C.S. Chem. Comm.*, 1974, 362.
[341] B. Franck, V. Scharf, and M. Schrameyer, *Angew. Chem. Internat. Edn.*, 1974, **13**, 136.

(70)

been further exemplified.[193] The synthesis and reactions of some α,β-epoxydiazomethylketones have been reported.[342] Improved syntheses of mucondialdehyde[343] and t-butylmalondialdehyde[344] have been described. (S)-(+)-4-Methylheptan-3-one, the principal alarm pheromone of the ant *Atta texana*, has been synthesized[345] in high optical purity. A synthesis of muscone is described above.[246] Silica catalyses a reaction between acetone and methanol in the presence of oxygen to produce methyl vinyl ketone.[346]

Carbonyl Umpolung and Related Techniques.—*Sulphur-based Examples.* Seebach has written an interesting review[347] of cases in which the normal reactivity of the carbonyl group has been reversed, making it susceptible to electrophilic attack; this reversal has been described in a number of ways, Seebach now recommending the term 'umpolung' (polarity reversal). Perhaps the best example of such umpolung is the well studied 1,3-dithian anion (71); one of the more recent examples of its use is to be seen in the synthesis pestalotin referred[79] to above. The advantages and disadvantages of the 1,3,5-trithian[348] (72), methylthiomethyl NN-dimethyldithiocarbamate[349] (73), and methylene bis(NN-dimethyldithiocarbamate)[350] (74) anions have been

(71)   (72)   (73)   (74)

(75)

[342] B. Zwanenburg and L. Thijs, *Tetrahedron Letters*, 1974, 2459; N. F. Woolsey and M. H. Khalil, *ibid.*, p. 4309; see also B. M. Trost and P. J. Whitman, *J. Amer. Chem, Soc.*, 1974, **96**, 7421.
[343] G. Kossmehl and B. Bohn, *Chem. Ber.*, 1974, **107**, 710.
[344] C. Reichardt and E.-U. Würthwein, *Chem. Ber.*, 1974, **107**, 3454.
[345] R. G. Riley and R. M. Silverstein, *Tetrahedron*, 1974, **30**, 1171.
[346] M. Okada and Y. Asami, *Chem. Letters*, 1973, 333.
[347] D. Seebach and M. Kolb, *Chem. and Ind.*, 1974, 687.
[348] D. Seebach, E. J. Corey, and A. K. Beck, *Chem. Ber.*, 1974, **107**, 367.
[349] I. Hori, T. Hayashi, and H. Midorikawa, *Synthesis*, 1974, 705.
[350] T. Nakai and M. Okawara, *Chem. Letters*, 1974, 731.

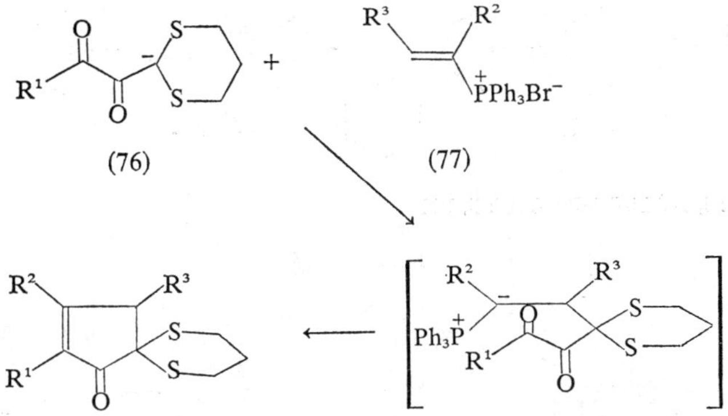

Functional Groups other than Alkanes, Acetylenes, Allenes, and Olefins 141

Scheme 112

delineated. The 1,3-dithiolan[351] (75) effects overall C-formylation of, *inter alia*, electron-rich aromatic systems; it is less reactive and more selective than the Vilsmeier reagent.

Both 1,3-dithian and methyl methylthiomethyl sulphoxide anions react with immonium salts, leading to α-aminoaldehydes;[352] immonium salts with α-hydrogens suffer competitive elimination to enamines. Addition of the 1,3-dithian anion (76) to the vinylphosphonium salt (77) is followed by an intramolecular Wittig reaction, and the overall process constitutes a new route to functionalized cyclopentenones (Scheme 112).[353] Application of the known cleavage[354] of cyclic α-diketone monothioacetals has allowed the preparation of functionalized cyclopentane derivatives of potential utility in prostaglandin synthesis.[355]

A homologative transformation[356] of carbonyl compounds into αβ-unsaturated ketones is illustrated in Scheme 113.

Methyl methylthiomethyl sulphoxide reacts with αω-dihalogenoalkanes in the presence of base to give cyclic ketones, including cyclobutanones[357] from 1,3-dihalogenoalkanes; a related sequence[358] of dialkylation yields symmetrical ketones. A highly efficient synthesis (Scheme 114) of rethrolones, the alcohol components of the pyrethrins, has been described.[359]

α-Lithiothioacetals react with trialkylboranes in the first step of an overall conversion of aldehydes into ketones[360] (Scheme 115).

[351] K. Hiratani, T. Nakai, and M. Okawara, *Bull. Chem. Soc. Japan*, 1973, **36**, 3510.
[352] L. Duhamel, P. Duhamel, and N. Mancelle, *Bull. Soc. chim. France*, 1974, 331.
[353] I. Kawamoto, S. Muramatsu, and Y. Yura, *Tetrahedron Letters*, 1974, 4223.
[354] J. A. Marshall and D. E. Seitz, *J. Org. Chem.*, 1974, **39**, 1814.
[355] E. Cossement, R. Binamé, and L. Ghosez, *Tetrahedron Letters*, 1974, 997.
[356] D. Seebach, M. Kolb, and B.-T. Gröbel, *Tetrahedron Letters*, 1974, 3171.
[357] K. Ogura, M. Yamashita, M. Suzuki, and G. Tsuchihashi, *Tetrahedron Letters*, 1974, 3653.
[358] G. Schill and P. R. Jones, *Synthesis*, 1974, 117.
[359] R. F. Romanet and R. H. Schlessinger, *J. Amer. Chem. Soc.*, 1974, **96**, 3701.
[360] S. Yamamoto, M. Shiono, and T. Mukaiyama, *Chem. Letters*, 1973, 961.

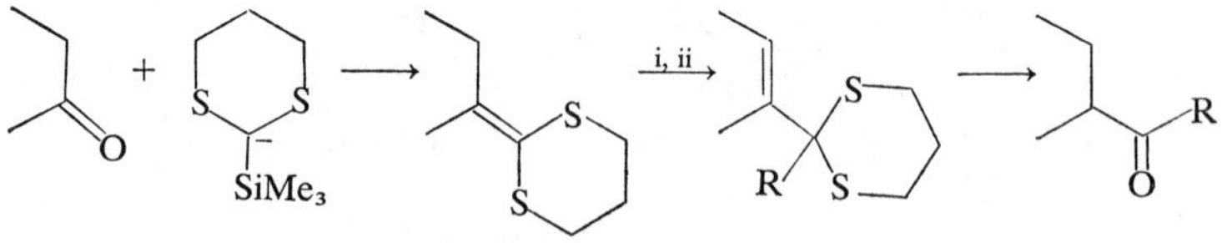

Reagents: i, Bu^{n}Li-HMPA-THF; ii, RX

Scheme 113

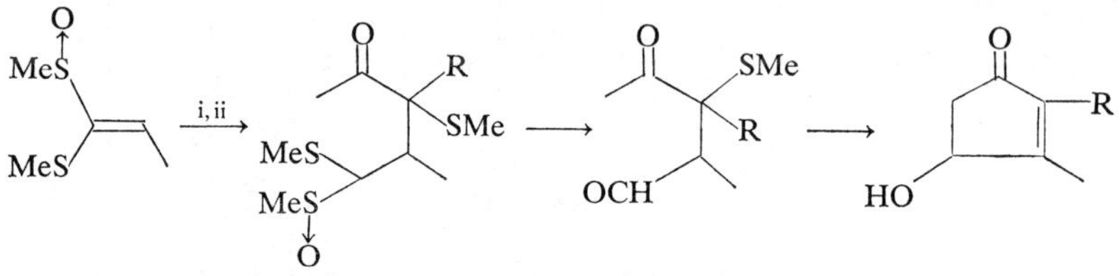

Reagents: i, MeSCHCOCH$_3$; ii, RX

Scheme 114

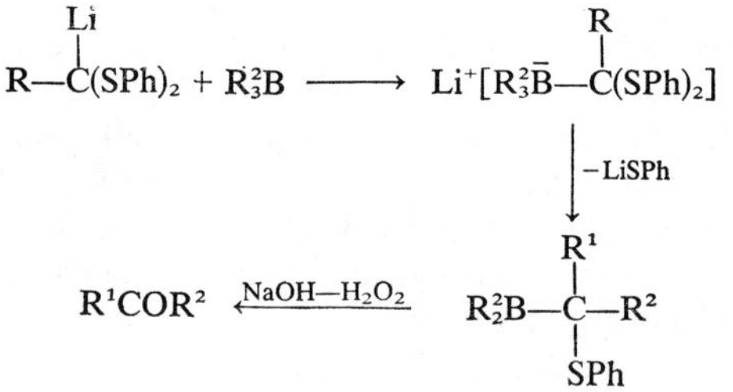

$$R-\underset{|}{\overset{Li}{C}}(SPh)_2 + R_3^2B \longrightarrow Li^+[R_3^2\bar{B}-\underset{|}{\overset{R}{C}}(SPh)_2]$$

$$\downarrow -LiSPh$$

$$R^1COR^2 \xleftarrow{NaOH-H_2O_2} R_2^2B-\underset{|}{\overset{R^1}{\underset{SPh}{C}}}-R^2$$

Scheme 115

A variant[361] on the use of keten thioacetal derivatives as two-carbon Michael acceptors[362] has been described. The masked cyanoacetaldhyde (78) has been prepared;[363] no details of its alkylation are given.

Other Cases. Baldwin[364] has reported on the utility of α-methoxyvinyl-lithium as a simple equivalent of the acetyl anion (Scheme 116), which adds exclusively 1,2 to αβ-unsaturated systems; while attempted extension to ββ-disubstituted enol ethers failed, related studies[365] on β-monosubstituted cases have been reported.

[361] T. Oishi, H. Takechi, and Y. Ban, *Tetrahedron Letters*, 1974, 3757.
[362] J. L. Herrmann, G. R. Kieczykowski, R. F. Romanet, P. J. Wepplo, and R. H. Schlessinger, *Tetrahedron Letters*, 1973, 4711; J. L. Herrmann, G. R. Kieczykowski, R. F. Romanet, and R. H. Schlessinger, *ibid.*, p. 4715.
[363] T. H. Jones and P. J. Kropp, *Synthetic Comm.*, 1974, **4**, 331.
[364] J. E. Baldwin, G. A. Höfle, and O. W. Lever, *J. Amer. Chem. Soc.*, 1974, **96**, 7125.
[365] J. Hartmann, M. Stähle, and M. Schlosser, *Synthesis*, 1974, 888.

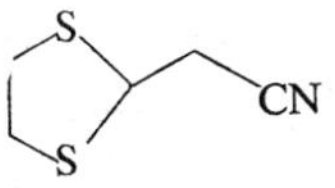

Functional Groups other than Alkanes, Acetylenes, Allenes, and Olefins

(78)

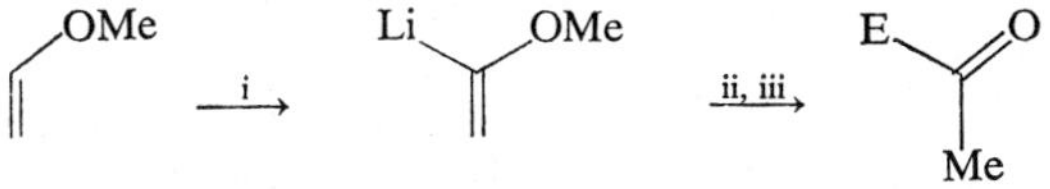

Reagents: i, Bu^{t}Li, −65 °C; ii, Electrophile; iii, H$_3$O$^+$

Scheme 116

The vinyl-lithium (79) functions[366] as the equivalent of the α-anion of acrolein in simple aldol-type condensations (Scheme 117), and has found use in stereospecific olefin synthesis. The scope and utility of π-(2-methoxyallyl)-nickel bromide,[312] a reagent which allows the introduction of the acetonyl functional group, is illustrated above (Scheme 106).

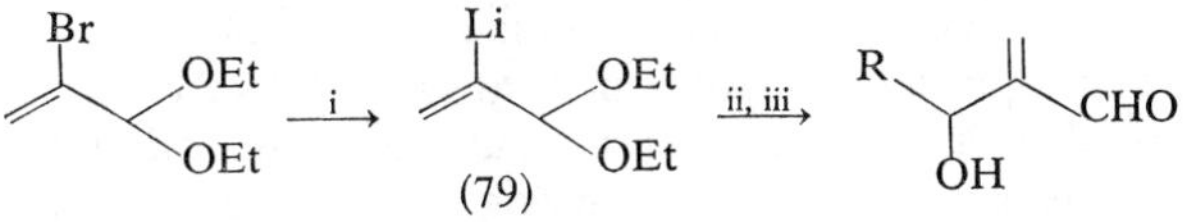

Reagents: i, Bu^{n}Li, −70 °C; ii, RCHO; iii, H$_3$O$^+$

Scheme 117

Vinylsilanes can be readily transformed[367] into carbonyl derivatives (Scheme 118); this synthetic equivalence has been used in a general route[368] to carbonyl compounds, in the conjugate addition of acyl anion equivalents via the silylated vinyl copper species[369] (80), and in the application of the allyl halide (81)[370] to a useful regiospecific annelation[371] sequence (Scheme 118).

Two groups have shown that the anions of aldehyde cyanohydrins will add conjugatively to αβ-unsaturated carbonyl compounds. Stetter[372] has given full details of the conjugate addition of aromatic and heterocyclic aldehydes in such a sequence. He has also made the process general for all

---

[366] J.-C. Depezay and Y. Le Merrer, *Tetrahedron Letters*, 1974, 2751, 2755.
[367] G. Stork and E. Colvin, *J. Amer. Chem. Soc.*, 1971, **93**, 2080.
[368] B.-T. Gröbel and D. Seebach, *Angew. Chem. Internat. Edn.*, 1974, **13**, 83.
[369] R. K. Boeckman and K. J. Bruza, *Tetrahedron Letters*, 1974, 3365; see also ref. 410.
[370] G. Stork and M. E. Jung, *J. Amer. Chem. Soc.*, 1974, **96**, 3682; G. Stork, M. E. Jung, E. Colvin, and Y. Noel, *ibid.*, p. 3684.
[371] For other regiospecific annelations, see P. L. Stotter and K. A. Hill, *J. Amer. Chem. Soc.*, 1974, **96**, 6524; H. Stetter and K. Elfert, *Synthesis*, 1974, 36.
[372] H. Stetter and M. Schreckenberg, *Chem. Ber.*, 1974, **107**, 210, 2453.

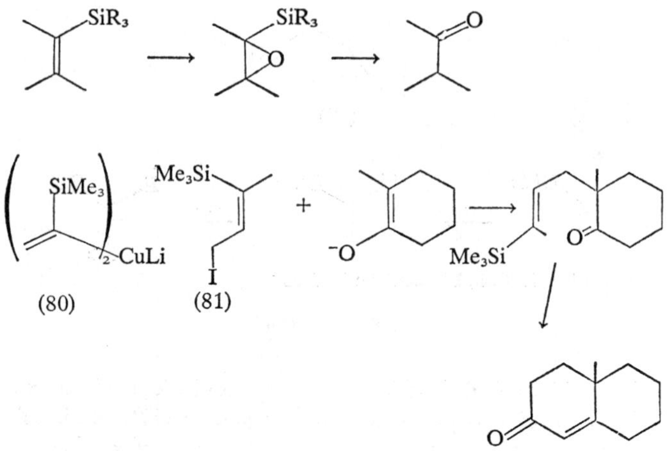

Scheme 118

aldehyde types, by employing thiazolium salts,[373] whose catalytic activity in the acyloin condensation is known, in the presence of a base (Scheme 119).

Stork[374] has reported that the anions (82) of protected cyanohydrins of αβ-unsaturated aldehydes will carry out conjugate addition; saturated alkoxynitriles give 1,2-adducts in varying amounts, sometimes exclusively (Scheme 120).

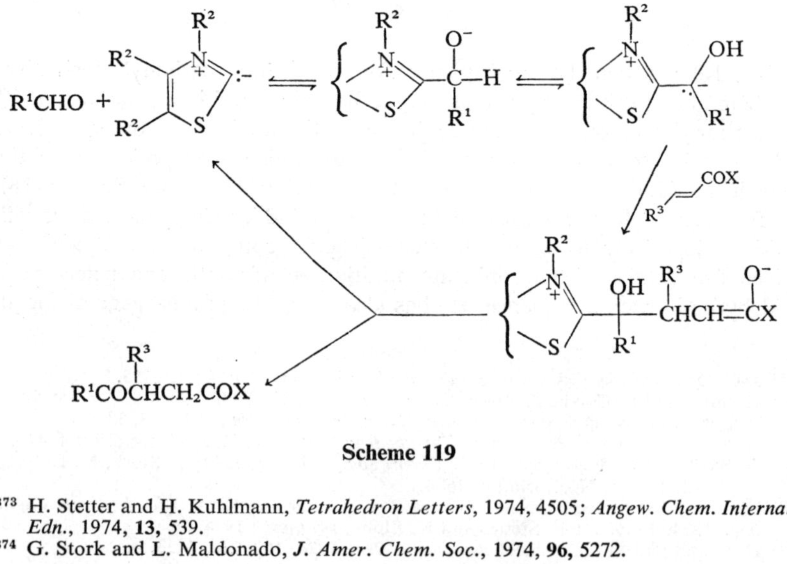

Scheme 119

---

[373] H. Stetter and H. Kuhlmann, *Tetrahedron Letters*, 1974, 4505; *Angew. Chem. Internat. Edn.*, 1974, **13**, 539.

[374] G. Stork and L. Maldonado, *J. Amer. Chem. Soc.*, 1974, **96**, 5272.

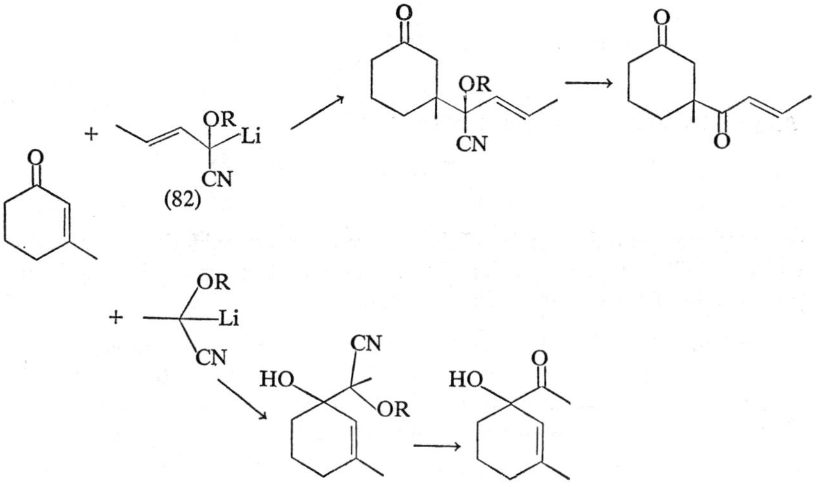

Scheme 120

Seebach has described studies on the alkylation of benzylthiol[375] and thio-allyl alcohol[376] dianions; the latter reacts preferentially at the γ-position with a range of electrophiles (Scheme 121).

The equivalence of lithio-aldimines with acyl anions has been extensively tabulated.[243] While direct alkylation of such species with a primary alkyl halide is a satisfactory route to unsymmetrical ketones, alkylation with secondary halides fails, as elimination is the major process. This can be overcome[377] by the use of dialkylchloroboranes as alkylating agents (Scheme 122).

The potential synthetic equivalence of nitriles and lower homologous

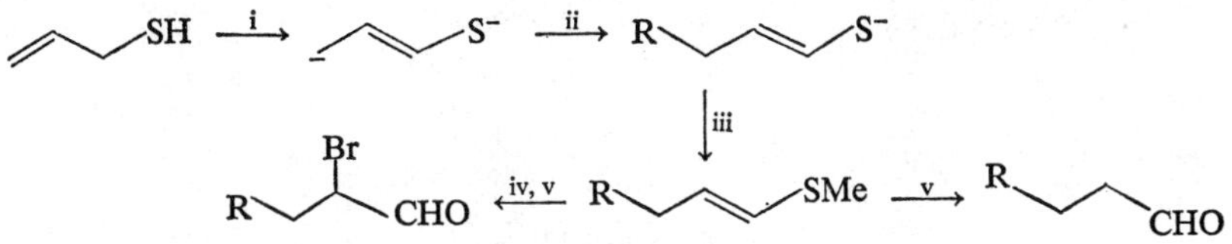

Reagents: i, 2 Bu^{n}Li–TMEDA, 0 °C; ii, RBr; iii, MeI; iv, Br$_2$; v, hydrolysis

Scheme 121

---

[375] D. Seebach and K.-H. Geiss, *Angew. Chem. Internat. Edn.*, 1974, **13**, 202.
[376] K.-H. Geiss, B. Seuring, R. Pieter, and D. Seebach, *Angew. Chem. Internat. Edn.*, 1974, **13**, 479.
[377] Y. Yamamoto, K. Kondo, and I. Moritani, *Tetrahedron Letters*, 1974, 793.

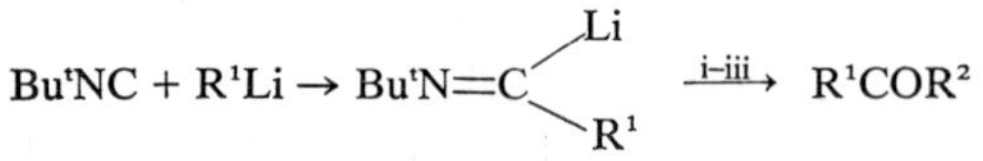

$$\text{Bu}^t\text{NC} + \text{R}^1\text{Li} \longrightarrow \text{Bu}^t\text{N}=\text{C}\genfrac{}{}{0pt}{}{\text{Li}}{\text{R}^1} \xrightarrow{\text{i-iii}} \text{R}^1\text{COR}^2$$

Reagents: i, R^2_2BCl; ii, $(CF_3CO)_2O$; iii, $NaOH-H_2O_2$

Scheme 122

carbonyl compounds has been demonstrated[239] by oxidative decyanation (Scheme 77). Nitroethylene has found use[378] as a keten equivalent in the preparation of prostaglandin intermediates.

Alkylation.—Potassium hydride is recommended[379] for the rapid, quantitative formation of the potassium enolates of ketones; little or no self-condensation or reduction is observed. Direct alkylation, with reactive alkylating agents, of the anions of $\alpha\beta$-unsaturated aldehydes, generated with potassamide in liquid ammonia, gave the expected α-alkyl-$\beta\gamma$-unsaturated products in fair yield.[380] Geranylacetone (83) can be selectively alkylated (Scheme 123) without protection of the carbonyl group[381] via the intermediacy of π-allyl palladium complexes.

The alkylation[382] of metallated aldimines with $\alpha\omega$-dihalogenoalkanes leads to ω-halogenoaldehydes or $\alpha\omega$-dialdehydes. The alkylation and reductive

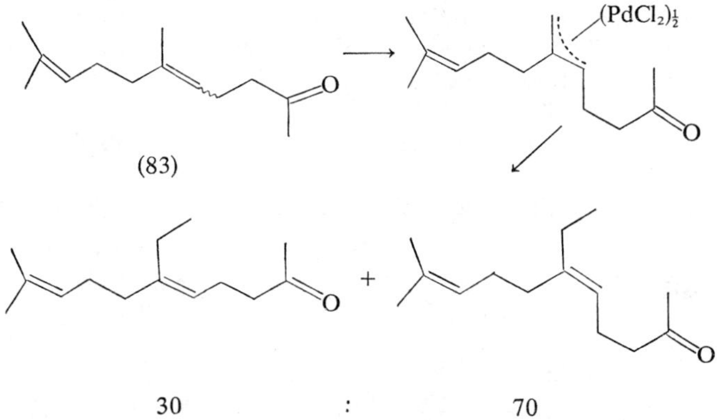

Scheme 123

[378] S. Ranganathan, D. Ranganathan, and A. K. Mehrotra, *J. Amer. Chem. Soc.*, 1974, **96**, 5261.
[379] C. A. Brown, *J. Org. Chem.*, 1974, **39**, 1324.
[380] S. A. G. de Graaf, P. E. R. Oosterhoff, and A. van der Gen, *Tetrahedron Letters*, 1974, 1653.
[381] B. M. Trost, T. J. Dietsche, and T. J. Fullerton, *J. Org. Chem.*, 1974, **39**, 737.
[382] Th. Cuvigny, J. F. Le Borgne, M. Larchevêque, and H. Normant, *J. Organometallic Chem.*, 1974, **70**, C5.

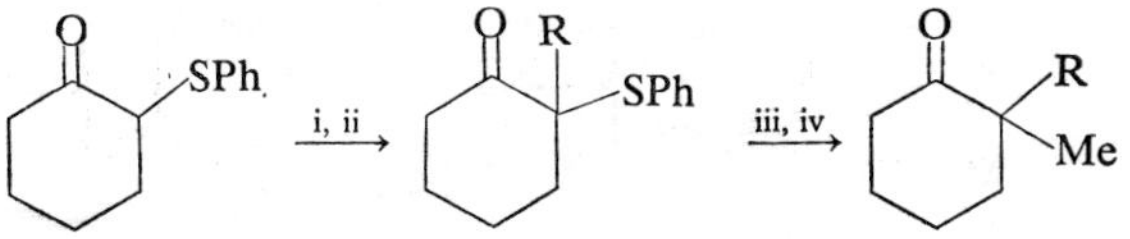

Functional Groups other than Alkanes, Acetylenes, Allenes, and Olefins

Reagents: i, LiNPr$_2^i$; ii, RX; iii, Li–NH$_3$; iv, MeI

Scheme 124

alkylation (Scheme 124) of α-phenylthio-ketones and -aldehydes, now available by direct enolate sulphenylation, results in regiospecific geminal α-dialkylation;[383] the phenylthio group increases the thermodynamic and kinetic acidity of the α-hydrogen by at least 3 pK units.

Further examples[384,385] of the orientational stability of copper-lithium enolates, generated by conjugate addition of lithium organocuprates to αβ-unsaturated enones, have been presented; the use of dimethoxyethane as a solvent for further alkylation is highly recommended.

A variety of regiospecific alkylation techniques based on cyclopropane rings have been described. β-Methyl-αβ-unsaturated ketones are available by alkylative ring fission[386] of the dichlorocarbene adducts (84) (Scheme 125). This sequence has been used to synthesize (±)-muscone from cyclododecanone.

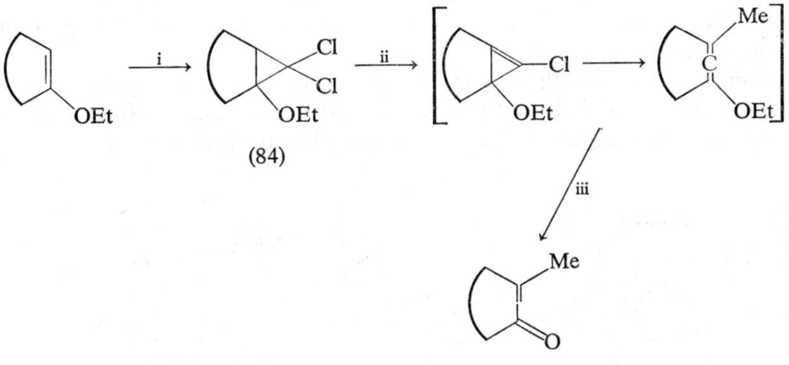

Reagents: i, :CCl$_2$; ii, 2MeLi, −95 °C; iii, H$_3$O$^+$

Scheme 125

---

[383] R. M. Coates, H. D. Pigott, and J. Ollinger, *Tetrahedron Letters*, 1974, 3955.
[384] R. M. Coates and L. O. Sandefur, *J. Org. Chem.*, 1974, **39**, 275.
[385] G. H. Posner, C. E. Whitten, J. J. Sterling, and D. J. Brunelle, *Tetrahedron Letters*, 1974, 2591.
[386] T. Hiyama, T. Mishima, K. Kitatani, and H. Nozaki, *Tetrahedron Letters*, 1974, 3297.

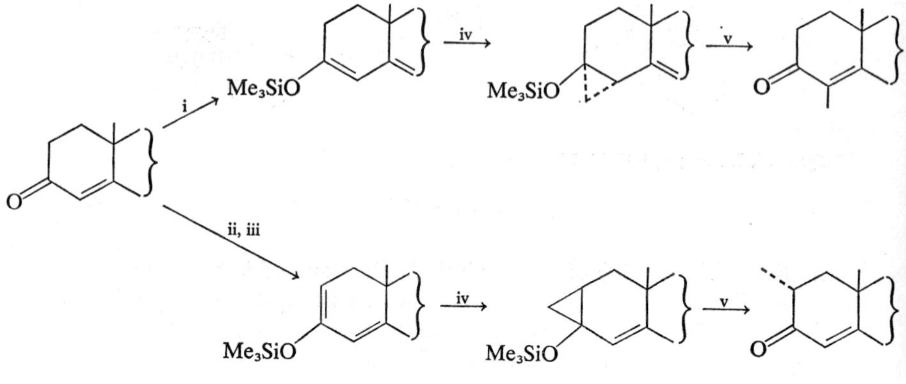

Reagents: i, Et$_3$N–Me$_3$SiCl–DMF; ii, LiNPr$_2^i$; iii, Me$_3$SiCl; iv, CH$_2$I$_2$–Zn–Ag; v, MeOH–H$^+$

Scheme 126

Conia, extending his studies on the cyclopropanation of specifically generated enol ethers and their subsequent reactivity, has reported the highly selective α- or α'-methylation of αβ-unsaturated ketones,[387] as illustrated in Scheme 126; he has also reported the related preparation of α'-methyl-α-cyclopropyl ketones.[388] When cisoid or labile α-ethylenic ketones are employed in such a sequence, a different course[389] of ring-opening occurs, and leads to rapid and general routes[390] to cyclobutanones and cyclopentanones (Scheme 127).

Simmons–Smith addition to cyclic silyl enol ethers in concentrated solution results in zinc iodide induced isomerization[391] of the initially formed cyclopropyl ethers to protected 2-methylenecycloalkanols (Scheme 128).

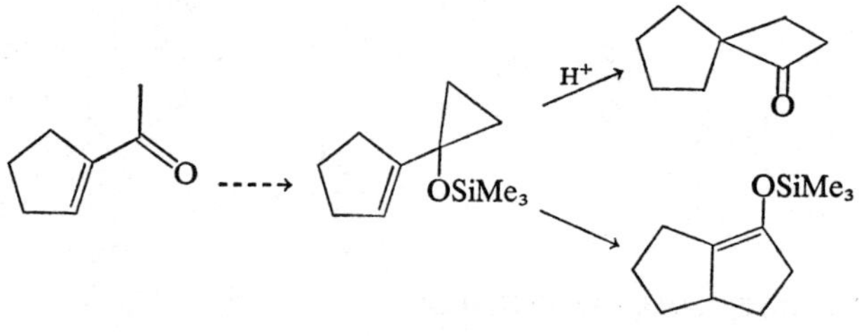

Scheme 127

[387] C. Girard and J. M. Conia, *Tetrahedron Letters*, 1974, 3327.
[388] C. Girard and J. M. Conia, *Tetrahedron Letters*, 1974, 3333.
[389] J. Salaun, B. Garnier, and J. M. Conia, *Tetrahedron*, 1974, **30**, 1413.
[390] C. Girard, P. Amice, J. P. Barnier, and J. M. Conia, *Tetrahedron Letters*, 1974, 3329.
[391] S. Murai, T. Aya, T. Renge, I. Ryu, and N. Sonoda, *J. Org. Chem.*, 1974, **39**, 858.

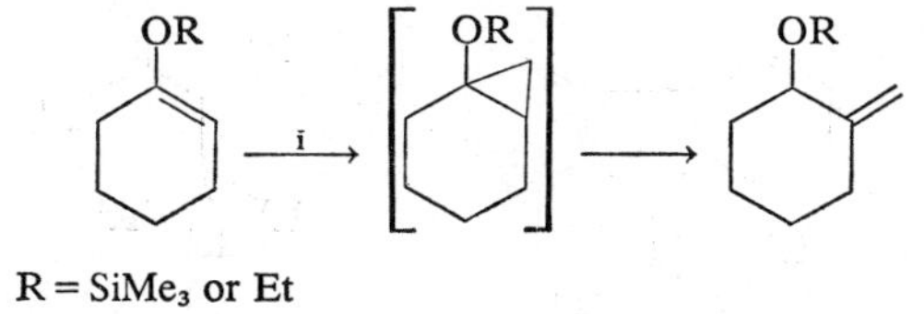

Functional Groups other than Alkanes, Acetylenes, Allenes, and Olefins 149

R = SiMe₃ or Et

Reagent: i, CH₂I₂–Zn–Cu (concentrated solution)

Scheme 128

Two independent reports[392,393] on the utility of metallated allyl ethers as homoenolate anion equivalents have been published, which illustrate the effective β-alkylation[394] of a saturated aldehyde (Scheme 129).

The sterically hindered allylamine (85) undergoes base-induced isomerization[395] and alkylation at the γ-position; the overall result again is effective β-alkylation (Scheme 130).

Peracid oxidation of silyl enol ethers[396] produces α-hydroxy-ketones in high yield; such species can also be obtained by direct reaction of the corresponding enolate anions with a molybdenum peroxide complex.[397]

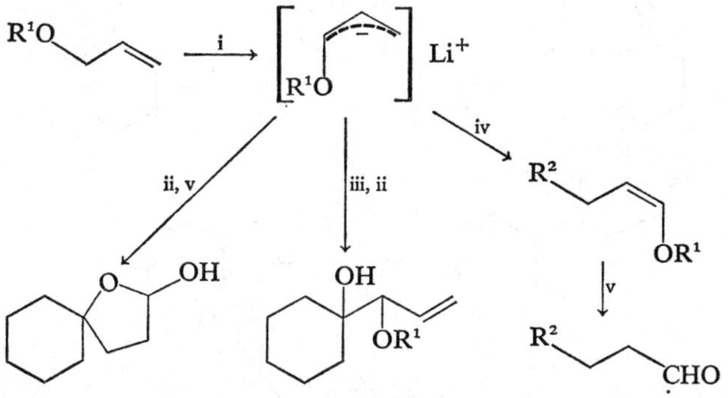

Reagents: i, Bu°Li; ii, cyclohexanone; iii, ZnCl₂; iv, R²X–HMPA; v, H₃O⁺

Scheme 129

---

[392] D. A. Evans, G. C. Andrews, and B. Buckwalter, *J. Amer. Chem. Soc.*, 1974, **96**, 5560.
[393] W. C. Still and T. L. Macdonald, *J. Amer. Chem. Soc.*, 1974, **96**, 5561.
[394] See also ref. 376.
[395] M. Julia, A. Schouteeten, and M. Ballarge, *Tetrahedron Letters*, 1974, 3433; see also T. A. Bryson and R. B. Gammill, *ibid.*, p. 3963.
[396] G. M. Rubottom, M. A. Vasquez, and D. R. Pelegrina, *Tetrahedron Letters*, 1974, 4319.
[397] E. Vedejs, *J. Amer. Chem. Soc.*, 1974, **96**, 5944.

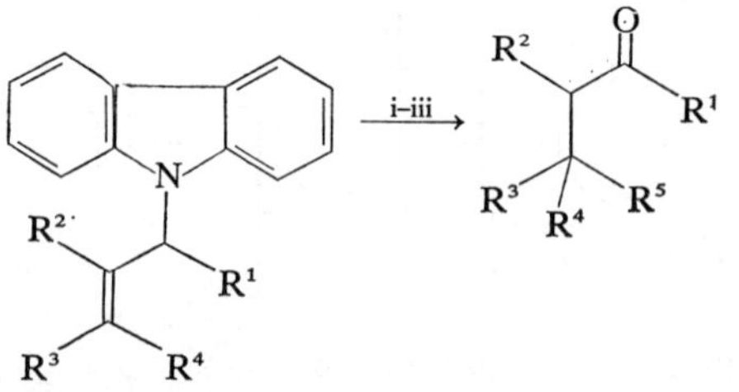

(85) R = H or Me

Reagents: i, Bu^{n}Li–TMEDA; ii, R^{5}X; iii, H$_3$O$^+$

Scheme 130

Introduction of αβ-Unsaturation.—The *syn* 1,2-elimination of selenoxide-as a method for the synthesis of enones has been extended[398] to the preparation of β-dicarbonyl enones (Scheme 131). The low yields of enone obtained from 2-phenylseleno-cyclohexanone, -cycloheptanone, and -cyclooctanone are due to competing Pummerer rearrangement; this can be obviated, and yields enhanced, by prior ketalization (Scheme 131).

Three independent reports[399–401] of the direct addition of benzeneselenenyl

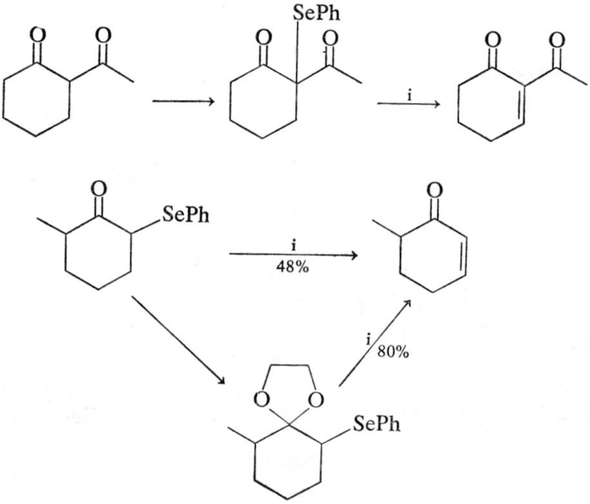

Reagent: i, H$_2$O$_2$, heat

Scheme 131

[398] H. J. Reich, J. M. Renga, and I. L. Reich, *J. Org. Chem.*, 1974, **39**, 2133.
[399] H. J. Reich, *J. Org. Chem.*, 1974, **39**, 428.
[400] K. B. Sharpless and R. F. Lauer, *J. Org. Chem.*, 1974, **39**, 429.
[401] D. L. J. Clive, *J.C.S. Chem. Comm.*, 1974, 100.

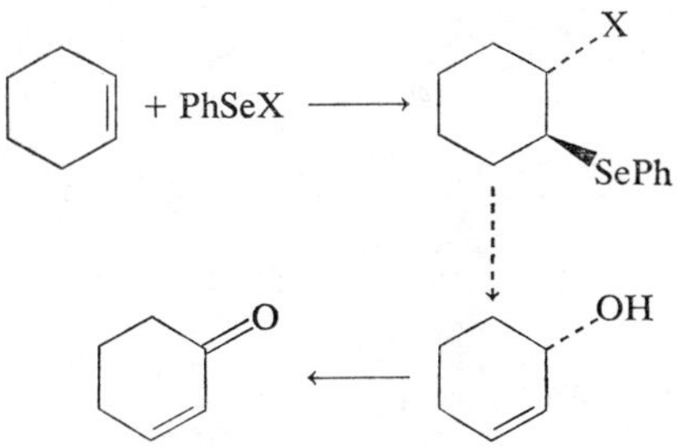

Functional Groups other than Alkanes, Acetylenes, Allenes, and Olefins 151

Scheme 132

reagents to unactivated olefins (Scheme 132) have appeared. The addition is rapid, and shows *trans* stereochemistry, but no regioselectivity is observed with terminal olefins; the products are readily convertible into $\alpha\beta$-unsaturated carbonyl compounds.

The dianion (86) is alkylated exclusively at the γ-position by a variety of alkyl halides; the resulting α-phenylsulphinyl ketones undergo ready elimination of benzenesulphinic acid, and this process represents a new general route[402] to alkyl vinyl ketones (Scheme 133). The dianion of phenylsulphinylacetone undergoes[403] a similar specific γ-alkylation; reductive removal of the phenylsulphinyl group leads to a range of alkylated acetone derivatives.

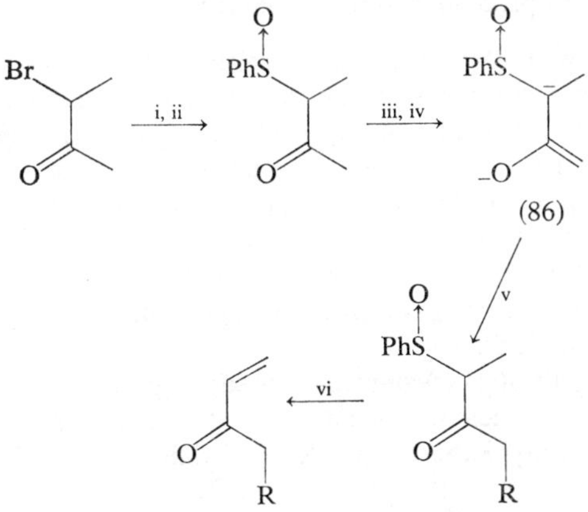

Reagents: i, NaSPh–EtOH; ii, oxidation; iii, NaH–HMPA–THF; iv, Bu^{n}Li–HMPA–THF; v, RX; vi, heat

Scheme 133

[402] P. A. Grieco, D. Boxler, and C. S. Pogonowski, *J.C.S. Chem. Comm.*, 1974, 497; see also B. M. Trost, W. P. Conway, P. E. Strege, and T. J. Dietsche, *J. Amer. Chem. Soc.*, 1974, **96**, 7165.
[403] I. Kuwajima and H. Iwasawa, *Tetrahedron Letters*, 1974, 107; see also ref. 312.

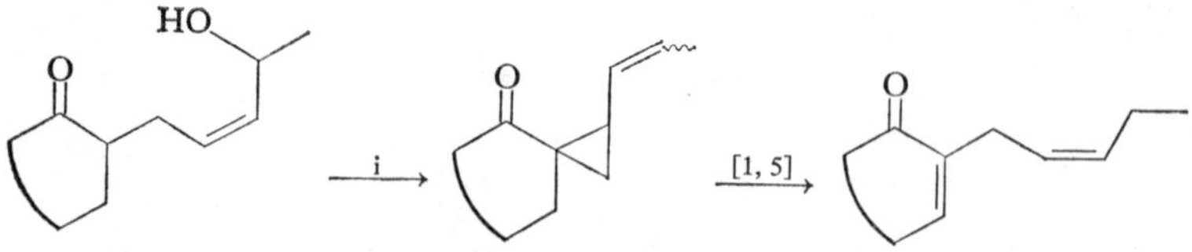

(87)

Reagent: i, dicyclohexylcarbodiimide—heat

Scheme 134

The alkylation of ketone and carboxylate enolates with benzyl bromomethyl sulphide, followed by oxidation and sulphoxide elimination, effects the introduction of an α-methylene group[404] in good yield; the reagent acts as a formaldehyde equivalent. Thermal cyclodehydration of unsaturated ε-ketols such as (87) is known to give vinyl spirannic ketones which undergo a clean thermal [1,5]sigmatropic rearrangement[405] to give *cis*-dienones (Scheme 134).

Lead tetra-acetate oxidative decarboxylation of γ-keto-acids efficiently produces[406] αβ-unsaturated ketones. A range of β-diketones give β-chloro-αβ-unsaturated ketones[407] in high yield on treatment with oxalyl chloride. The anodic oxidation of enol esters in acetic acid at carbon rod anodes has been reported[408] to produce mainly αβ-unsaturated ketones.

Conjugate Addition (Alkylation) and Reduction.—While vinylcuprates readily transfer[385] a vinyl group to cyclic αβ-unsaturated ketones, the ethynyl group cannot be transferred directly, due to strong copper–acetylene bonding. A partial solution to this problem was found in the use of ethynylalanes, but this is successful[409] only with acyclic or cisoid enones and fails with cyclic *S-trans* cases. Corey[410] has reported that the stannylated copper reagent (88) adds readily to such cyclic enones to give a product which can be converted by oxidation into the corresponding ethynyl derivatives, and thence to methyl ketones[411] (Scheme 135).

Lithium phenylthio(alkyl)cuprates are easily prepared,[412] and efficiently transfer alkyl groups, including tertiary ones (Scheme 136). Nickel and copper acetylacetonate catalyse the conjugate methylation[413,414] of αβ-unsaturated

[404] H. J. Reich and J. M. Renga, *J.C.S. Chem. Comm.*, 1974, 135; see also ref. 136.
[405] Y. Bahurel, L. Cottier, and G. Descotes, *Synthesis*, 1974, 118.
[406] J. E. McMurry and L. C. Blasczak, *J. Org. Chem.*, 1974, **39**, 2217.
[407] R. D. Clark and C. H. Heathcock, *Synthesis*, 1974, 47.
[408] T. Shono, Y. Matsumura, and Y. Nakagawa, *J. Amer. Chem. Soc.*, 1974, **96**, 3532.
[409] J. Hooz and R. B. Layton, *J. Amer. Chem. Soc.*, 1974, **96**, 7320.
[410] E. J. Corey and R. H. Wollenberg, *J. Amer. Chem. Soc.*, 1974, **96**, 5581.
[411] See also ref. 369.
[412] G. H. Posner, D. J. Brunelle, and L. Sinoway, *Synthesis*, 1974, 662.
[413] E. A. Jeffrey, A. Meisters, and T. Mole, *J. Organometallic Chem.*, 1974, **74**, 365.
[414] E. C. Ashby and G. Heinsohn, *J. Org. Chem.*, 1974, **39**, 3297.

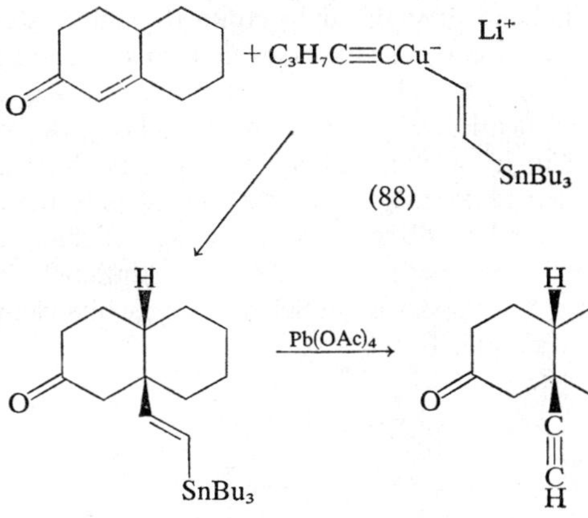

Scheme 135

ketones with trimethylaluminium and lithium tetramethylaluminate. A photochemical route[415] for the conjugate introduction of an aryl group into a cyclic $\alpha\beta$-unsaturated ketone system has been described (Scheme 137). The vapour-phase introduction of vinyl ketones in Michael additions is recommended[416] as an alternative to *in situ* generation from Mannich base

PhSCu + RLi ⟶ [PhSCuR]⁻Li⁺

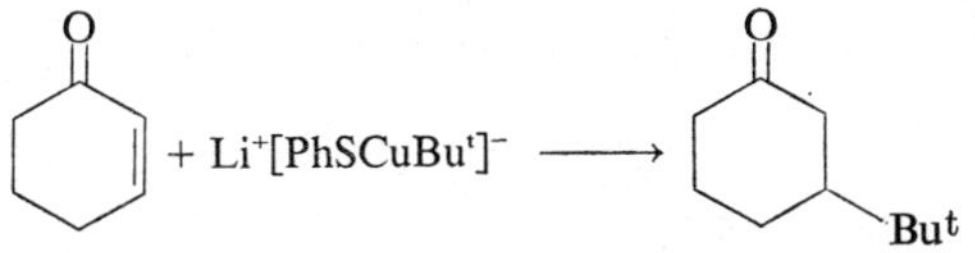

Scheme 136

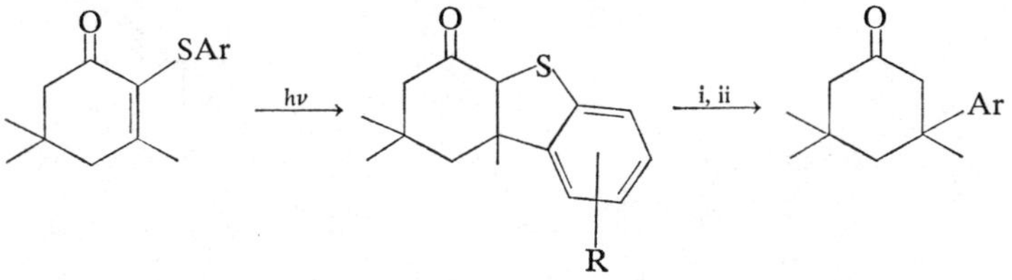

Reagents: i, Ni; ii, CrO₃

Scheme 137

[415] A. G. Schultz, *J. Org. Chem.*, 1974, **39**, 3185.
[416] C. D. De Boer, *J. Org. Chem.*, 1974, **39**, 2426.

methiodides. The silylated methyl vinyl ketone (89) successfully traps[417,418] even readily equilibrated enolate anions, which do not need to be 'copper enolates' (Scheme 138), provided careful attention is given to details, including thorough purification of the solvent, dimethoxyethane; the original procedure is indeed regiospecific, earlier difficulties having been caused by the 'adventitious presence of traces of protic impurities in the medium'.

Asymmetric induction is observed[419] in some Michael condensations catalysed by partially resolved 2-(hydroxymethyl)quinuclidine. A simple annelation procedure[420] is illustrated in Scheme 139; while yields are modest, the technique is very simple.

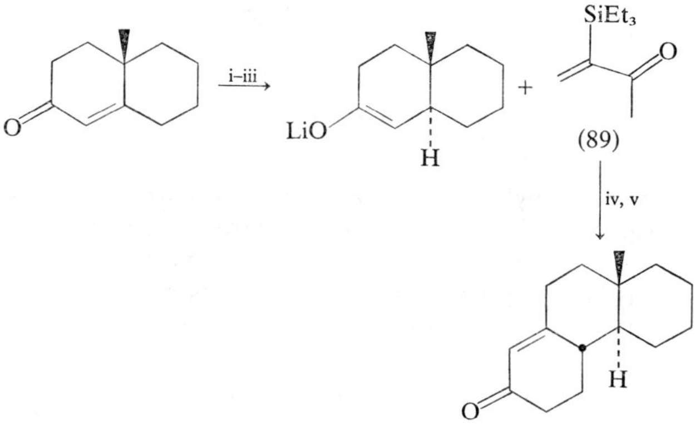

Reagents: i, Li–NH₃–Bu'OH (1 equiv); ii, Me₃SiCl, isolation; iii, LiMe; iv, NH₄Cl–H₂O; v, NaOMe–MeOH

Scheme 138

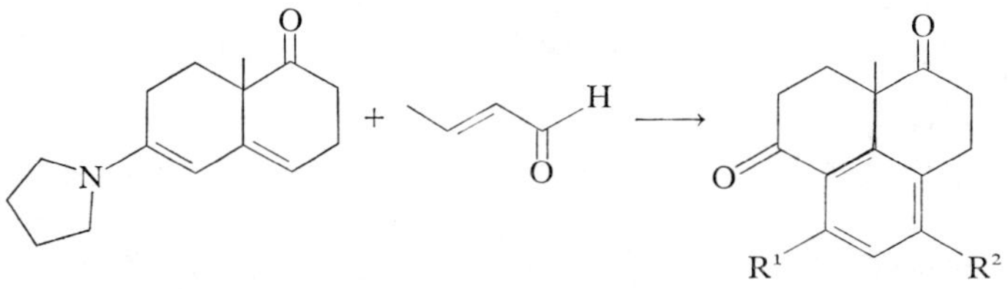

$R^1, R^2 = Me, H$

Scheme 139

[417] R. K. Boeckman, *J. Amer. Chem. Soc.*, 1974, **96**, 6179.
[418] G. Stork and J. Singh, *J. Amer. Chem. Soc.*, 1974, **96**, 6181.
[419] B. Långström and G. Bergson, *Acta Chem. Scand.*, 1973, **27**, 3118.
[420] P. Houdewind, J. C. Lapierre Armand, and U. K. Pandit, *Tetrahedron Letters*, 1974, 591.

Functional Groups other than Alkanes, Acetylenes, Allenes, and Olefins   155

Robinson–Mannich annelation[421] of cyclohexane-1,3-dione gives products dependent on the reaction conditions (Scheme 140).

Other examples of conjugate addition are discussed above in the section on umpolung (p. 140).

Several descriptions of 'ate' complexes of copper(I) hydride have appeared, and their differing properties have been exemplified. The lithium hydridocuprate (90) conjugatively reduces $\alpha\beta$-unsaturated carbonyl compounds; the addition of hexamethylphosphoramide facilitates[422] reduction in those cases where the β-carbon is highly substituted. The related species (91) also reduces saturated ketones to alcohols, and a variety of halides, mesylates, and tosylates to hydrocarbons.[423] The potassium hydridocuprate (92) efficiently reduces, *inter alia*, vinyl, aryl, and alkyl halides to the parent hydrocarbons.[424] The lithium dihydridocuprate (93) has been prepared.[425] A theoretical and mechanistic study[426] of the regioselectivity of reduction of cyclopentenone and cyclohexenone has revealed that the very 'soft' hydride donor (94) is best for maximum 1,4-reduction.

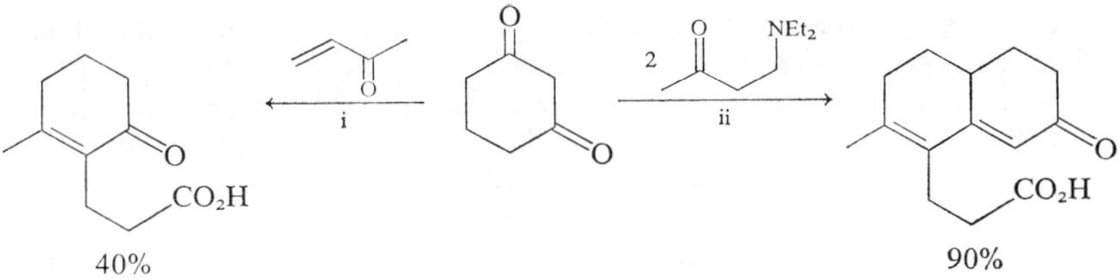

Reagents: i, Triton B; ii, py–PhH, heat

Scheme 140

Li⁺[CuHR]⁻	LiCuHBu^n	KCuH$_2$
(90)	(91)	(92)
R = 1-pentyne, OBu^t, SPh		

LiCuH$_2$	LiAlH(SBu^t)$_3$
(93)	(94)

[421] K. Balasubramanian, J. P. John, and S. Swaminathan, *Synthesis*, 1974, 51.
[422] R. K. Boeckman and R. Michalak, *J. Amer. Chem. Soc.*, 1974, **96**, 1623.
[423] S. Masamune, G. S. Bates, and P. E. Georghiou, *J. Amer. Chem. Soc.*, 1974, **96**, 3686.
[424] T. Yoshida and E. Negishi, *J.C.S. Chem. Comm.*, 1974, 762.
[425] E. C. Ashby, T. F. Korenowski, and R. D. Schwartz, *J.C.S. Chem. Comm.*, 1974, 157.
[426] J. Durand, N. T. Anh, and J. Huet, *Tetrahedron Letters*, 1974, 2397.

αβ-Unsaturated carbonyl compounds are selectively hydrogenated by CO and water in an autoclave;[427] a catalytic amount of a rhodium carbonyl complex is necessary (Scheme 141). Soluble iridium-based catalysts for the 1,4-reduction of conjugated enones have been described,[428] as have chromium(II) reagents.[429] Aqueous solutions of titanium(III) ions reduce the double bond of enedicarbonyl compounds[430] at low temperatures; only enediesters are not reduced, due to their high reduction potentials.

$$R^1CH{=}CHCOR^2 + CO + H_2O \xrightarrow[THF]{Rh_6(CO)_{16}} R^1CH_2CH_2COR^2 + CO_2$$

Scheme 141

Reduction and Reductive Coupling.—Lithium dimesitylborohydride bis(dimethoxyethane) (95) is a new crystalline reagent for the selective reduction[431] of ketones; essentially complete equatorial attack is seen in all cases studied and leads to axial alcohols (Scheme 142): highly hindered ketones such as camphor are resistant to such reduction under normal conditions.

Full comparative details have been given on the stereochemistry of reduction of a wide variety of ketones by dialkylboranes;[432] exceptionally high steric control is frequently observed. The catalytic properties of some dimethyl sulphoxide complexes of iridium hydrides have been described in detail;[433] their use favours production of axial alcohols.

The chirally modified aluminohydride[434] (96) reduces aldehydes, ketones, and ozonides to optically active carbinols in optical yields of up to 75%; since the enantiomer of (96) is equally accessible, both alcohol enantiomers are available at will, as illustrated in the synthesis of pestalotin[79] referred to

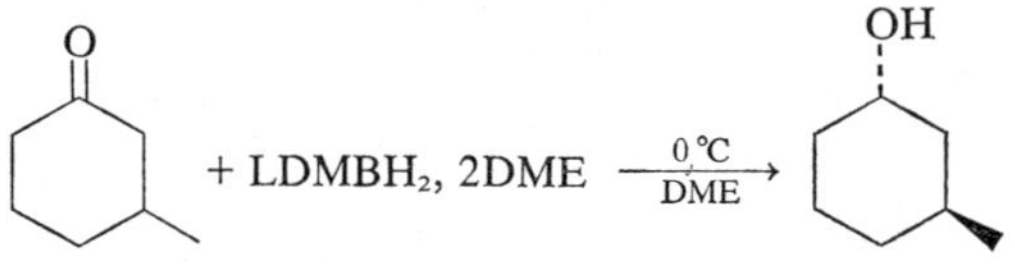

Scheme 142

[427] T. Kitamura, N. Sakamoto, and T. Joh, *Chem. Letters*, 1973, 379.
[428] H. B. Henbest and J. Trocha-Grimshaw, *J.C.S. Perkin I*, 1974, 601.
[429] H. O. House and E. F. Kinloch, *J. Org. Chem.*, 1974, **39**, 1173.
[430] L. C. Blaszczak and J. E. McMurry, *J. Org. Chem.*, 1974, **39**, 258.
[431] J. Hooz, S. Akiyama, F. J. Cedar, M. J. Bennett, and R. M. Tuggle, *J. Amer. Chem. Soc.*, 1974, **96**, 274.
[432] H. C. Brown and V. Varma, *J. Org. Chem.*, 1974, **39**, 1631.
[433] Y. M. Y. Haddad, H. B. Henbest, and J. Trocha-Grimshaw, *J.C.S. Perkin I*, 1974, 592; Y. M. Y. Haddad, H. B. Henbest, J. Husbands, T. R. B. Mitchell, and J. Trocha-Grimshaw, *ibid.*, p. 596.
[434] D. Seebach and H. Daum, *Chem. Ber.*, 1974, **107**, 1748.

(96)

(97)

(98) R = Me, Et Pr^i, or PhCH$_2$

earlier. A (−)-ephedrine-based chiral aluminohydride (97)[435] gives very good optical yields on reduction of phenyl alkyl ketones to the corresponding alkanols. Meyers[436] has recommended the oxazoline modified aluminohydride (98) as an asymmetric reducing agent and the parent chiral oxazoline can be recovered; he has also provided a useful comparative table of various asymmetric reducing agents.

More examples[437] of the utility of chiral rhodium complexes in catalysing the asymmetric hydrosilylation of ketones have been given. The asymmetric reduction of ketones with chiral organoaluminium[438] and dialkylzinc[439] reagents has been studied further. The stereochemical course of hydride-reduction of some α-chiral β-aminoketones has been investigated.[440]

Low-valent titanium,[441] presumably titanium(II), induces the reductive coupling of carbonyl compounds to yield products of pinacol condensation and olefins,[442] the relative proportions of which are dependent on the reaction conditions (Scheme 143); the reagent does indeed convert 1,2-diols into olefins. Such coupling is seen to advantage in the efficient synthesis of β-carotene, which is also illustrated[443] in Scheme 143.

[435] I. Jacquet and J. P. Vigneron, *Tetrahedron Letters*, 1974, 2065.
[436] A. I. Meyers and P. M. Kendall, *Tetrahedron Letters*, 1974, 1337.
[437] I. Ojima, T. Kogure, and Y. Nagai, *Chemistry Letters*, 1973, 541.
[438] G. Giacomelli, R. Menicagli, and L. Lardicci, *J. Org. Chem.*, 1974, **39**, 1757.
[439] G. Giacomelli, L. Lardicci, and R. Santi, *J. Org. Chem.*, 1974, **39**, 2736; L. Lardicci and G. Giacomelli, *J.C.S. Perkin I*, 1974, 337.
[440] L. Angiolini and M. Tramontini, *J. Org. Chem.*, 1974, **39**, 2056.
[441] J. E. McMurry, *Accounts Chem. Res.*, 1974, **7**, 281.
[442] T. Mukaiyama, T. Sato, and J. Hanna, *Chem. Letters*, 1973, 1041; see also T. Mukaiyama, M. Hayashi, and K. Narasaka, *ibid.*, p. 291.
[443] J. E. McMurry and M. P. Fleming, *J. Amer. Chem. Soc.*, 1974, **96**, 4708.

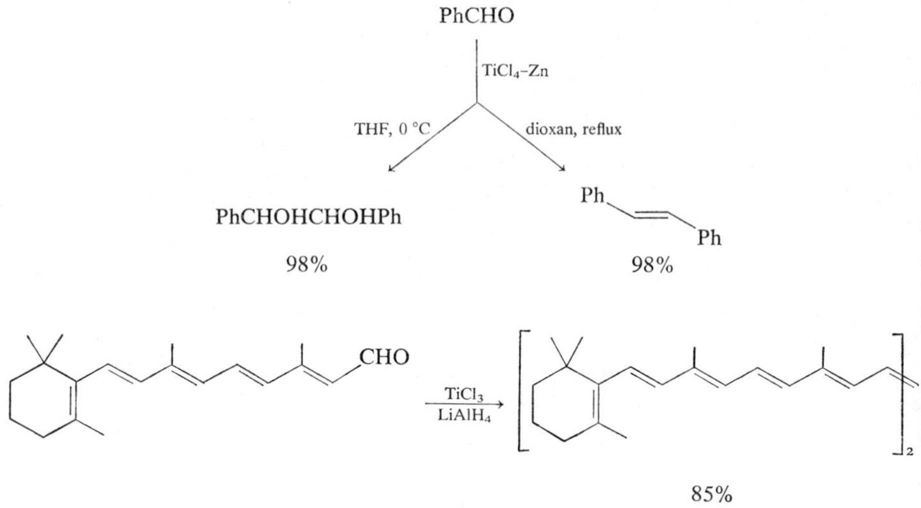

PhCHO

TiCl₄–Zn

THF, 0 °C ↙   ↘ dioxan, reflux

PhCHOHCHOHPh   Ph_____/Ph
98%   98%

[structure: β-ionylidene polyene aldehyde] —TiCl₃/LiAlH₄→ [coupled dimer]₂

85%

Scheme 143

Two groups[444,445] have independently described routes to phenylseleno-(99) and phenylthio-(100) alkyl-lithium compounds; these react with carbonyl compounds, with effective coupling of the carbonyl carbon atoms, to give the range of products shown in Scheme 144.

Aldehydes and ketones are reduced to hydrocarbons by silanes in the presence of alkyl iodides, which, though not consumed to a significant extent, are essential for reasonable conversion; yields are said to be good.[446] The scope and advantages of the reduction of carbonyl compounds to alcohols and alcohol derivatives with silanes in acidic media have been discussed.[447] Triethylaminoaluminium hydride[448] is a stable, hydrocarbon-soluble reducing agent for ketones.

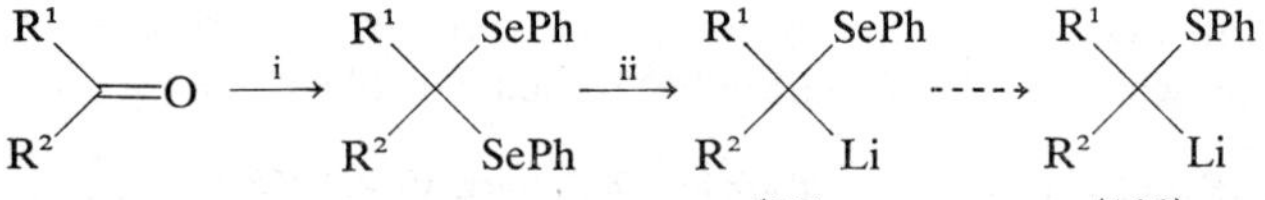

$R^1R^2C=O$ —i→ $R^1R^2C(SePh)_2$ —ii→ $R^1R^2C(SePh)(Li)$ ----→ $R^1R^2C(SPh)(Li)$
                                          (99)   (100)

[444] W. Dumont, P. Bayet, and A. Krief, *Angew. Chem. Internat. Edn.*, 1974, **13**, 804.
[445] D. Seebach and A. K. Beck, *Angew. Chem. Internat. Edn.*, 1974, **13**, 806.
[446] J. L. Levenson and L. Kaplan, *J.C.S. Chem. Comm.*, 1974, 23.
[447] M. P. Doyle, D. J. DeBruyn, S. J. Donnelly, D. A. Kooistra, A. A. Odubela, C. T. West, and S. M. Zonnebelt, *J. Org. Chem.*, 1974, **39**, 2740.
[448] S. Cacchi, B. Giannoli, and D. Misiti, *Synthesis*, 1974, 728.

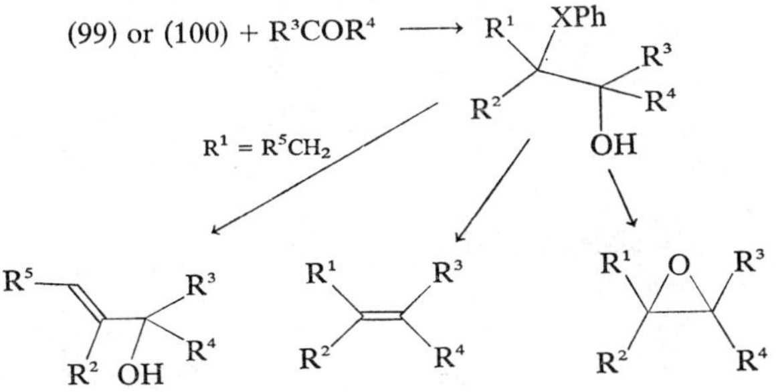

(99) or (100) + R³COR⁴ ⟶

Reagents: i, PhSeH–H⁺; ii, Bu^{n}Li

Scheme 144

Condensations.—*Directed Aldol Condensations.* Direct regiospecific aldol condensations can be performed with surprising ease. Stork[449] has reported that enolate anions condense smoothly and regiospecifically with formaldehyde to give α-hydroxymethylketones, precursors of, *inter alia*, α-methyleneketones. The method of choice of generation of specific anions is *via* conjugate addition to or reduction of αβ-unsaturated enones, with intermediate trapping as silyl enol ethers and subsequent liberation. He has also reported that methyl ketones react regiospecifically[450] *via* their kinetic enolates, generated at low temperature, with aldehydes as electrophiles (Scheme 145).

Trimethylsilyl enol ethers[451] and enol acetates[452] react with carbonyl compounds, or the corresponding acetals,[453] in the presence of Lewis acids

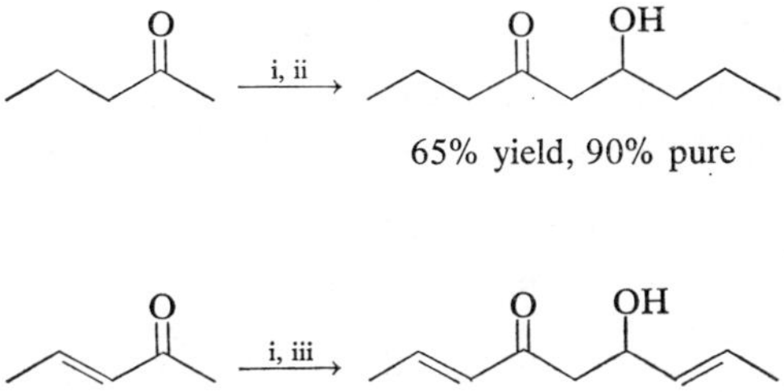

65% yield, 90% pure

Reagents: i, LiNPr$^i_2$; ii, ∼CHO; iii, ∼CHO

Scheme 145

[449] G. Stork and J. d'Angelo, *J. Amer. Chem. Soc.*, 1974, **96**, 7114.
[450] G. Stork, G. A. Kraus, and G. A. Garcia, *J. Org. Chem.*, 1974, **39**, 3459.
[451] T. Mukaiyama, K. Narasaka, and K. Banno, *Chem. Letters*, 1973, 1011; T. Mukaiyama, K. Banno, and K. Narasaka, *J. Amer. Chem. Soc.*, 1974, **96**, 7503.
[452] T. Mukaiyama, T. Izawa, and K. Saigo, *Chem. Letters*, 1974, 323; see also ref. 59.
[453] T. Mukaiyama and M. Hayashi, *Chem. Letters*, 1974, 15; see also ref. 106.

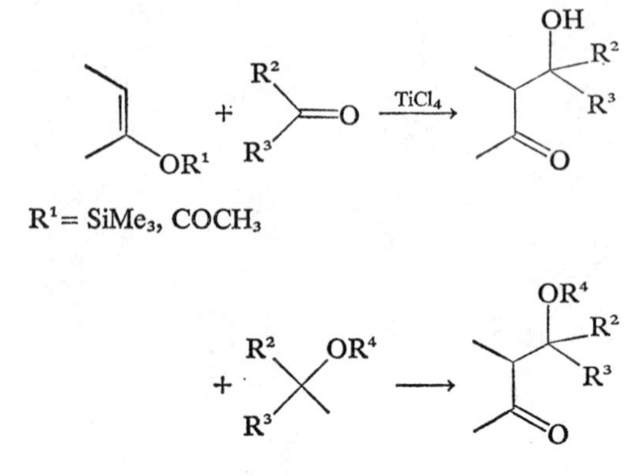

R^2____ + R^2____=O $\xrightarrow{TiCl_4}$ (β-hydroxy ketone structure)
    OR^1 R^3

R^1 = SiMe$_3$, COCH$_3$

R^2, OR^4 + R^3 → (β-alkoxy ketone structure)

Scheme 146

such as titanium(IV) chloride to give β-hydroxy- or β-alkoxy-ketones, respectively (Scheme 146).

Isomeric Mannich bases derived from unsymmetrical ketones can be synthesized regioselectively[454] by suitable selection of reaction conditions (Scheme 147); such observed differences probably reflect the influence of the conditions on the relative rates of enolization and alkylation. Solvent and ion-pair effects on the self-condensation of aliphatic aldehydes have been studied.[455]

cis-Jasmone (101) has been synthesized by regiospecific condensation[456] of the 1,4-diketone (102); the alternative aldol product (103) is not formed under the kinetic conditions employed, but can be isomerized to *cis*-jasmone under more vigorous treatment (Scheme 148).

Organometallics. Ashby has summarized his work on the mechanism of addition of Grignard reagents to ketones.[457] Although the orientation of

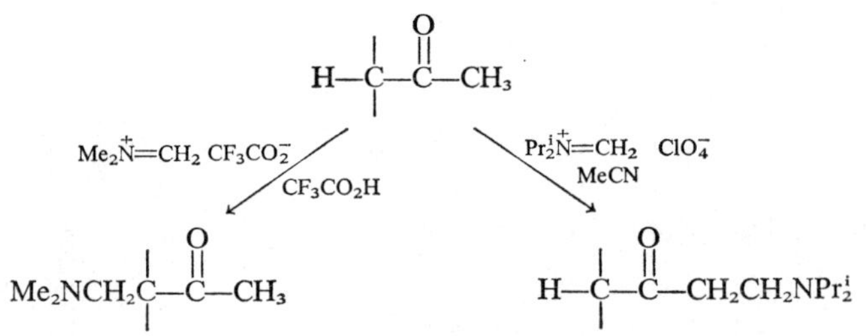

$$H-\underset{|}{C}-\underset{\|}{\overset{O}{C}}-CH_3$$

Me$_2\overset{+}{N}$=CH$_2$ CF$_3$CO$_2^-$ / CF$_3$CO$_2$H   Pr$_2^i\overset{+}{N}$=CH$_2$ ClO$_4^-$ / MeCN

Me$_2$NCH$_2\underset{|}{C}-\underset{\|}{\overset{O}{C}}-CH_3$ $H-\underset{|}{C}-\underset{\|}{\overset{O}{C}}-CH_2CH_2NPr_2^i$

Scheme 147

[454] Y. Jasor, M.-J. Luche, M. Gaudry, and A. Marquet, *J.C.S. Chem. Comm.*, 1974, 253.
[455] G. Casnati, A. Pochini, G. Salerno, and R. Ungaro, *Tetrahedron Letters*, 1974, 959.
[456] P. M. McCurry and R. K. Singh, *J. Org. Chem.*, 1974, **39**, 2316, 2317.
[457] E. C. Ashby, J. Laemmle, and H. M. Neumann, *Accounts Chem. Res.*, 1974, **7**, 272.

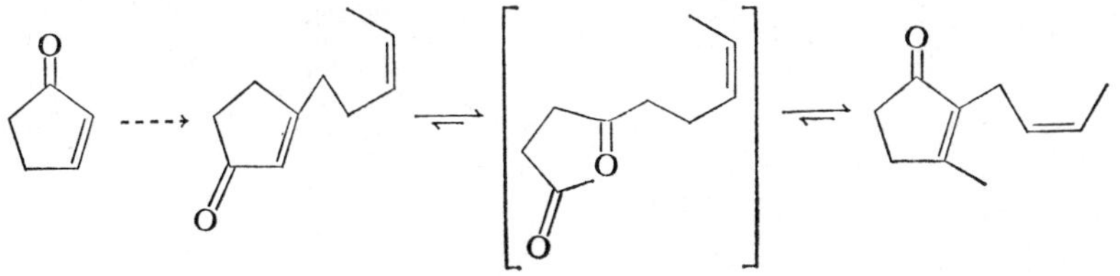

Functional Groups other than Alkanes, Aceytlenes, Allenes, and Olefins 161

(103)   (102)   (101)

Scheme 148

addition of ambident allylic nucleophiles is normally determined by the degree of steric hindrance at the carbonyl group, *gem*-dichloroallyl-lithium adds in what seems to be electronically governed orientational behaviour,[458] the main criterion being the degree of 'hardness' of the carbonyl group. 2-Methylbutenyl-lithium (104) and *cis*-crotyl-lithium both react with carbonyl compounds at the γ-carbon to give branched products in moderate to high selectivity, unless the carbonyl group is highly hindered, when α-attack and linear products are observed (Scheme 149).[459] A study of the stereochemistry of addition of 'ate' complexes such as lithium tetramethylaluminate has been published.[460] The chiral oxazoline described on p. 157 for the induction of asymmetry in hydride reduction of ketones performs a similar function in the addition of Grignard reagents.[461] Chiral chromium tricarbonyl

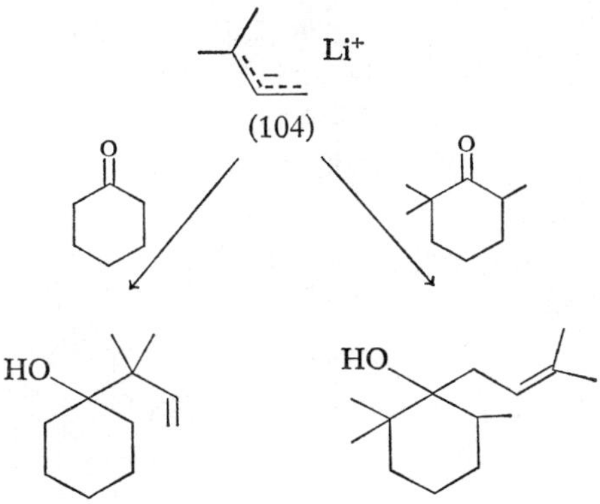

Scheme 149

[458] D. Seyferth, G. J. Murphy, and R. A. Woodruff, *J. Amer. Chem. Soc.*, 1974, **96**, 5011.
[459] V. Rautenstrauch, *Helv. Chim. Acta*, 1974, **57**, 496.
[460] E. C. Ashby, L. Chao, and J. Laemmle, *J. Org. Chem.*, 1974, **39**, 3258.
[461] A. I. Meyers and M. E. Ford, *Tetrahedron Letters*, 1974, 1341.

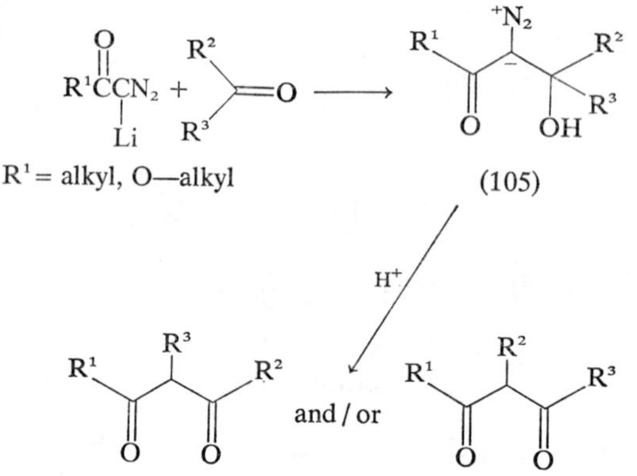

Scheme 150

complexes[462] of some aromatic ketones undergo asymmetric attack by Grignard reagents to give optically active carbinols.

α-Diazo-β-hydroxycarbonyl compounds (105) rearrange on treatment with mineral acid to β-dicarbonyl species[463] (Scheme 150); in the case of cycloalkanone precursors, this constitutes another[464] method of ketone ring expansion.

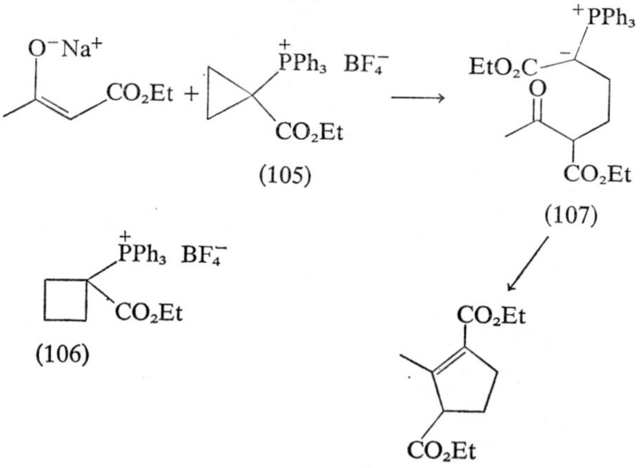

Scheme 151

---

[462] A. Meyer and G. Jaouen, *J.C.S. Chem. Comm.*, 1974, 787.
[463] U. Schöllkopf, B. Banhidai, H. Frasnelli, R. Meyer, and H. Beckhaus, *Annalen*, 1974, 1767; cf. E. Wenkert and C. A. McPherson, *J. Amer. Chem. Soc.*, 1972, **94**, 8084.
[464] See also refs. 314, 315, 317, 318.

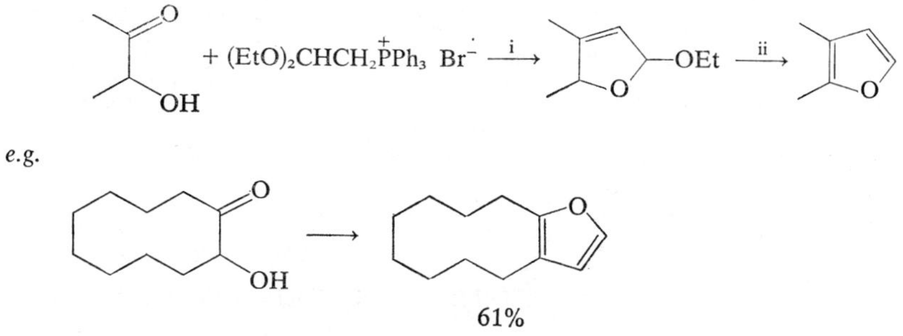

Functional Groups other than Alkanes, Acetylenes, Allenes, and Olefins   163

[Scheme showing: α-hydroxy ketone + (EtO)₂CHCH₂P⁺Ph₃ Br⁻ →(i) dihydrofuran-OEt →(ii) furan]

e.g.

[2-hydroxycyclodecanone → fused furan, 61%]

Reagents: i, NaH; ii, TsOH

Scheme 152

Wittig Types. Wittig–Horner–Emmons olefin synthesis has been reviewed.[465] The cyclopropylphosphonium salt (105) is highly susceptible to nucleophilic ring-opening reactions, due to the *gem* electron-withdrawing groups; such a propensity is used to advantage[466] in the cycloalkenylation procedure illustrated in Scheme 151. Attempts to use the homologous salt (106) were unrewarding, and no stabilized ylide similar to (107) was detected.

α-Hydroxy-ketones react with β-ethoxyvinylphosphonium salts (generated *in situ*) to give furans[467] in good yield (Scheme 152).

Wittig reaction of n-hexanal with the stabilized ylide (108) generates not only all four geometrical isomers of (109), the expected α-condensation product, but also both geometrical isomers of (110), the unprecedented γ-condensation product; either ester can be made to predominate by appropriate choice of reaction conditions. One application of such α-condensation is seen in a synthesis[468] of (111) (Scheme 153), selective epoxidation of which gives a dehydro analogue of the C_{18} Cecropia juvenile hormone.

While adamantanone reacts with methylenetriphenylphosphorane to give the expected methylene olefin, the corresponding thione gives the thiirane.[469]

General. Trost[470] has published details of the stereoselectivity and regiospecificity of spiroannelations[471] and cyclopentanone annelations[472] using his cyclopropyl sulphide-based reagents.

Cycloaddition[473] of the allyl anion (112) to electrophilic olefins, formally

[465] J. Boutagy and R. Thomas, *Chem. Rev.*, 1974, **74**, 87.
[466] P. L. Fuchs, *J. Amer. Chem. Soc.*, 1974, **96**, 1607.
[467] M. E. Garst and T. A. Spencer, *J. Org. Chem.*, 1974, **39**, 584; see also C. J. Harris, J. J. Cleary, and T. M. Harris, *ibid.*, p. 72.
[468] E. J. Corey and B. W. Erickson, *J. Org. Chem.*, 1974, **39**, 821.
[469] A. P. Krapcho, M. P. Silvon, and S. D. Flanders, *Tetrahedron Letters*, 1974, 3817.
[470] B. M. Trost, *Accounts Chem. Res.*, 1974, **7**, 85.
[471] B. M. Trost and D. E. Keeley, *J. Amer. Chem. Soc.*, 1974, **96**, 1252.
[472] B. M. Trost and S. Kurozumi, *Tetrahedron Letters*, 1974, 1929.
[473] J. P. Marino and W. B. Mesbergen, *J. Amer. Chem. Soc.*, 1974, **96**, 4050.

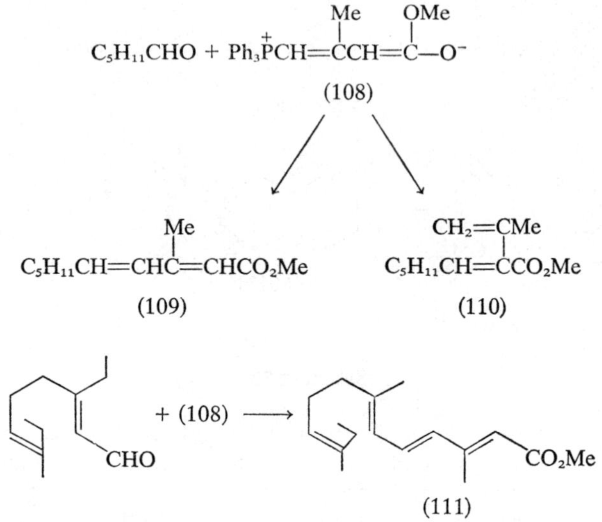

$$C_5H_{11}CHO + Ph_3\overset{+}{P}CH=\overset{\underset{|}{Me}}{C}CH=\overset{\underset{|}{OMe}}{C}-O^-$$

(108)

$$C_5H_{11}CH=CH\overset{\underset{|}{Me}}{C}=CHCO_2Me$$

(109)

$$C_5H_{11}CH=\overset{\underset{|}{CH_2=CMe}}{C}CO_2Me$$

(110)

Scheme 153

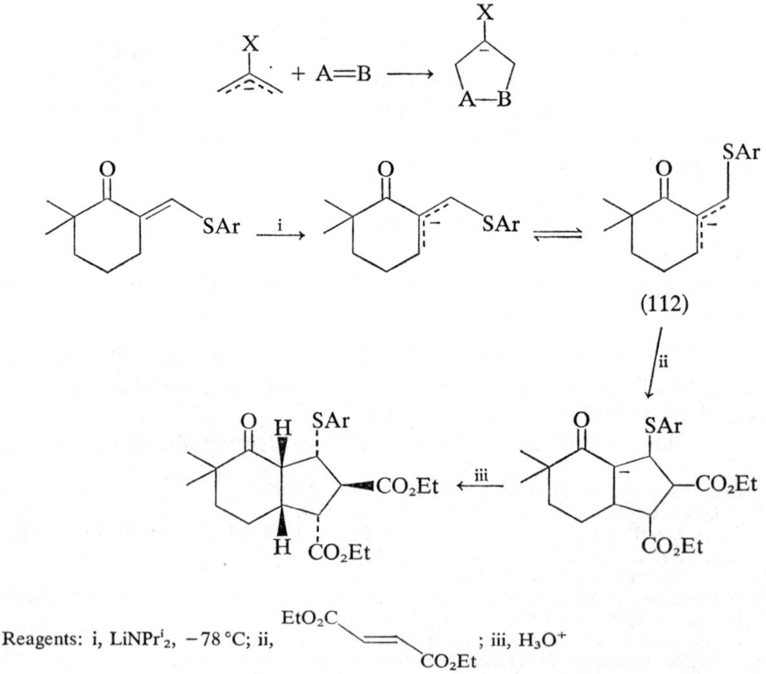

Reagents: i, LiNPr$^i_2$, −78 °C; ii, EtO$_2$C−CH=CH−CO$_2$Et ; iii, H$_3$O$^+$

Scheme 154

a $[_\pi 4_s + _\pi 2_s]$ process, gives a product in which all five chiral centres are formed stereoselectively (Scheme 154).

Halogen Derivatives and Oxyallyl Zwitterions.—Halogenation of silyl enol ethers provides a regiospecific route[474] to α-halogeno-ketones and -aldehydes (Scheme 155); related nitrosation[475] gives good yields of α-nitroso- or α-oximino-carbonyl compounds, trapped in the case of aldehydes as the corresponding glyoximes: these reactions are also applicable to ester and lactone silyl enol ethers.

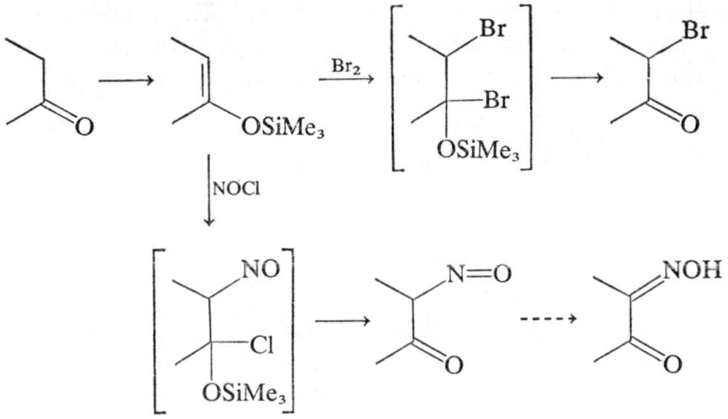

Scheme 155

Diazomethylketones react with N-halogenosuccinimides in poly(hydrogen fluoride)–pyridine solution to give α-fluoro-α-halogenomethylketones.[476] A variety of Grignard reagents react with polychlorinated enamines by substitution of the halogen *gem* to nitrogen; hydrolysis leads to αα-dichloromethylketones[477] in good yield (Scheme 156).

Trimethylsilyl cyclopropyl ethers act as relatively stable synthetic equivalents[478] of the frequently labile cyclopropanols in the reaction of the latter with bromine to give β-bromoketones (Scheme 157); the cyclopropyl ethers are easily prepared by Simmons–Smith methylenation of silyl enol ethers. Efficient syntheses of some polyhalogenated cross-conjugated diene aldehydes have been reported.[479] A series of polyhalogenoacetones constitutes[480] the major volatile fraction of the halogenated metabolites of the red seaweed, *Asparagopsis taxiformis*.

---

[474] R. H. Reuss and A. Hassner, *J. Org. Chem.*, 1974, **39**, 1785.
[475] J. K. Rasmussen and A. Hassner, *J. Org. Chem.*, 1974, **39**, 2558; for other nitrosation procedures, see M. F. Chen and S. F. MacDonald, *Canad. J. Chem.*, 1974, **52**, 1760; M. Kataoka and M. Ohno, *Bull. Chem. Soc. Japan*, 1973, **46**, 3474.
[476] G. A. Olah and J. Welch, *Synthesis*, 1974, 896.
[477] J. Ficini and A. Duréault, *Bull. Soc. chim. France*, 1974, 1533.
[478] S. Murai, Y. Seki, and N. Sonoda, *J.C.S. Chem. Comm.*, 1974, 1032.
[479] F. Pochat, *Bull. Soc. chim. France*, 1974, 1373.
[480] W. Fenical, *Tetrahedron Letters*, 1974, 4463.

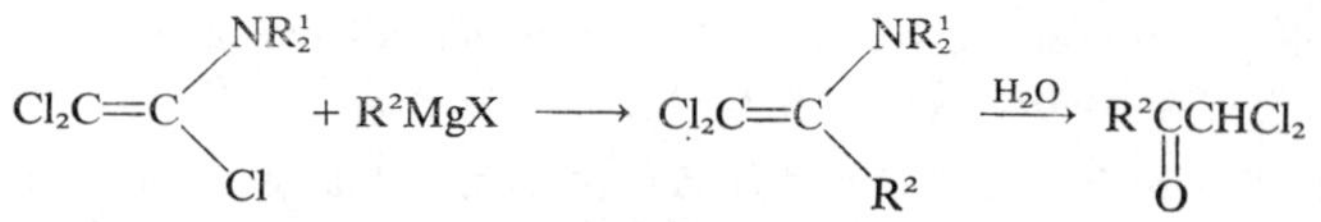

$$Cl_2C=C\begin{smallmatrix}NR^1_2\\Cl\end{smallmatrix} + R^2MgX \longrightarrow Cl_2C=C\begin{smallmatrix}NR^1_2\\R^2\end{smallmatrix} \xrightarrow{H_2O} R^2CCHCl_2$$
$$\parallel$$
$$O$$

Scheme 156

The separate observations that α-halogenoketones are readily substituted by pyridine, and that pyridinium salts undergo facile 1,4-reduction with sodium dithionite, have led[481] to an efficient procedure for the reductive dehalogenation of α-halogenoketones (Scheme 158). A Zn–Cu couple can serve as a convenient alternative[482] to the Zn–Ag couple[483] for reductive dehalogenation of β-chlorovinyl ketones[407] to conjugated enones.

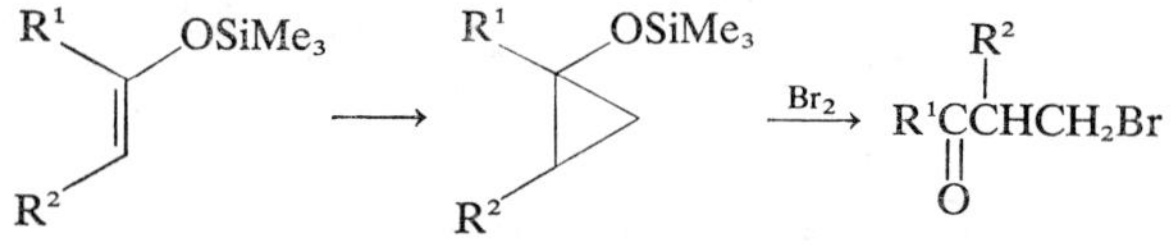

Scheme 157

In an extension of his studies on α-ketomethylene free radicals, Ghera[484] has reported that the radical generated by action of a Zn–Cu couple on phenacyl bromide adds to conjugated enynes and dienes, as illustrated in Scheme 159.

The reaction of αα'-dibromoketones with organoboranes in the presence of base gives products dependent[485] on the type and amount of base used (Scheme 160); the monoalkylcycloalkanones formed arise from exclusive attack at the less hindered side of unsymmetrical ketones.

Cycloaddition of oxyallyl zwitterions to dienes suggests a promising route to cycloheptenones and bridged derivatives, the main hindrance being the

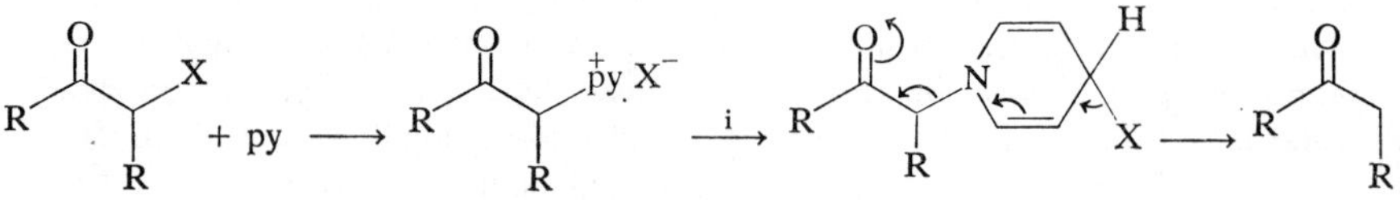

Reagent: i, $Na_2S_2O_4$–H_2O

Scheme 158

[481] T.-L. Ho and C. M. Wong, *J. Org. Chem.*, 1974, **39**, 562.
[482] R. M. Blankenship, K. A. Burdett, and J. S. Swenton, *J. Org. Chem.*, 1974, **39**, 2300.
[483] R. D. Clark and C. H. Heathcock, *J. Org. Chem.*, 1973, **38**, 3659.
[484] E. Ghera and S. Shoua, *Tetrahedron Letters*, 1974, 3843.
[485] R. H. Prager and J. M. Tippett, *Austral. J. Chem.*, 1974, **27**, 1457, 1467.

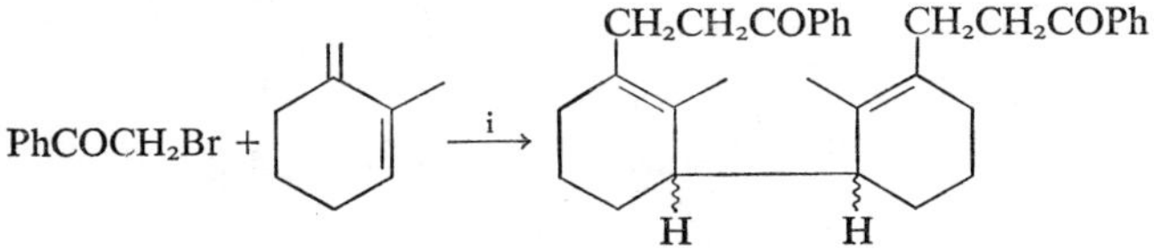

Functional Groups other than Alkanes, Acetylenes, Allenes, and Olefins 167

PhCOCH$_2$Br + [cyclohexene] →^{i} [bicyclic product with CH$_2$CH$_2$COPh groups]

Reagent: i, Zn–Cu–DMSO

Scheme 159

relative lack of good routes to the required zwitterions. Chan[486] has now reported that fluoride ion specifically attacks the silicon atom of the oxiran (113); this allows facile entry into the allene oxide–oxyallyl zwitterion–cyclopropanone set of valence bond isomers (Scheme 161).

2-(*N*-Alkylimino)cyclobutanones[487] are formed when αα'-dibromoketones react with Cu in the presence of an isocyanide, presumably *via* insertion of the isocyanide into an intermediate cyclopropanone or its equivalent. *N*-Methylformamide acts as a polar and weakly protic solvent for the Zn–Cu couple reductive dimerization of αα'-dibromoketones, leading to 1,4-diketones (Scheme 162); the mechanism is envisaged[488] as shown, involving combination of a zinc oxyallyl species and a zinc enolate.

The reductive dehalogenation of αα'-dibromoketones with sodium iodide in the presence of pyrroles and Cu leads[489] to a variety of 6,7-dehydrotropinones (Scheme 163).

General Properties and Reactions.—Unambiguous experimental proof[490] of the existence of *front* octants for ketones has been presented. The ketone

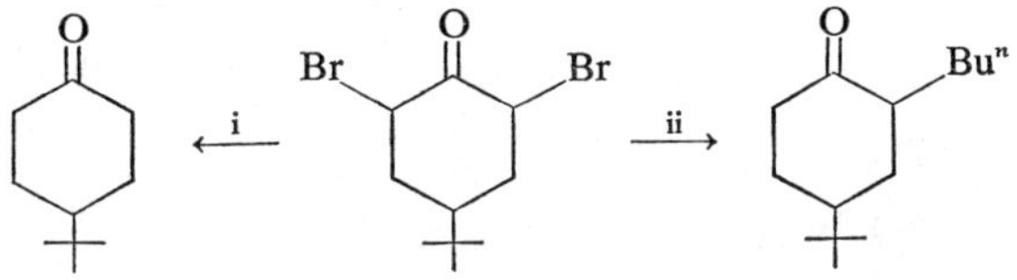

[4-t-butylcyclohexanone] ←^{i}— [2,6-dibromo-4-t-butylcyclohexanone] —^{ii}→ [2-n-butyl-4-t-butylcyclohexanone]

60%   100%

Reagents: i, 2 KOPr^i–4 Bu$^n_3$B; ii, 2 KOCEt$_3$–4 Bu$^n_3$B

Scheme 160

[486] T. Chan, M. P. Li, W. Mychajlowskij, and D. N. Harpp, *Tetrahedron Letters*, 1974, 3511.
[487] Y. Ito, M. Asada, K. Yonezawa, and T. Saegusa, *Synthetic Comm.*, 1974, **4**, 87.
[488] C. Chassin, E. A. Schmidt, and H. M. R. Hoffmann, *J. Amer. Chem. Soc.*, 1974, **96**, 606.
[489] G. Fierz, R. Chidgey, and H. M. R. Hoffmann, *J. Amer. Chem. Soc.*, 1974, **96**, 5466.
[490] D. A. Lightner and D. E. Jackman, *J.C.S. Chem. Comm.*, 1974, 344; D. A. Lightner and T. C. Chang, *J. Amer. Chem. Soc.*, 1974, **96**, 3015.

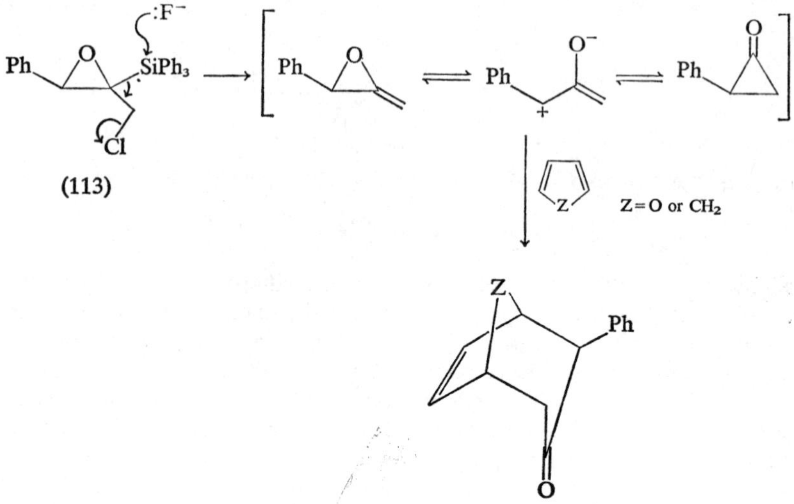

Scheme 161

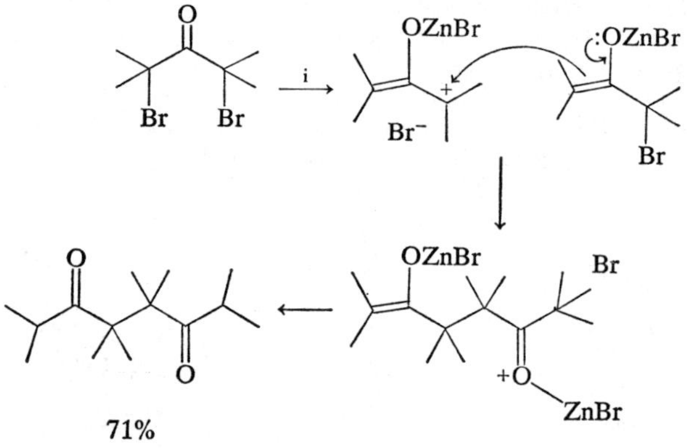

(113)

Z = O or CH₂

Scheme 162

Reagents: i, Zn–Cu–HCONHMe

71%

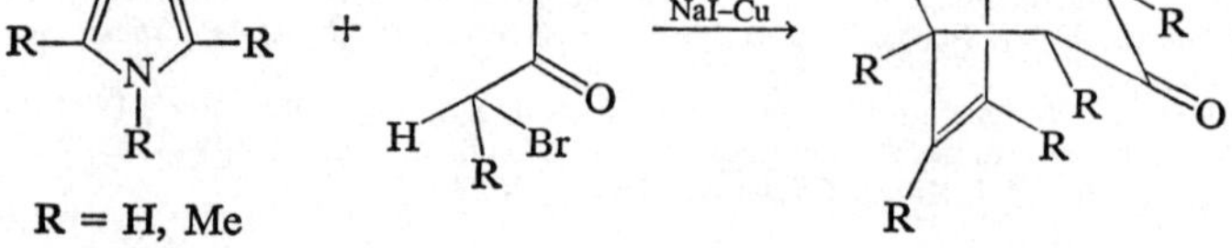

Scheme 163

R = H, Me

Functional Groups other than Alkanes, Acetylenes, Allenes, and Olefins 169

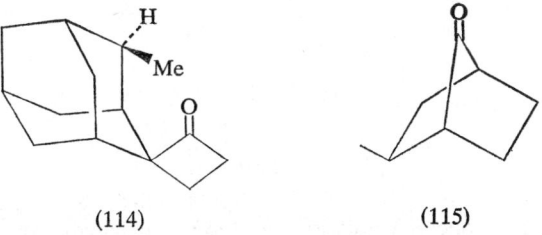

(114)   (115)

(114) has the lone dissymmetric perturber (the methyl group) projecting into the lower right or upper left front octant, and indeed shows a strong negative effect. However, the ketone (115) also shows a negative Cotton effect; this too is ascribed[491] to the methyl group being present in a front octant, implying that it is not a plane which defines the front octant, but rather it is a curve bent back behind the carbonyl group.

The $n \rightarrow \pi^*$ transitions of some fluoro- and iodo-ketones have been studied[492] on the basis of the number of intervening bonds between functional groups.

An *ab initio* study[493] of nucleophilic addition to the carbonyl group has produced results in qualitative agreement with those derived from solid-state data. A similar study[494] of the protonation of the carbonyl group has given a basicity order of amide > carboxylic acid > acid fluoride $\simeq$ acetaldehyde > keten > formaldehyde, an order which closely follows the degree of negative charge on oxygen. Gas-phase ionic reactions related to acid-catalysed carbonyl reactions in solution have been studied[495] by ion cyclotron resonance.

Cyclohexanone and acetophenones form complexes with $AgBF_4$ in dichloromethane, as shown[496] by the enhanced solubility of the silver salts. ^{13}C N.m.r. spectroscopic evidence indicated that the carbonyl group is acting as a N-donor, giving a rather weak Ag—O bond (116); no evidence of *syn-anti* isomerism could be detected, even at low temperatures.

The resolution of ketones and determination of their specific rotation *via* formation of the corresponding dioxolanes with chiral 1,2-diols have been described.[497]

o-Nitrophenylethylene glycol is[498] a photosensitive protecting group for

[491] D. A. Lightner and D. E. Jackman, *J. Amer. Chem. Soc.*, 1974, **96**, 1938.
[492] E. E. Ernstbrunner and J. Hudec, *J. Amer. Chem. Soc.*, 1974, **96**, 7106.
[493] H. B. Burgi, J. M. Lehn, and G. Wipff, *J. Amer. Chem. Soc.*, 1974, **96**, 1956.
[494] A. C. Hopkinson and I. G. Csizmadia, *Canad. J. Chem.*, 1974, **52**, 546; see also A. Levi, G. Modena, and G. Scorrano, *J. Amer. Chem. Soc.*, 1974, **96**, 6585; D. G. Lee and M. H. Sadar, *ibid.*, p. 2862.
[495] J. K. Pau, J. K. Kim, and M. C. Caserio, *J.C.S. Chem. Comm.*, 1974, 120, 121.
[496] D. R. Crist, Z.-H. Hsieh, G. J. Jordan, F. P. Schinco, and C. A. Maciorowski, *J. Amer. Chem. Soc.*, 1974, **96**, 4932.
[497] J. Y. Conan, A. Natat, F. Guinot, and G. Lamaty, *Bull. Soc. chim. France*, 1974, 1400, 1405; *Tetrahedron Letters*, 1974, 1667; J. Brugidou, H. Christol, and R. Sales, *Bull. Soc. chim. France*, 1974, 2027, 2033.
[498] J. Hébert and D. Gravel, *Canad. J. Chem.*, 1974, **52**, 187.

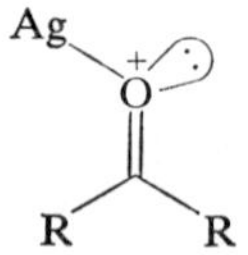

(116)

aldehydes and ketones; irradiation at 350 nm efficiently regenerates the carbonyl compound. DMF–dialkylsulphate adducts[499] and Amberlyst 15[500] have been separately recommended for the preparation of acetals.

Flash thermolysis of allyl sulphides at 900 K generates the corresponding thiocarbonyl compounds (Scheme 164); by this means, both thioacrolein and thiobenzaldehyde have been prepared[501] and characterized spectroscopically (at 77 K). ^{13}C N.m.r. chemical shifts of carbonyl and thiocarbonyl groups have been correlated.[502] Various aldehydes and ketones are smoothly converted into the corresponding vinyl sulphides[503a] by treatment with thiols in the presence of titanium(IV) chloride; $\alpha\beta$-unsaturated enones undergo conjugate addition, to give [503b] 3-alkylthiocarbonyl compounds.

In work based on the conceptually useful analogy between mass spectrometric and anodic chemistry, it has been found[504] that anodic oxidation of simple ketones in acetonitrile leads to γ-hydrogen abstraction, with subsequent solvent capture and/or rearrangement and capture (Scheme 165); no β-cleavage is observed.

A theoretical model[505] for the Baeyer–Villiger rearrangement has been described. Deslongchamps[506] has provided full details of his work on the ozonolysis of acetals. Kinetic and product studies of the base-catalysed decomposition of some α-hydroperoxyketones has revealed[507] that the reaction proceeds largely through 1,2-dioxetan intermediates (Scheme 166),

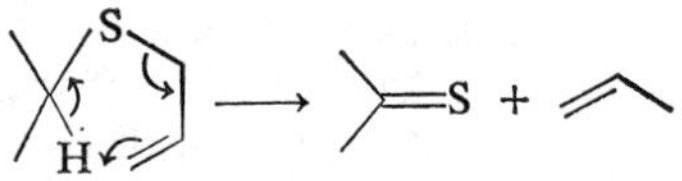

Scheme 164

[499] W. Kantlehner, H.-D. Gutbrod, and P. Gross, *Annalen*, 1974, 690.
[500] S. A. Patwardhan and S. Dev, *Synthesis*, 1974, 348.
[501] H. G. Giles, R. A. Marty, and P. DeMayo, *J.C.S. Chem. Comm.*, 1974, 409.
[502] H.-O. Kalinowski and H. Kessler, *Angew. Chem. Internat. Edn.*, 1974, **13**, 90.
[503] (*a*) T. Mukaiyama and K. Saigo, *Chem. Letters*, 1973, 4799; (*b*) T. Mukaiyama, T. Izawa, K. Saigo, and H. Takei, *ibid.*, p. 355.
[504] J. Y. Becker, L. R. Byrd, and L. L. Miller, *J. Amer. Chem. Soc.*, 1974, **96**, 4718.
[505] V. A. Stoute, M. A. Winnik, and I. G. Csizmadia, *J. Amer. Chem. Soc.*, 1974, **96**, 6388.
[506] P. Deslongchamps, P. Atlani, D. Fréhel, A. Malaval, and C. Moreau, *Canad. J. Chem.*, 1974, **52**, 3691.
[507] W. H. Richardson, V. F. Hodge, D. L. Stiggall, M. B. Yelvington, and F. C. Montgomery, *J. Amer. Chem. Soc.*, 1974, **96**, 6652.

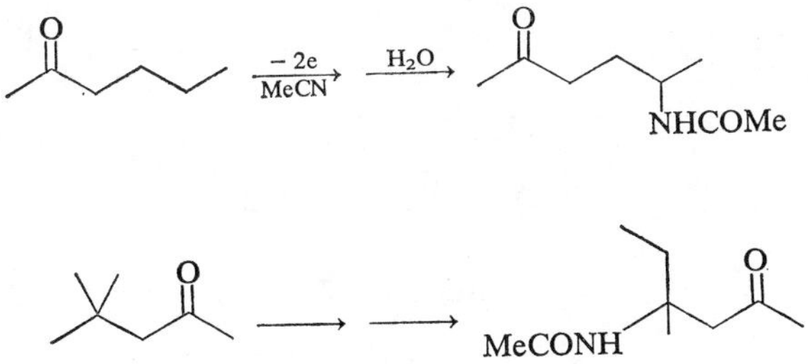

Functional Groups other than Alkanes, Acetylenes, Allenes, and Olefins

[Scheme showing ketone → (−2e, MeCN) → (H₂O) → product with NHCOMe group]

[Scheme showing ketone → intermediate → product with MeCONH group]

Scheme 165

$$R^1\underset{\underset{OOH}{|}}{\overset{\overset{O}{\|}}{C}}CR^2R^3 \xrightarrow{\text{base}} \left[R^1-\underset{\underset{O-O}{|}}{\overset{\overset{OH}{|}}{C}}-CR^2R^3 \right] \longrightarrow \begin{array}{c} R^1CO_2H \\ + \\ R^2COR^3 \end{array}$$

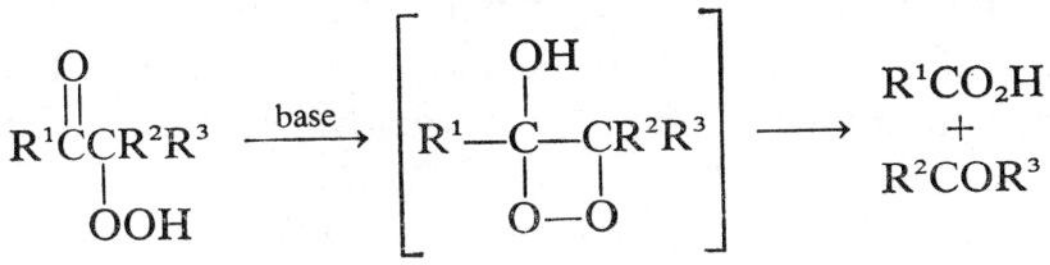

Scheme 166

and that the excited state carbonyl production and chemiluminescence are associated with such a cyclic path.

Conjugated dienones and diene esters are efficiently epoxidized at the $\gamma\delta$-double bond[508] by molecular oxygen when heated in solvents with readily abstractable hydrogen atoms, such as cumene; these conditions do not affect isolated double bonds, nor singly conjugated ones. 1,4-Diketones are produced in modest yield[509] by the dimerization of monoketones with nickel peroxide.

Further examples of the reductive amination[510] of carbonyl compounds using the hydridotetracarbonylferrate(-II) anion have been described. Carbonyl compounds, including non-enolizable aldehydes, are converted into the corresponding methylene olefins[511] by either of the aluminium reagents (117) and (118) (Scheme 167). Details have been given of the carbonyl C-dimethylation[512] of ketones with trimethylaluminium.

The use of trimethylsilyl cyanide[513] permits an efficient synthesis of β-aminomethyl alcohols, including those from ketones which do not form

---

[508] H. Hart and P. B. Lavrik, *J. Org. Chem.*, 1974, **39**, 1793.
[509] E. G. E. Hawkins and R. Large, *J.C.S. Perkin I*, 1974, 280.
[510] (*a*) Y. Watanabe, M. Yamashita, T. Mitsudo, M. Tanaka, and Y. Takegami, *Tetrahedron Letters*, 1974, 1879; (*b*) Y. Watanabe, T. Mitsudo, M. Yamashita, S. C. Shim, and Y. Takegami, *Chem. Letters*, 1974, 1265; (*c*) G. P. Boldrini, M. Panunzio, and A. Umani-Ronchi, *Synthesis*, 1974, 733.
[511] A. Bongini, D. Savoia, and A. Umani-Ronchi, *J. Organometallic Chem.*, 1974, **72**, C4.
[512] A. Meisters and T. Mole, *Austral. J. Chem.*, 1974, **27**, 1655; see also E. A. Jeffrey, A. Meisters, and T. Mole, *ibid.*, p. 2569.
[513] D. A. Evans, G. L. Carroll, and L. K. Truesdale, *J. Org. Chem.*, 1974, **39**, 914.

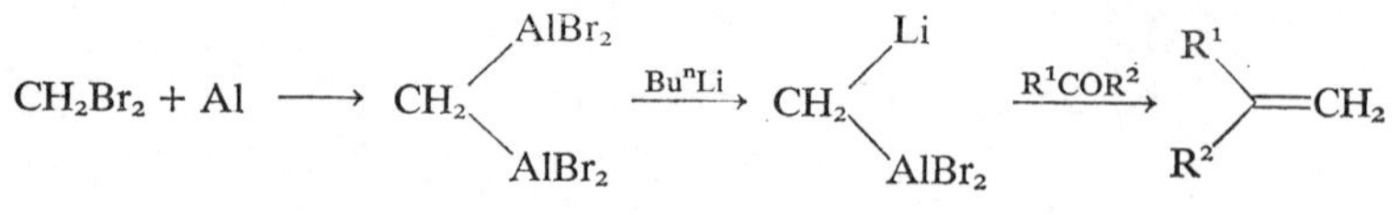

CH₂Br₂ + Al ⟶ CH₂(AlBr₂)(AlBr₂) —Bu^nLi→ CH₂(Li)(AlBr₂) —R¹COR²→ R¹R²C=CH₂

(117)   (118)

Scheme 167

stable cyanohydrins, and those from conjugated enones, where clean 1,2-addition is observed. Phosphorus pentachloride in pyridine converts α-CH₂ ketones directly into acetylenes[514] in modest yield; vinyl chlorides are the major contaminants. A full account has been given[515] of the photochemical ring expansion of α-cyanoketones to lactam derivatives. The conversions of carbonyl compounds into homologous acids,[8–10,12] and of aldehydes into nitriles,[225–228] were described previously. Further examples of the reductive cleavage[516] of the central σ-bond of 1,4-dicarbonyl compounds have been presented. Acetylacetone forms two isomeric enol acetates (119) and (120), both in the (S)-cis-conformation, in approximately equal proportions (Scheme 168); the stability of the isomer (119) is attributed[517] to some degree

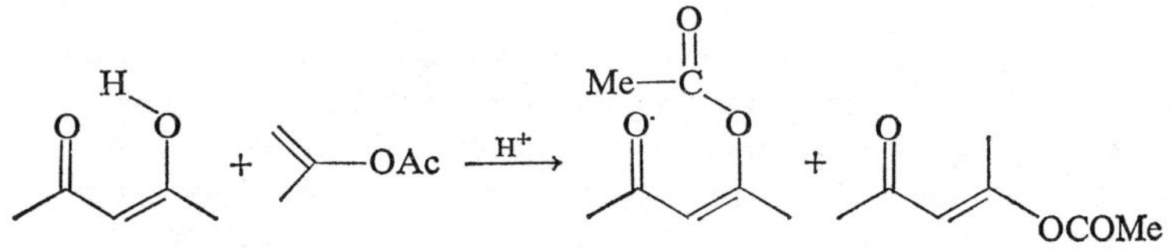

(119)   (120)

Scheme 168

of neighbouring-group interaction, a rationale consistent with ¹H n.m.r. data. Cargill has reviewed[518] his work on acid-catalysed rearrangements of βγ-unsaturated ketones.

9 Alcohols

Preparation.—Various aspects of chiral and stereoselective reductions of carbonyl compounds were discussed on p. 156, as were organometallic

[514] C. M. Wong and T.-L. Ho, *Synthetic Comm.*, 1974, **4**, 25.
[515] G. K. Chip and T. R. Lynch, *Canad. J. Chem.*, 1974, **52**, 2249.
[516] J. Grimshaw and R. J. Haslett, *J.C.S. Chem. Comm.*, 1974, 174.
[517] D. V. C. Awang, *Canad. J. Chem.*, 1973, **51**, 3752.
[518] R. L. Cargill, T. E. Jackson, N. P. Peet, and D. M. Pond, *Accounts Chem. Res.*, 1974, **7**, 106.

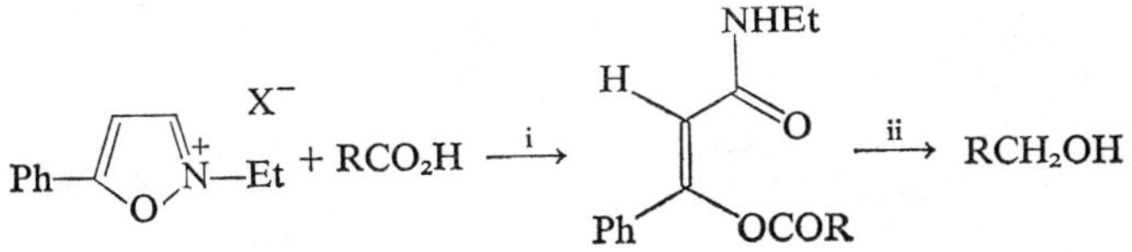

Functional Groups other than Alkanes, Acetylenes, Allenes, and Olefins 173

$$\text{Ph}\underset{O}{\overset{\overset{+}{N}-\text{Et}}{\diagdown}} X^- + \text{RCO}_2\text{H} \xrightarrow{i} \underset{\text{Ph}}{\overset{H}{\diagdown}}C=C\underset{\text{OCOR}}{\overset{\text{C(O)NHEt}}{\diagdown}} \xrightarrow{ii} \text{RCH}_2\text{OH}$$

Reagents: i, Et$_3$N; ii, NaBH$_4$–H$_2$O

Scheme 169

condensations with carbonyl compounds on p. 160. A number of routes to ^{2}H-[519–522] and ^{3}H-[523] labelled alcohols have been described.

Carboxylic acids react with isoxazolium salts to give enol ethers (Scheme 169), which can be reduced by aqueous sodium borohydride to primary alcohols; this method is of value when lithium aluminium hydride cannot be used, e.g., in peptide studies.[524]

A novel synthesis[525] of secondary and tertiary alcohols from alkynes is illustrated in Scheme 170; migration occurs with retention, and the methyl group of the intermediate 'ate' complex (121) shows very low migratory aptitude.

Tertiary alcohols are produced in high yield by the sequence show in Scheme 171, all steps of which can be performed[526] in a single vessel. The

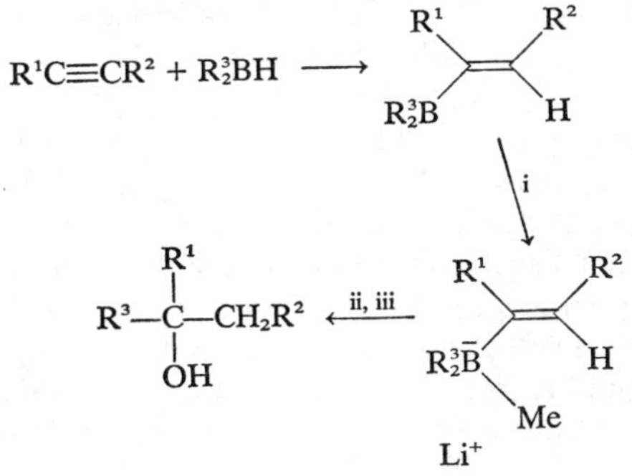

$$R^1C\equiv CR^2 + R_2^3BH \longrightarrow \underset{R_2^3B}{\overset{R^1}{\diagdown}}C=C\underset{H}{\overset{R^2}{\diagdown}}$$

$$\underset{\text{OH}}{\overset{R^1}{\underset{|}{R^3-C-CH_2R^2}}} \xleftarrow{ii, iii} \underset{R_2^3B}{\overset{R^1}{\diagdown}}C=C\underset{\underset{Li^+}{\diagdown Me}}{\overset{R^2}{\diagdown H}}$$

(121)

Reagents: i, MeLi; ii, HCl or MeSO$_3$H; ii, NaOH–H$_2$O$_2$

Scheme 170

[519] Y. Sasson and J. Blum, *J.C.S. Chem. Comm.*, 1974, 309.
[520] R. Shanker, *Chem. and Ind.*, 1974, 76.
[521] C. Sreekumar and C. N. Pillai, *Synthesis*, 1974, 498.
[522] S. L. Regen, *J. Org. Chem.*, 1974, **39**, 260.
[523] R. H. Cornforth, *Tetrahedron*, 1974, **30**, 3933.
[524] P. L. Hall and R. B. Perfetti, *J. Org. Chem.*, 1974, **39**, 111.
[525] G. Zweifel and R. P. Fisher, *Synthesis*, 1974, 339.
[526] R. J. Hughes, A. Pelter, and K. Smith, *J.C.S. Chem. Comm.*, 1974, 863.

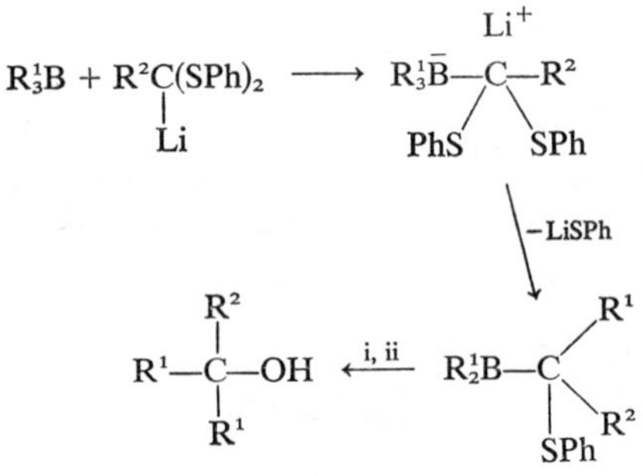

Scheme 171

$$R^1_3B + R^2C(SPh)_2Li \longrightarrow R^1_3\bar{B}-\underset{PhS}{\overset{Li^+}{\underset{|}{C}}}-R^2$$
$$\underset{SPh}{} \downarrow -LiSPh$$
$$R^1-\underset{R^1}{\overset{R^2}{\underset{|}{C}}}-OH \xleftarrow{i, ii} R^1_2B-\underset{SPh}{\overset{R^1}{C}}-R^2$$

Reagents: i, $HgCl_2$; ii, $NaOH-H_2O_2$

superiority[527] of the BH_3–Me_2S complex has been discussed, and a simplified hydroboration–oxidation procedure[528] has been described. Full details of the hydroborating selectivity of 9-borabicyclo[3,3,1]nonane[529] and of thexylborane[530] have been given.

The direct electrophilic addition of organoselenium species to olefins, which allows the preparation of, *inter alia*, allylic alcohols, was described previously (p. 150), as was another organoselenium based route (p. 158) to such compounds. The selective oxymercuration of 1,5-dienes leads[531] to $\gamma\delta$-unsaturated alcohols. Full details[532] of the directed oxymercuration of allylic alcohols to *cis*-1,2-diols have been given. An electrochemical route[533] for the conversion of olefins into methyl alkyl carbonates, and thence alkanols, has been reported. Sutherland[534] has described a modified Prins reaction suitable for the transformation of conjugated dienes into 1,5-diol diformates, as illustrated in Scheme 172.

Further studies[535] have been reported on the regioselectivity of ring-opening of allylic oxirans to allylic and homoallylic alcohols with organometallic reagents. Diethylaluminium 2,2,6,6-tetramethylpiperidide (122) functions[536] as a mild reagent for the regiospecific isomerization of epoxides to allylic alcohols (Scheme 173); a mechanistic rationale for the regiospecificity is proposed.

[527] C. F. Lane, *J. Org. Chem.*, 1974, **39**, 1437.
[528] V. Hach, *Synthesis*, 1974, 340.
[529] H. C. Brown, E. F. Knights, and C. G. Scouten, *J. Amer. Chem. Soc.*, 1974, **96**, 7765.
[530] E. Negishi and H. C. Brown, *Synthesis*, 1974, 77.
[531] V. Gomez Aranda, J. Barluenga, M. Yus, and G. Asensio, *Synthesis*, 1974, 806.
[532] L. E. Overman and C. B. Campbell, *J. Org. Chem.*, 1974, **39**, 1474.
[533] R. Brettle and J. R. Sutton, *J.C.S. Chem. Comm.*, 1974, 449.
[534] J. J. S. Bajorek, R. Battaglia, G. Pratt, and J. K. Sutherland, *J.C.S. Perkin I*, 1974, 1243.
[535] C. B. Rose and S. K. Taylor, *J. Org. Chem.*, 1974, **39**, 578.
[536] A. Yasuda, S. Tanaka, K. Oshima, H. Yamamoto, and H. Nozaki, *J. Amer. Chem. Soc.*, 1974, **96**, 6513.

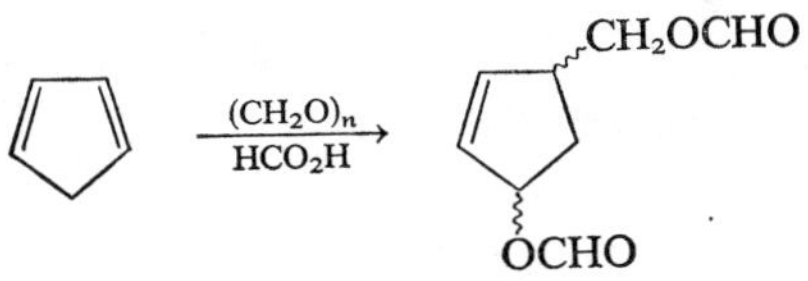

Functional Groups other than Alkanes, Acetylenes, Allenes, and Olefins 175

Scheme 172

Schlosser[537] has indicated the utility of metallated functionalized alkenes for the preparation of configurationally pure, specifically substituted alkenols, as exemplified in Scheme 174.

Ketones are cleanly homologated to primary allylic alcohols by acid-catalysed rearrangement[538] of the derived vinyl carbinols (Scheme 175).

Evans, in an application of his studies[278] on the [2,3]sigmatropic rearrangement of allylic sulphoxides, has described[539] the preparation of the diol (123) (Scheme 176), of obvious synthetic potential.

A re-investigation[540] of the catalysed decomposition of alkyl hydroperoxides has suggested that unsaturated primary alcohols can be obtained with good regiospecificity (Scheme 177).

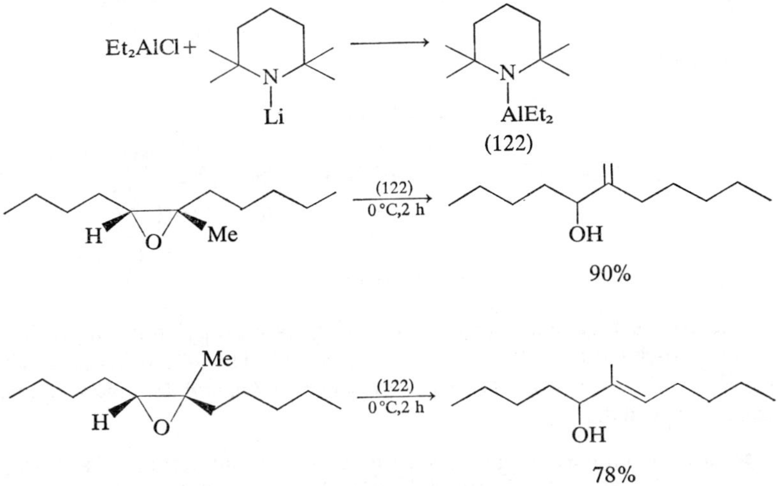

Scheme 173

[537] J. Hartmann, R. Muthukrishnan, and M. Schlosser, *Helv. Chim. Acta*, 1974, **57**, 2261; M. Schlosser and E. Hammer, *ibid.*, p. 2547.
[538] J. H. Babler and D. O. Olsen, *Tetrahedron Letters*, 1974, 351.
[539] D. A. Evans, T. C. Crawford, T. T. Fujimoto, and R. C. Thomas, *J. Org. Chem.*, 1974, **39**, 3176.
[540] Ž. Čeković and M. M. Green, *J. Amer. Chem. Soc.*, 1974, **96**, 3000.

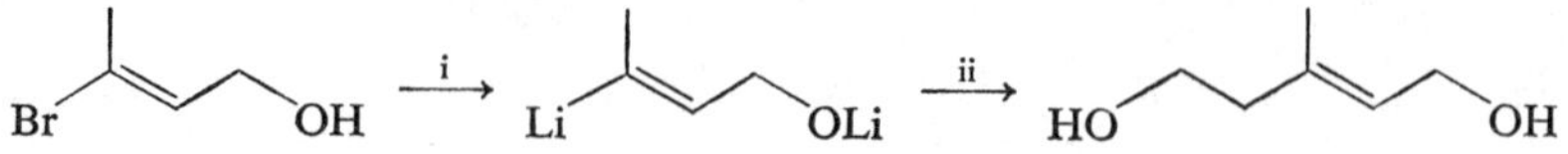

Br/\/\OH —i→ Li/\/\OLi —ii→ HO/\/\/\OH

Reagents: i, Bu^{s}Li; ii, $\overset{O}{\triangle}$

Scheme 174

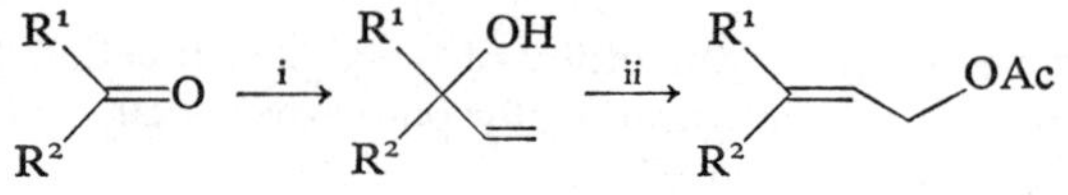

$R^1R^2C=O$ —i→ $R^1R^2C(OH)CH=CH_2$ —ii→ $R^1R^2C=CHCH_2OAc$

Reagents: i, CH$_2$=CHLi; ii, Ac$_2$O–AcOH–H$^+$

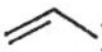

Scheme 175

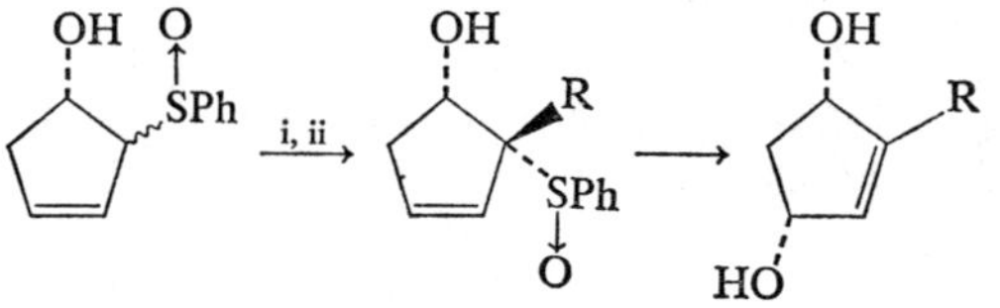

(cyclopentenol with SPh and O substituents) —i, ii→ (cyclopentenol with R and SPh(O) substituents) → (cyclopentenediol with R) (123)

Reagents: LiNEt$_2$; ii, RX

Scheme 176

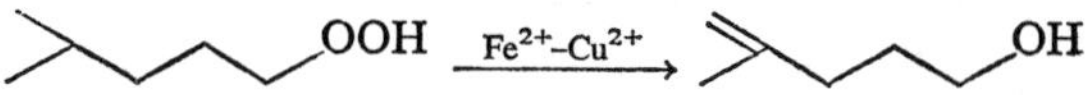

$\diagup\diagdown$OOH —Fe^{2+}–Cu^{2+}→ $\diagup\diagdown$OH

76% of total

Scheme 177

Some routes to allenic alcohols involving cyclopropyl rearrangement[541] and propargyl reduction[542] have been described. The photochemical cleavage[543] of ethers to alcohols in the presence of trialkylaluminium species has been reported.

Properties.—Complexing agents, such as copper hexafluoroacetylacetonate, are of considerable utility in the determination of absolute configuration of alcohols, glycols, and related systems by the exciton chirality method,[544]

---

[541] B. Ragonnet, M. Santelli, and M. Bertrand, *Bull. Soc. chim. France*, 1973, 3119; G. Léandri, H. Monti, and M. Bertrand, *ibid.*, 1974, 1919.
[542] A. Claesson, L.-I. Olsson, and C. Bogentoft, *Acta Chem. Scand.*, 1973, **27**, 2941; 1974, **28**, 765.
[543] J. Furukawa, K. Omura, O. Yamamoto, and K. Ishikawa, *J.C.S. Chem. Comm*, 1974, 77.
[544] J. Dillon and K. Nakanishi, *J. Amer. Chem. Soc.*, 1974, **96**, 4055, 4057, 4059.

Functional Groups other than Alkanes, Acetylenes, Allenes, and Olefins 177

which has been demonstrated to be operative[545] even at considerable separation of the two interacting chromophores; it has also been applied[546] to the chirality determination of conjugated enone, ester, and lactone benzoates. The absolute configurations of secondary chiral alcohols (and amines) can be ascertained[547] by a study of the ^{1}H n.m.r. spectra of the derived diastereoisomeric esters (and amides) of α-phenylbutyric and hydratropic acids. Significant preparative-scale resolution of racemic alcohols can be achieved[548] *via* their hippurate esters, by selective hydrolysis catalysed by α-chymotrypsin.

The average solution conformations of all 18 possible isomers of the six-carbon aliphatic alcohols have been elucidated[549] by ^{1}H and ^{13}C n.m.r. spectroscopy. The thermochemistry of some aliphatic alcohols has been studied[550] using pulsed ion cyclotron resonance spectroscopy.

Reactions.—Various aspects of alcohol oxidation, including the selective mono-oxidation of diols and the oxidation of alkoxides with singlet oxygen, are discussed on p. 123.

The preparation of some specific primary and secondary alkyl peroxyl radicals has been described.[551] Allylic trichloroacetimidates, readily prepared from allylic alcohols and trichloroacetonitrile, undergo a facile thermal [3,3]sigmatropic rearrangement to allylic amides, effecting the alcohol-into-amine conversion[552] shown in Scheme 178; the rearrangement can also be catalysed at low temperature by mercury(II) ions, when a two-step iminomercuration-deoxymercuration is proposed.

A number of reagent combinations for the conversion of alcohols into

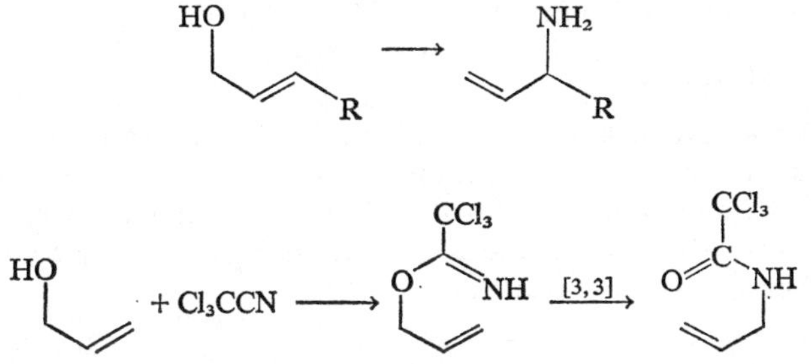

Scheme 178

[545] S. L. Chen, N. Harada, and K. Nakanishi, *J. Amer. Chem. Soc.*, 1974, **96**, 7352.
[546] M. Koreeda, N. Harada, and K. Nakanishi, *J. Amer. Chem. Soc.*, 1974, **96**, 266.
[547] G. Helmchen, *Tetrahedron Letters*, 1974, 1527.
[548] Y. Y. Lin, D. N. Palmer, and J. B. Jones, *Canad. J. Chem.*, 1974, **52**, 469.
[549] K. L. Williamson, D. R. Clutter, R. Emch, M. Alexander, A. E. Burroughs, C. Chua, and M. E. Bogel, *J. Amer. Chem. Soc.*, 1974, **96**, 1471.
[550] R. T. McIver and J. S. Miller, *J. Amer. Chem. Soc.*, 1974, **96**, 4323.
[551] T. J. Kemp and M. J. Welbourn, *Tetrahedron Letters*, 1974, 87.
[552] L. E. Overman, *J. Amer. Chem. Soc.*, 1974, **96**, 597.

$$\begin{array}{c} Ph \diagdown \diagup OC(CF_3)_2Ph \\ S \\ Ph \diagup \diagdown OC(CF_3)_2Ph \end{array}$$

(124)

halides have been described, including boron trichloride,[553] aqueous hydrochloric acid[554] with phase transfer catalysis, alkali halides[555] in poly(hydrogen fluoride)–pyridine solution, and the DMF–phosphorus pentachloride[556] Vilsmeier reagent. Dichlorine heptoxide in CCl_4 converts alcohols[557a] and amines[557b] into the respective perchlorates. The sulphurane (124) efficiently transforms 1,2-diols capable of assuming *trans* periplanar geometry into oxirans; longer chain diols are also converted into the corresponding cyclic ethers.[558]

Aliphatic alcohols and phenols can be reduced to hydrocarbons[559] by hydrogenation of the carbodi-imide-derived isoureas (Scheme 179).

When methanol or trimethylphosphate is heated in polyphosphoric acid, a complex mixture of hydrocarbons is formed[560] in 36—39% yield; to explain the initial coupling to a two-carbon molecule, five-coordinate carbon or carbene intermediates are suggested. Full details have been given[561] of the reductive C-methylation of tertiary and benzylic alcohols with trimethylaluminium.

The lithium aluminium hydride reduction of propargylic alcohols is well known, and its mechanism has recently been studied further;[562] the first cases of addition of butyl-lithium to such alcohols have now been reported,[563] and some examples are illustrated in Scheme 180.

10 Amines

Preparation.—Several applications of the hydridotetracarbonylferrate(−II) anion as an imine reducing agent are referred to above;[510] one such application[510b] is seen in a method for the NN-dimethylation of primary amines (Scheme 181). The preparations of some sterically hindered secondary amines[564] have been described.

[553] H. Yazawa, H. Nakamura, K. Tanaka, and K. Kariyone, *Tetrahedron Letters*, 1974, 3991.
[554] D. Landini, F. Montanari, and F. Rolla, *Synthesis*, 1974, 37.
[555] G. A. Olah and J. Welch, *Synthesis*, 1974, 653.
[556] D. R. Hepburn and H. R. Hudson, *Chem. and Ind.*, 1974, 664.
[557] (*a*) K. Baum and C. D. Beard, *J. Amer. Chem. Soc.*, 1974, **96**, 3233. (*b*) C. D. Beard and K. Baum, *ibid.*, p. 3237.
[558] J. C. Martin, J. A. Franz, and R. J. Arhart, *J. Amer. Chem. Soc.*, 1974, **96**, 4604.
[559] E. Vowinkel and C. Wolff, *Chem. Ber.*, 1974, **107**, 907, 1739; E. Vowinkel and H.-J. Baese, *ibid.*, p. 1213; E. Vowinkel and I. Büthe, *ibid.*, p. 1353.
[560] D. E. Pearson, *J.C.S. Chem. Comm.*, 1974, 397.
[561] D. W. Harney, A. Meisters, and T. Mole, *Austral. J. Chem.*, 1974, **27**, 1639.
[562] B. Grant and C. Djerassi, *J. Org. Chem.*, 1974, **39**, 968
[563] L.-I. Olsson and A. Claesson, *Tetrahedron Letters*, 1974, 2161.
[564] J. C. Stowell and S. J. Padegimas, *Synthesis*, 1974, 127.

Functional Groups other than Alkanes, Acetylenes, Allenes, and Olefins 179

$$R^1OH + R^2N{=}C{=}NR^2 \longrightarrow R^2NHC{=}NR^2 \xrightarrow{i} R^1H + (R^2NH)_2CO$$
$$\hspace{5cm} |$$
$$R^1 = \text{alkyl or aryl} \hspace{2cm} OR^1$$

Reagent: i, H_2–Pd–C.

Scheme 179

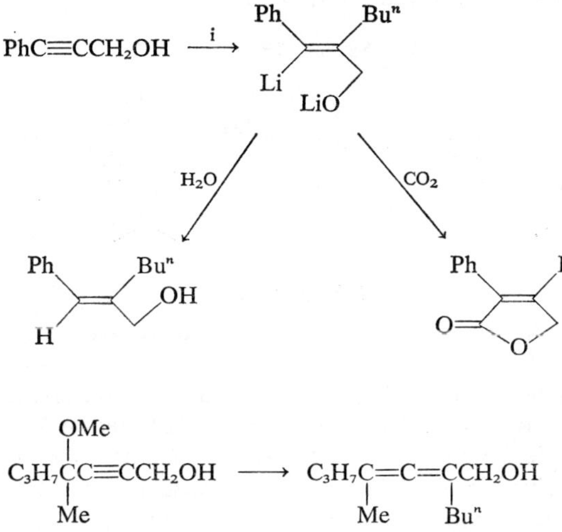

PhC≡CCH$_2$OH $\xrightarrow{i}$ [Ph, Bu^n, Li, LiO intermediate]

H_2O → Ph,Bu^n,H,OH alkene

CO_2 → Ph,Bu^n lactone (O=C-O ring)

$$\underset{\underset{Me}{|}}{C_3H_7}\overset{\overset{OMe}{|}}{C}C{\equiv}CCH_2OH \longrightarrow \underset{\underset{Me}{|}}{C_3H_7}C{=}C{=}\underset{\underset{Bu^n}{|}}{C}CH_2OH$$

Reagent: i, Bu^{n}Li–TMEDA–Et$_2$O

Scheme 180

Amine hydrohalides react with ethylene carbonate[565] to give salts of β-hydroxyethylamines. The protected hydroxylamine (125) undergoes ready alkylation, and subsequent cleavage gives[566] the corresponding monoalkylated hydroxylamine (Scheme 182).
Conjugated dienes react with diethylamine in the presence of sodium naphthalenide to give $\beta\gamma$-unsaturated amines;[567] the example shown (Scheme

$$R{-}NH_2 + (CH_2O)_n \xrightarrow{KHFe(CO)_4} R{-}NMe_2$$

R = alkyl or aryl

Scheme 181

[565] T. Yoshino, S. Inaba, H. Komura, and Y. Ishido, *Bull. Chem. Soc. Japan*, 1974, **47**, 405.
[566] Y. Isowa and H. Kurita, *Bull. Chem. Soc. Japan*, 1974, **47**, 720.
[567] T. Fujita, K. Suga, and S. Watanabe, *Austral. J. Chem.*, 1974, **27**, 531; *cf.* R. J. Schlott, J. C. Falk, and K. W. Narducy, *J. Org. Chem.*, 1972, **37**, 4243.

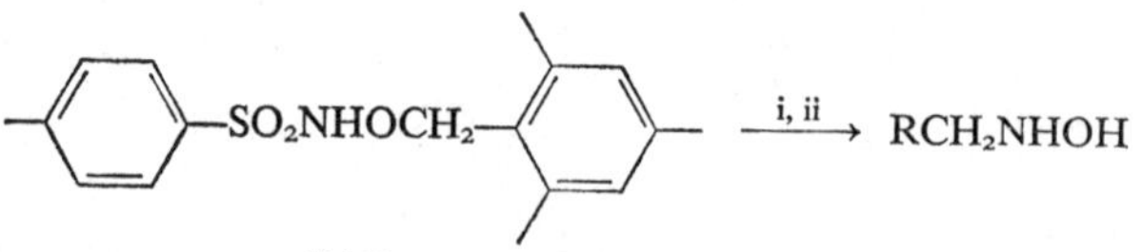

180 *Aliphatic Chemistry*

Ar—SO$_2$NHOCH$_2$—Ar' $\xrightarrow{i, ii}$ RCH$_2$NHOH

(125)

Reagents: i, RCH$_2$X–NaOMe; ii, HBr–AcOH

Scheme 182

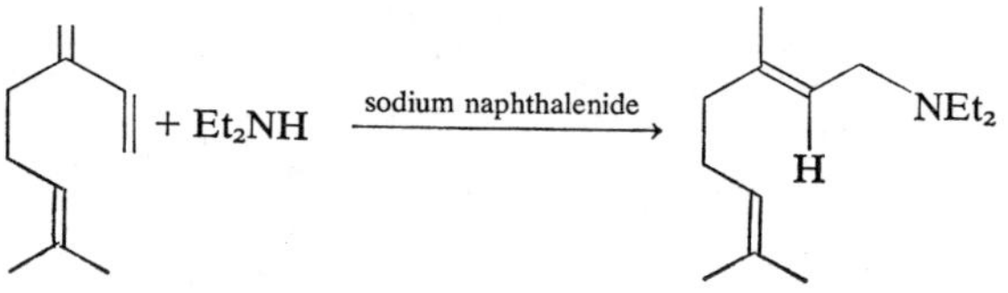

myrcene + Et$_2$NH $\xrightarrow{\text{sodium naphthalenide}}$ product-NEt$_2$

Scheme 183

183) was converted further into geranyl acetate by treatment with acetic anhydride.

Studies continue[568] on the alkylation of ammonia and amines *via* metallation of the benzophenone-derived imines, as illustrated in Scheme 184.

Simple high-yield preparations of nitrogen analogues of crown ethers have been described.[569] Aryl iodides couple with bis(trimethylsilyl)amido-copper(I) to give, after methanolysis, substituted anilines[570] (Scheme 185).

A method for the mono-methylation[571] of primary aromatic and hetero-aromatic amines is shown in Scheme 186.

Benzylic and allylic alcohols undergo a palladium-catalysed reaction with amines (Scheme 187);[572] N-substituted pyrroles can be prepared in this

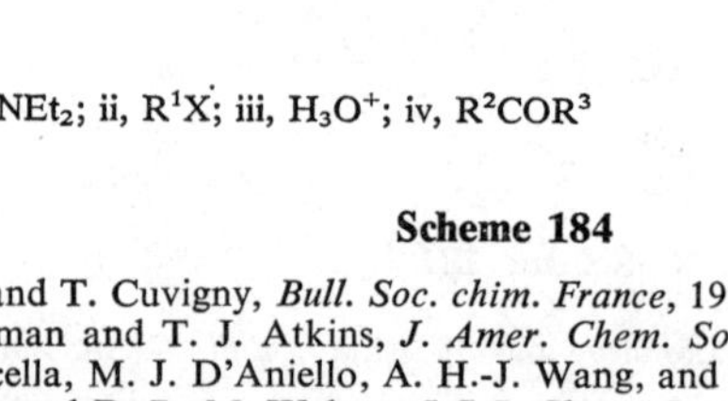

Ph$_2$C=NMe $\xrightarrow{i}$ Ph$_2$C=NCH$_2$Li $\xrightarrow{ii, iii}$ R^{1}CH$_2$NH$_2$

$\downarrow$ iv, iii

R^{2}R^{3}C(OH)—CH$_2$NH$_2$

Reagents: i, LiNEt$_2$; ii, R^{1}X; iii, H$_3$O$^+$; iv, R^{2}COR^3

Scheme 184

[568] P. Hullot and T. Cuvigny, *Bull. Soc. chim. France*, 1973, 2985, 2989.
[569] J. E. Richman and T. J. Atkins, *J. Amer. Chem. Soc.*, 1974, **96**, 2268; F. Wagner, M. T. Mocella, M. J. D'Aniello, A. H.-J. Wang, and E. K. Barefield, *ibid.*, p. 2625.
[570] F. D. King and D. R. M. Walton, *J.C.S. Chem. Comm.*, 1974, 256.
[571] R. A. Crochet and C. DeW. Blanton, *Synthesis*, 1974, 55.
[572] S.-I. Murahashi, T. Shimamura, and I. Moritani, *J.C.S. Chem. Comm.*, 1974, 931.

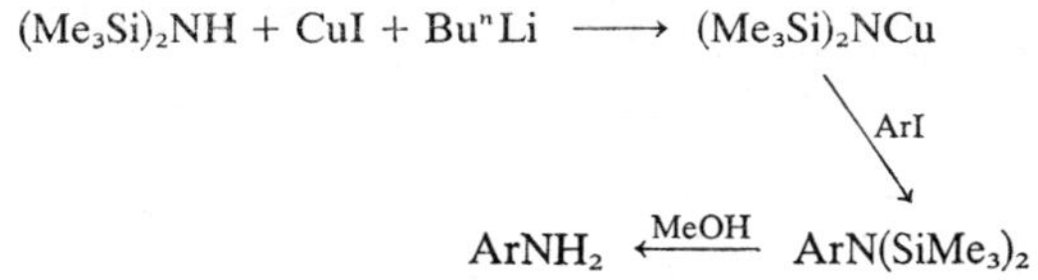

$(Me_3Si)_2NH + CuI + Bu^nLi \longrightarrow (Me_3Si)_2NCu$

$\searrow ArI$

$ArNH_2 \xleftarrow{MeOH} ArN(SiMe_3)_2$

Scheme 185

$ArNH_2 + HC(OEt)_3 \xrightarrow{heat} ArN{=}CHOEt \xrightarrow{NaBH_4} ArNHMe$

Scheme 186

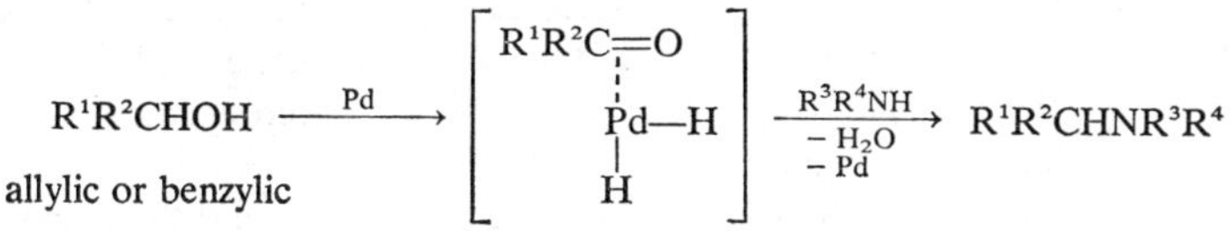

$R^1R^2CHOH \xrightarrow{Pd}$ allylic or benzylic $\left[\begin{array}{c} R^1R^2C{=}O \\ | \\ Pd{-}H \\ | \\ H \end{array} \right] \xrightarrow[-Pd]{R^3R^4NH \atop -H_2O} R^1R^2CHNR^3R^4$

Scheme 187

manner, using but-2-ene-1,4-diol. The conversion of allylic alcohols into transposed allylic amines is illustrated in Scheme 178.

Silver nitrite[573] and 2,2,2-trichloroethyl chloroformate[574] are separately recommended for the demethylation of tertiary methyl amines, as is phenyl isocyanide dichloride[575] for the dealkylation of cyclic tertiary amines; the products in each case can be readily converted into secondary amines. The silyl quaternary ammonium salt (126) undergoes a novel ylide rearrangement[576] (Scheme 188) to give triphenylsilylmethyl amines.

O-Mesitylenesulphonylhydroxylamine is recommended[577] as a potent electrophilic aminating agent in the preparation of amines *via* hydroboration. A hydroboration route to allylic amines is described above.[192]

The chiral reducing ability of lithium aluminium hydride-monosaccharide complexes has been extended[578] to the reduction of oximes and derivatives; optically active amines are obtained in up to 56% optical purity. β-Arylethylamines are generated conveniently and efficiently by reduction[579] of β-nitrostyrene precursors with sodium bis-(2-methoxyethoxy)aluminium hydride. Titanium(II) ions effect[580] the reduction of aromatic nitro compounds

---

[573] L. Bernardi and G. Bosisio, *J.C.S. Chem. Comm.*, 1974, 690.
[574] T. A. Montzka, J. D. Matiskella, and R. A. Partyka, *Tetrahedron Letters*, 1974, 1325.
[575] G. Leclerc, B. Rouot, and C. G. Wermuth, *Tetrahedron Letters*, 1974, 3765.
[576] Y. Sato, Y. Ban, and H. Shirai, *J.C.S. Chem. Comm.*, 1974, 182.
[577] Y. Tamura, J. Minamikawa, S. Fujii, and M. Ikeda, *Synthesis*, 1974, 196.
[578] S. R. Landor, O. O. Sonola, and A. R. Tatchell, *J.C.S. Perkin I*, 1974, 1902.
[579] J. R. Butterick and A. M. Unrau, *J.C.S. Chem. Comm.*, 1974, 307.
[580] T.-L. Ho and C. M. Wong, *Synthesis*, 1974, 45.

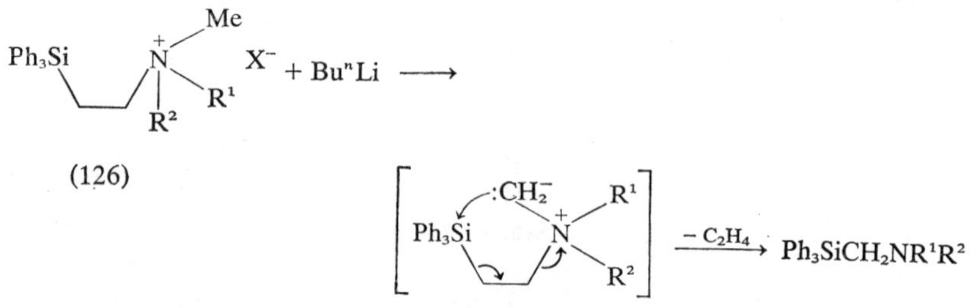

Ph$_3$Si−N$^+$(Me)(R^1)(R^2) X$^-$ + Bu^{n}Li ⟶

(126)

$$\left[Ph_3Si \cdots :CH_2^- \cdots N^+ R^1 R^2 \right] \xrightarrow{-C_2H_4} Ph_3SiCH_2NR^1R^2$$

Scheme 188

to anilines. Nitroxyl radicals are reduced to amine hydriodides[446] by silanes in the presence of iodine or methyl iodide. A mechanistic study[581] has been reported on the acid-catalysed deoxygenation of heterocyclic and tertiary amine *N*-oxides with dimethyl sulphoxide. Several classes of organic azides react readily with sodium hydride to give moderate yields of the corresponding amine[582] and quantitative yields of nitrogen. The addition of nitrosyl chloride to Schiff bases produces ideal intermediates[583] for the synthesis of α-functionalized *N*-nitrosoamines (Scheme 189).

The preparation and properties of 2,2-bis(methylsulphonyl)vinylamines[584] (127), and those of acylated enediamines[585] (128), have been described.

Properties and Reactions.—While in most cases correlation of the absolute configurations of chiral amines with results from the Horeau method is

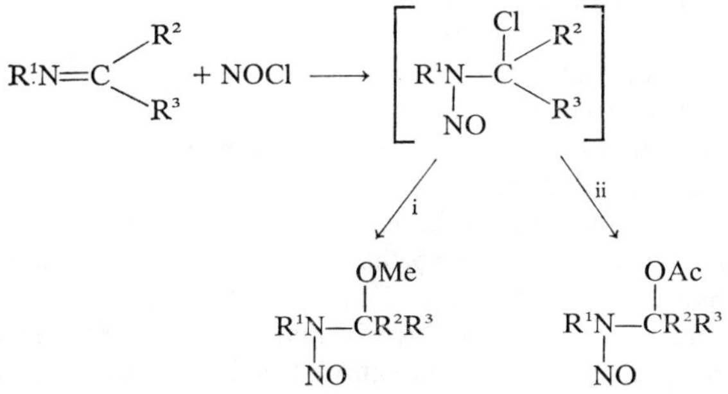

$$R^1N=C(R^2)(R^3) + NOCl \longrightarrow \left[R^1N-C(Cl)(NO)(R^2)(R^3) \right]$$

i ↙   ↘ ii

R^{1}N(NO)—CR^{2}R^3(OMe)   R^{1}N(NO)—CR^{2}R^3(OAc)

Reagents: i, MeOH–Et$_3$N; ii, AcOH–Et$_3$N

Scheme 189

[581] M. E. C. Biffin, C. Bocksteiner, J. Miller, and D. B. Paul, *Austral. J. Chem.*, 1974, **27**, 789.
[582] Y.-J. Lee and W. D. Closson, *Tetrahedron Letters*, 1974, 381.
[583] M. Wiessler, *Angew. Chem. Internat. Edn.*, 1974, **13**, 743.
[584] A. R. Friedman and D. R. Graber, *J. Org. Chem.*, 1974, **39**, 1432.
[585] P. Duhamel, L. Duhamel, and V. Truxillo, *Tetrahedron Letters*, 1974, 51.

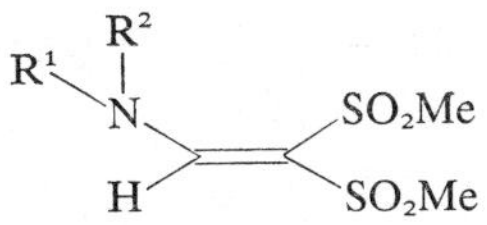

Functional Groups other than Alkanes, Acetylenes, Allenes, and Olefins 183

$$\underset{H}{\overset{R^1\underset{|}{\diagdown}\overset{R^2}{N}}{\diagup}}C=C\underset{SO_2Me}{\overset{SO_2Me}{\diagup}}$$
(127)

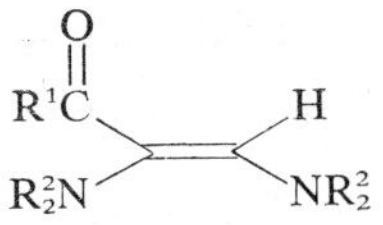

$$\underset{R_2^2N}{\overset{R^1C\overset{O}{\diagdown}\overset{\parallel}{}}{\diagup}}C=C\underset{NR_2^2}{\overset{H}{\diagup}}$$
(128)

valid, it has been recommended[586] that such assignments should always be substantiated by direct chemical correlation or by unambiguous c.d. measurements. Bis-2,4-dinitrophenyl derivatives of diamines show[587] induced Cotton effects, allowing chirality determination; those with negative C—N chirality show a negative Cotton effect for the longest wavelength band.

Lithium naphthalenide can replace[588] sodium–liquid ammonia in the specific reductive cleavage of aliphatic and aromatic trimethylammonium iodides to hydrocarbons and trimethylamine. Specific syntheses of some N-fluoro compounds[589] have been described, as has the preparation of amine perchlorates.[557b] Full details[590] have been given of the preparation and reactions of stable quaternary ammonium salts of the type (129). A new convenient route[591] to primary sulphamides (130) has been described.

$$R_3\overset{+}{N}-X\ BF_4^-$$
(129) X = CN, CO_2R^1, or COR^2

$$RNHSO_2NH_2$$
(130)

A simple high-yield procedure[592] for the large scale synthesis of symmetrical t-azoalkanes is outlined in Scheme 190. Nitroso compounds react with NN-dichloroamines in the presence of base in a new directed synthesis[593] of unsymmetrical azoxyalkanes.

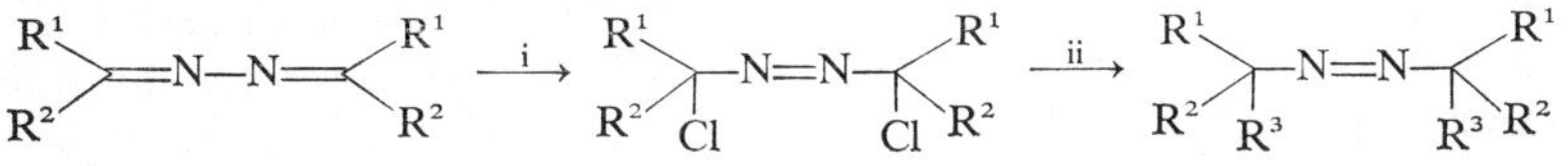

$$\underset{R^2}{\overset{R^1}{\diagdown}}C=N-N\underset{R^2}{\overset{R^1}{\diagup}} \xrightarrow{i} \underset{R^2}{\overset{R^1}{\diagdown}}\underset{Cl}{\overset{}{C}}-N=N-\underset{Cl}{\overset{}{C}}\underset{R^2}{\overset{R^1}{\diagup}} \xrightarrow{ii} \underset{R^2}{\overset{R^1}{\diagdown}}\underset{R^3}{\overset{}{C}}-N=N-\underset{R^3}{\overset{}{C}}\underset{R^2}{\overset{R^1}{\diagup}}$$

Reagents: i, Cl_2; ii, AlR_3^3, R^3 = alkyl or aryl

Scheme 190

[586] H. E. Smith, A. W. Gordon, and A. F. Bridges, *J. Org. Chem.*, 1974, **39**, 2309.
[587] M. Kawai, U. Nagai, and T. Kobayashi, *Tetrahedron Letters*, 1974, 1881.
[588] I. Angres and H. E. Zieger, *J. Org. Chem.*, 1974, **39**, 1013.
[589] D. H. R. Barton, R. M. Hesse, M. M. Pechet, and H. T. Toh, *J.C.S. Perkin I*, 1974, 732.
[590] J. V. Paukstelis and M. Kim, *J. Org. Chem.*, 1974, **39**, 1494, 1499, 1503; G. Fodor, S. Abidi, and T. C. Carpenter, *ibid.*, p. 1507.
[591] J. D. Catt and W. L. Matier, *J. Org. Chem.*, 1974, **39**, 566.
[592] W. Duismann, H.-D. Beckhaus, and C. Rüchardt, *Annalen*, 1974, 1348.
[593] F. R. Sullivan, E. Luck, and P. Kovacic, *J. Org. Chem.*, 1974, **39**, 2967.

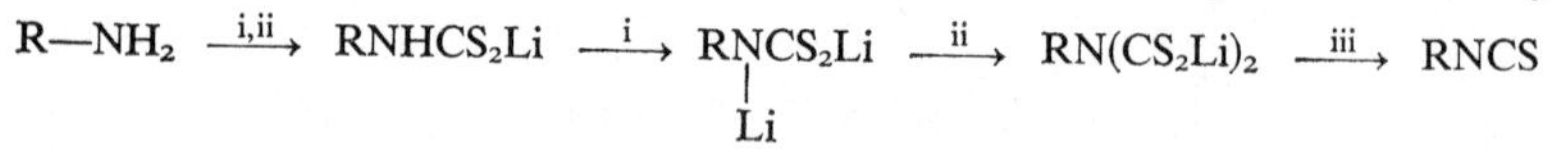

R—NH$_2$ $\xrightarrow{i,ii}$ RNHCS$_2$Li $\xrightarrow{i}$ RNCS$_2$Li $\xrightarrow{ii}$ RN(CS$_2$Li)$_2$ $\xrightarrow{iii}$ RNCS
$\qquad\qquad\qquad\qquad\qquad\quad$ |
$\qquad\qquad\qquad\qquad\qquad\;\;$ Li

Reagents: i, Bu^{n}Li; ii, CS$_2$; iii, heat

Scheme 191

Aromatic and aliphatic amines are converted into isothiocyanates by treatment[594] with n-butyl-lithium and carbon disulphide (Scheme 191).

11 Alkyl Halides

Preparation.—Reviews have been published on the utility of sulphur tetrafluoride[595] as a fluorinating agent, and on routes to monofluoro compounds.[596] Selenium tetrafluoride and its pyridine complex are convenient reagents[597] for the fluorination of alcohols and carbonyl compounds, as illustrated in Scheme 192.

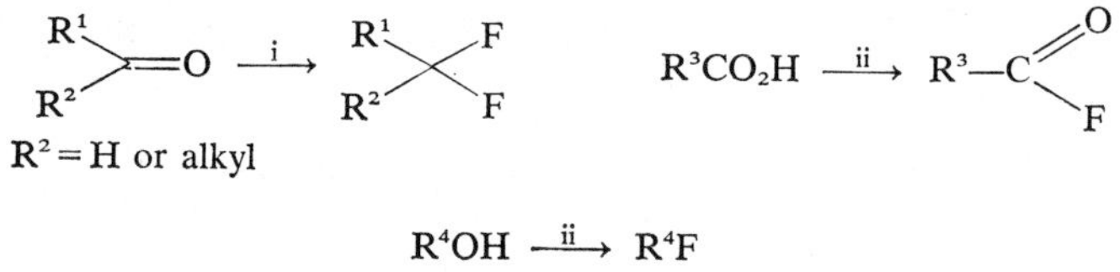

$$\underset{R^2 = H \text{ or alkyl}}{\overset{R^1}{\underset{R^2}{>}}=O} \xrightarrow{i} \overset{R^1}{\underset{R^2}{>}}\!\!\!<\!\!\!\overset{F}{\underset{F}{}} \qquad R^3CO_2H \xrightarrow{ii} R^3-C\overset{O}{\underset{F}{\diagdown}}$$

$$R^4OH \xrightarrow{ii} R^4F$$

Reagents: i, SeF$_4$; ii, py–SF$_4$

Scheme 192

Unsolvated 'naked' fluoride ion, prepared[598] by dissolving potassium fluoride in benzene or acetonitrile with the aid of a crown ether, functions as a potent nucleophile and base (Scheme 193); similar enhancements of reactivity of other halide ions[599] and cyanide ion[600] have been described. A related use of phase transfer catalysis has been reported[601] for the transformation of alkyl halides and mesylates into alkyl fluorides with aqueous potassium fluoride. Full details[602] have been published on the use of phenyl tetrafluorophosphorane for the conversion of alcohol silyl ethers into alkyl fluorides.

[594] S. Sakai, T. Aizawa, and T. Fujinami, *J. Org. Chem.*, 1974, **39**, 1970.
[595] G. A. Boswell, W. C. Ripka, R. M. Scribner, and C. W. Tullock, *Org. Reactions*, 1974, **21**, 1.
[596] C. M. Sharts and W. A. Sheppard, *Org. Reactions*, 1974, **21**, 125.
[597] G. A. Olah, M. Nojima, and I. Kerekes, *J. Amer. Chem. Soc.*, 1974, **96**, 925.
[598] C. L. Liotta and H. P. Harris, *J. Amer. Chem. Soc.*, 1974, **96**, 2250; cf. ref. 98.
[599] D. J. Sam and H. E. Simmons, *J. Amer. Chem. Soc.*, 1974, **96**, 2252; see also W. T. Ford, R. J. Hauri, and S. G. Smith, *ibid.*, p. 4316.
[600] F. L. Cook, C. W. Bowers, and C. L. Liotta, *J. Org. Chem.*, 1974, **39**, 3416.
[601] D. Landini, F. Montanari, and F. Rolla, *Synthesis*, 1974, 428.
[602] D. J. Costa, N. E. Boutin, and J. G. Riess, *Tetrahedron*, 1974, **30**, 3793.

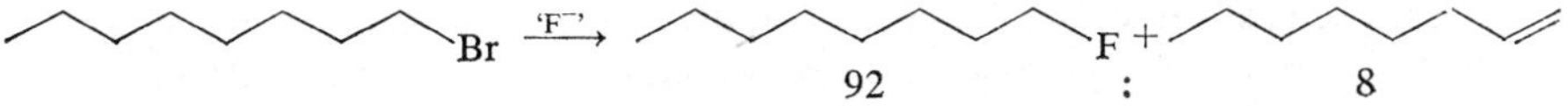

Functional Groups other than Alkanes, Acetylenes, Allenes, and Olefins 185

$$\text{\textasciitilde\textasciitilde\textasciitilde Br} \xrightarrow{\text{'F}^-\text{'}} \text{\textasciitilde\textasciitilde\textasciitilde F} + \text{\textasciitilde\textasciitilde}$$

$$92 \quad : \quad 8$$

Scheme 193

Some new methods for the conversion of alcohols into alkyl halides are discussed on p. 177. Phosphorodiamidites (131) are convenient and versatile intermediates[603] for the preparation of a range of halogenated species (Scheme 194).

Benzylic and tertiary alkyl chlorides are converted into the corresponding iodides[604] by reaction with sodium iodide in a non-polar solvent in the presence

$$\text{ROH} + (\text{Me}_2\text{N})_2\text{PCl} \longrightarrow (\text{Me}_2\text{N})_2\text{POR}$$
$$(131)$$

$$\text{CCl}_4 \swarrow \quad \downarrow \text{PhCCl}_3 \quad \searrow \text{Cl}_3\text{CCO}_2\text{Et}$$

$$\text{RCCl}_3 \quad \text{RCCl}_2\text{Ph} \quad \text{RCCl}_2\text{CO}_2\text{Et}$$

Scheme 194

of iron(III) chloride. Improved syntheses of di-iodomethane and chloro-iodomethane, by phase transfer catalysed iodide displacement on dichloromethane, have been described.[605]

Whereas organoaluminium and organoboron compounds undergo positional rearrangement slowly, internally metallated zirconium analogues[606] undergo such a process rapidly at room temperature, and rearrange to a product in which the zirconium is attached to the least hindered, accessible position of the olefin as a whole. These rearranged species react readily with electrophiles such as halogens (Scheme 195).

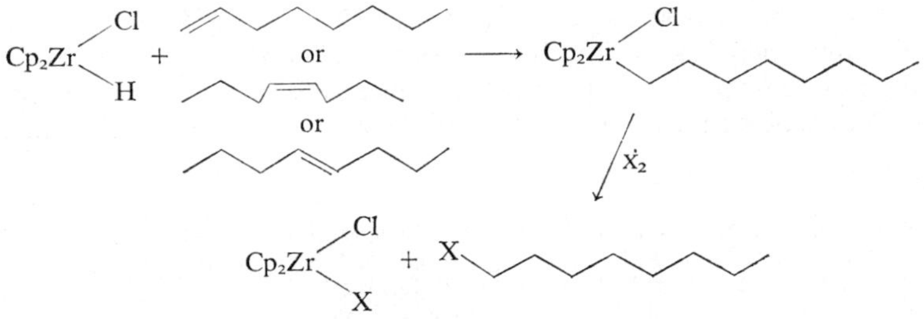

Scheme 195

[603] J. H. Hargis and W. D. Alley, *J. Amer. Chem. Soc.*, 1974, **96**, 5927.
[604] J. A. Miller and M. J. Nunn, *Tetrahedron Letters*, 1974, 2691.
[605] D. Landini and F. Rolla, *Chem. and Ind.*, 1974, 533.
[606] D. W. Hart and J. Schwartz, *J. Amer. Chem. Soc.*, 1974, **96**, 8115.

Full details[607] have been published on the *cis* chlorination of simple acyclic and six-membered cyclic olefins with antimony(v) chloride in chlorinated solvents. Organoboranes are efficiently cleaved by iron(III) or copper(II) chloride to the corresponding alkyl chlorides and up to two alkyl groups of the organoborane are utilized;[608] alkyl thiocyanates can be prepared similarly. Olefins react with copper(II) salts and potassium iodide or thiocyanate in alcohol solvents to afford[609] 1,2-alkoxyiodo- or 1,2-alkoxythiocyano-alkanes in good yield, and Markovnikov orientation and *trans* stereochemistry are observed in suitable cases; $\alpha\beta$-unsaturated carbonyl compounds give the products as dialkyl acetals (Scheme 196). A related process using t-butyl hypochlorite has been described.[610]

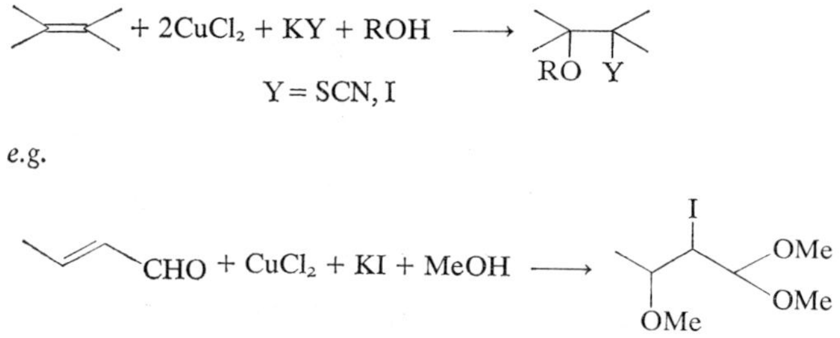

$$>=< + 2CuCl_2 + KY + ROH \longrightarrow >\!\!\!\underset{RO\;\;Y}{\overset{Y}{\rightthreetimes\!\!\!\leftthreetimes}}\!\!\!<$$

$$Y = SCN, I$$

e.g.

$$\diagdown\!\!\!\diagup\!\!\diagdown\!\!\text{CHO} + CuCl_2 + KI + MeOH \longrightarrow \diagdown\!\!\!\underset{OMe}{\overset{I}{\diagup\!\!\diagdown}}\!\!\!\underset{}{\overset{}{\diagup\!\!\diagdown}}\!\!\overset{OMe}{OMe}$$

Scheme 196

The chlorinating properties of antimony(v) chloride intercalated on graphite[611] include α-chlorination of alkyl bromides (Scheme 197).

Breslow[612] has described the halogenation of steroids, with substituent and proximity effects producing significant selectivity; in appropriate cases, halogenation–dehydrohalogenation introduces a 9(11) double bond (Scheme 198), a sequence used in a new synthesis of cortisone. Direct insertion[613] of dihalogenocarbenes into C—H bonds of cycloalkanes leads to (dihalogenomethyl)cycloalkanes. Further studies[614] of allylic bromination with *N*-bromosuccinimide have provided evidence supporting a bromine atom chain mechanism; the reagent serves merely as a reservoir of more bromine.

---

[607] S. Uemura, A. Onoe, and M. Okano, *Bull. Chem. Soc. Japan*, 1974, **47**, 692.
[608] A. Arase, Y. Masuda, and A. Suzuki, *Bull. Chem. Soc. Japan*, 1974, **47**, 2511.
[609] A. Onoe, S. Uemura, and M. Okano, *Bull. Chem. Soc. Japan*, 1974, **47**, 2818.
[610] C. Walling and R. T. Clark, *J. Org. Chem.*, 1974, **39**, 1962.
[611] J. L. Luche, J. Bertin, and H. B. Kagan, *Tetrahedron Letters*, 1974, 759; J. Bertin, J. L. Luche, H. B. Kagan, and R. Setton, *ibid.*, p. 763.
[612] R. Breslow, R. J. Corcoran, J. A. Dale, S. Liu, and P. Kalicky, *J. Amer. Chem. Soc.*, 1974, **96**, 1973; R. Breslow, R. J. Corcoran, and B. B. Snider, *ibid.*, p. 6791; R. Breslow, B. B. Snider, and R. J. Corcoran, *ibid.*, p. 6792.
[613] D. Seyferth and Y.-M. Cheng, *Synthesis*, 1974, 114.
[614] J. C. Day, M. J. Lindstrom, and P. S. Skell, *J. Amer. Chem. Soc.*, 1974, **96**, 5616.

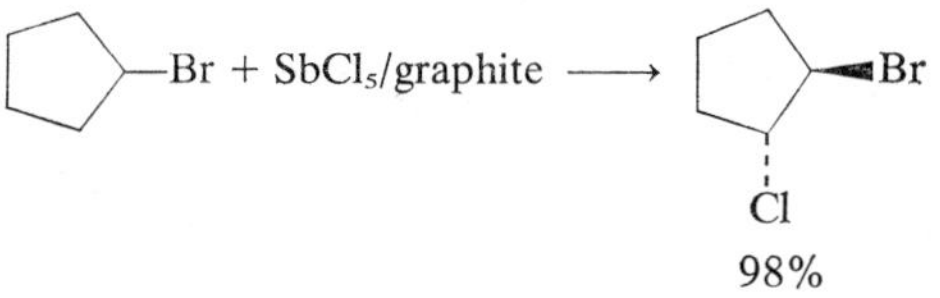

Functional Groups other than Alkanes, Acetylenes, Allenes, and Olefins 187

Scheme 197

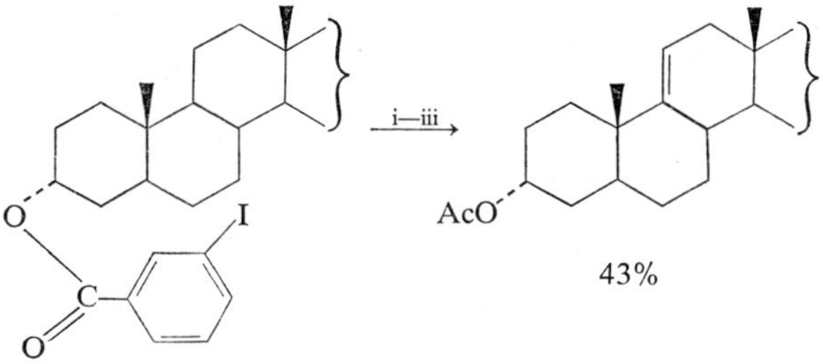

Reagents: i, PhICl₂, hv; ii, KOH–MeOH; iii, Ac₂O

Scheme 198

The readily accessible vinylsilane (132) acts as a precursor to stereo-specifically substituted[615] vinyl halides; a related route to 2-bromoalk-1-enes[616] has also been described (Scheme 199).

The stereospecifically pure vinylcopper (133) can be converted into the corresponding iodide with retention of configuration;[617] attempted bromination resulted in oxidation, but this was circumvented via the vinylmercury (134), again formed with configurational retention (Scheme 200).

The halides (135) are recommended[618] as halogenomethylating agents of broad utility (Scheme 201); their low volatility should decrease the carcinogenic hazards associated with use of more volatile chloromethylethers such as bis(chloromethyl)ether.

Alkyl (but not primary) and aryl Grignard reagents undergo a coupling reaction[619] with αω-dibromoalkanes, leading to a wide variety of monobromoalkanes; the presence of either dilithium tetrachlorocuprate or tetramethylethylenediamine is mandatory. Some routes to trichloromethyl compounds have been summarized.[620]

Halogenoalkyl and Halonium Ions.—*Ab initio* calculations[621] on the geometric structure and energy surface of the chloroethyl cation have given results in

---
[615] R. B. Miller and T. Reichenbach, *Tetrahedron Letters*, 1974, 543.
[616] R. K. Boeckman and D. M. Blum, *J. Org. Chem.*, 1974, **39**, 3307.
[617] J. F. Normant, C. Chuit, G. Cahiez, and J. Villieras, *Synthesis*, 1974, 803; *J. Organometallic Chem.*, 1974, **77**, 269.
[618] G. A. Olah, D. A. Beal, S. H. Yu, and J. A. Olah, *Synthesis*, 1974, 560.
[619] L. Friedman and A. Shani, *J. Amer. Chem. Soc.*, 1974, **96**, 7101.
[620] R. Kh. Freidlina and E. C. Chukovskaya, *Synthesis*, 1974, 477.
[621] W. J. Hehre and P. C. Hibberty, *J. Amer. Chem. Soc.*, 1974, **96**, 2665.

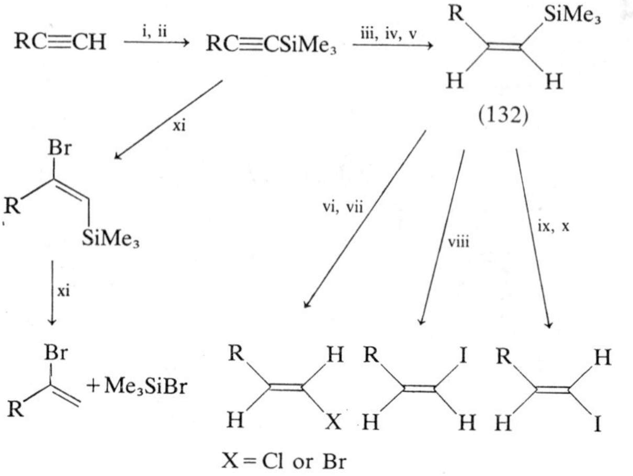

Reagents: i, EtMgBr; ii, Me$_3$SiCl; iii, (C$_6$H$_{11}$)$_2$BH; iv, Ac$_2$O–heat; v, NaOH–H$_2$O$_2$; vi, Cl$_2$ or Br$_2$; vii, NaOMe; viii, I$_2$; ix, I$_2$–CF$_3$CO$_2$Ag; x, KF–DMSO–H$_2$O; xi, HBr.

Scheme 199

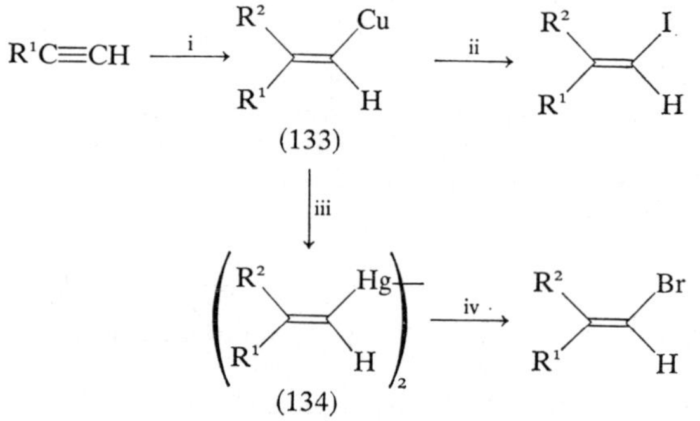

Reagents: i, R^{2}MgBr–CuBr; ii, I$_2$; iii, HgBr$_2$; iv, Br$_2$–py

Scheme 200

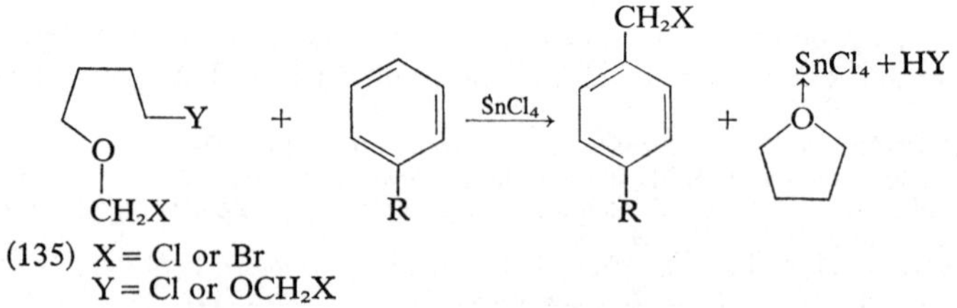

(135) X = Cl or Br
Y = Cl or OCH$_2$X

Scheme 201

Functional Groups other than Alkanes, Acetylenes, Allenes, and Olefins 189

Scheme 202

good agreement with those derived experimentally from super-acid data. Evidence has been presented[622] against strongly bridged bromonium ion intermediates[623] in bromocyclization reactions such as those of allylic amides. Gas-phase and solution stability studies[624] on cyclic halonium ions relative to acyl cations have been reported. Olah[625] has described the first bicyclic three-membered ring bromonium ion (136); the corresponding 1,2-dichloride precursor underwent a rapid 1,2-hydride shift to the chloro-carbonium ion (137) (Scheme 202).

Halogenoalkyl Radicals.—Further evidence[626] for a bridged transition state in the formation of β-bromoalkyl radicals has been presented. β-Chloro- and β-bromo-t-butyl radicals, (138) and (139), have been generated in an [^{2}H$_{16}$]adamantane matrix by X-irradiation of isobutyl chloride or bromide at 77 K. E.p.r. spectroscopic studies have shown that whereas radical (138) prefers[627] the same eclipsed conformation (141) as it does in solution, radical (139) prefers the staggered conformation (142); radical (139) is much less stable than (138). The analogous β-fluoro radical (140) also prefers an eclipsed conformation.

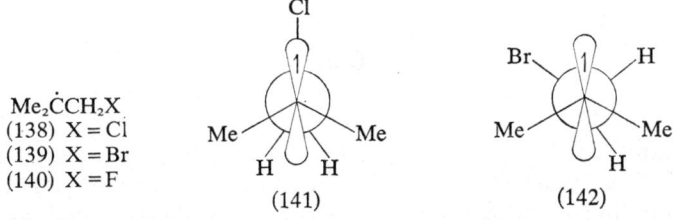

Me$_2$ĊCH$_2$X
(138) X = Cl
(139) X = Br
(140) X = F

[622] S. P. McManus and D. W. Ware, *Tetrahedron Letters*, 1974, 4271.
[623] G. A. Olah, P. W. Westerman, E. G. Melby, and Y. K. Mo, *J. Amer. Chem. Soc.*, 1974, **96**, 3565.
[624] R. D. Wieting, R. H. Staley, and J. L. Beauchamp, *J. Amer. Chem. Soc.*, 1974, **96**, 7552.
[625] G. A. Olah, G. Liang, and J. Staral, *J. Amer. Chem. Soc.*, 1974, **96**, 8112.
[626] E. N. Cain and R. K. Solly, *J.C.S. Chem. Comm.*, 1974, 148.
[627] R. V. Lloyd, D. E. Wood, and M. T. Rogers, *J. Amer. Chem. Soc.*, 1974, **96**, 7130.

$$RX + Zn \text{ powder} \xrightarrow{DMF} RZnX \xrightarrow{H_2O} RH$$

Scheme 203

Reduction to Hydrocarbons and Oxidative Coupling.—The ability of lithium n-butylhydridocopper to reduce a wide variety of alkyl halides is noted above.[423] Aryl iodides are reduced in good yield to the parent aromatic system[628] on heating with sodium hydride in tetrahydrofuran. A simple non-reductive procedure[629] for the dehalogenation of alkyl halides is shown in Scheme 203.

Alkyl halides can be oxidatively coupled to olefins[630] via selenide alkylation and selenoxide fragmentation (Scheme 204).

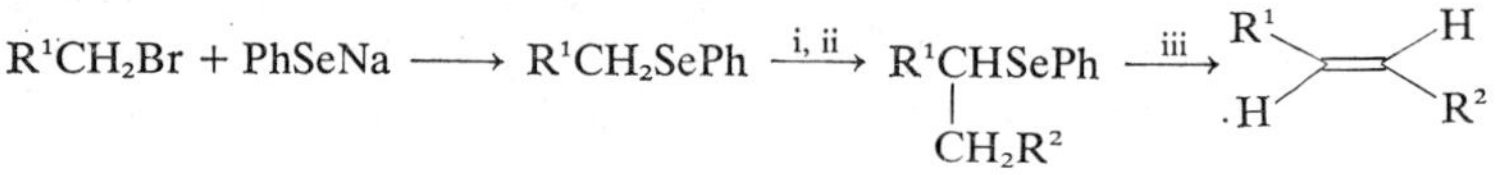

$$R^1CH_2Br + PhSeNa \longrightarrow R^1CH_2SePh \xrightarrow{i, ii} \underset{CH_2R^2}{R^1CHSePh} \xrightarrow{iii} \underset{H}{\overset{R^1}{\diagdown}}C=C\underset{R^2}{\overset{H}{\diagup}}$$

Reagents: i, LiNPr$_2^i$; ii, R^{2}CH$_2$Br; iii, H$_2$O$_2$, heat

Scheme 204

General Properties and Reactions.—N.m.r. spectroscopic studies[631] of 2-fluoroethanol have confirmed that the major conformer (95%) is gauche (143); no evidence of intramolecular F—H bonding could be detected.

The palladium-catalysed carboxylation of vinyl iodides to esters was described earlier.[94] A range of alkyl bromides undergo a facile mercury(I) and/or mercury(II) assisted solvolysis (Scheme 205),[632] providing simple high-yield routes to a range of alcohol derivatives.

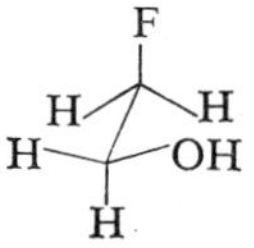

(143)

12 Ethers

Preparation and Reactions.—Copper(I) alkoxides, readily prepared by reaction of methylcopper with the parent alcohol, are useful[633] for the formation of, in particular, alkyl aryl ethers by alkoxide displacement on aryl

---

[628] R. B. Nelson and G. W. Gribble, *J. Org. Chem.*, 1974, **39**, 1425.
[629] G. Mehta and S. K. Kapoor, *J. Organometallic Chem.*, 1974, **66**, C33.
[630] R. H. Mitchell, *J.C.S. Chem. Comm.*, 1974, 990.
[631] R. C. Griffith and J. D. Roberts, *Tetrahedron Letters*, 1974, 3499.
[632] A. McKillop and M. E. Ford, *Tetrahedron*, 1974, **30**, 2467.
[633] G. M. Whitesides, J. S. Sadowski, and J. Lilburn, *J. Amer. Chem. Soc.*, 1974, **96**, 2829.

Functional Groups other than Alkanes, Acetylenes, Allenes, and Olefins 191

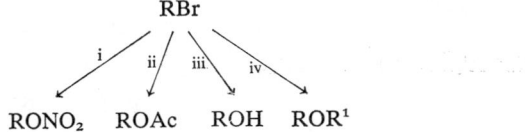

Reagents: i, Hg(NO₃)₂; ii, Hg(OAc)₂-AcOH; iii, Hg(ClO₄)₂-H₂O; iv, Hg(ClO₄)₂-R¹OH

Scheme 205

halides. The advantageous use[634] of phase transfer catalysis in the preparation of ethers of both simple and highly hindered phenols has been reported. The reagent combination sodium hydride–methyl iodide in tetrahydrofuran is recommended once again[635] for the rapid conversion of alcohols into methyl ethers. In an extension of their earlier studies, French authors[636] report that allylic alcohols react with all three classes of alcohol in the presence of copper powder to give allyl ethers in good yield (Scheme 206); an ionic mechanism is implicated.

The addition of t-butyl hypochlorite to a cold olefin–alcohol solution containing a few drops of boron trifluoride etherate leads to 2-chloro-t-butyl ethers;[610] a preparation[609] of 2-iodoalkyl ethers was illustrated earlier in Scheme 196.

A hydroboration route[281] to α-keto and *cis* allylic ethers is shown in Scheme 97. (Z)- and (E)-Allylic ethers are readily available from trialkylalkynylborates, by the process[637] shown in Scheme 207.

Borane in tetrahydrofuran reduces acetals to ethers,[638] generally in good yield and under mild conditions, in a process mechanistically related to the same reduction with lithium aluminium hydride–aluminium chloride; cyclic acetals give 2-hydroxyethers (Scheme 208). The selectivity of reduction of lactones to cyclic ethers with trichlorosilane has been investigated.[639]

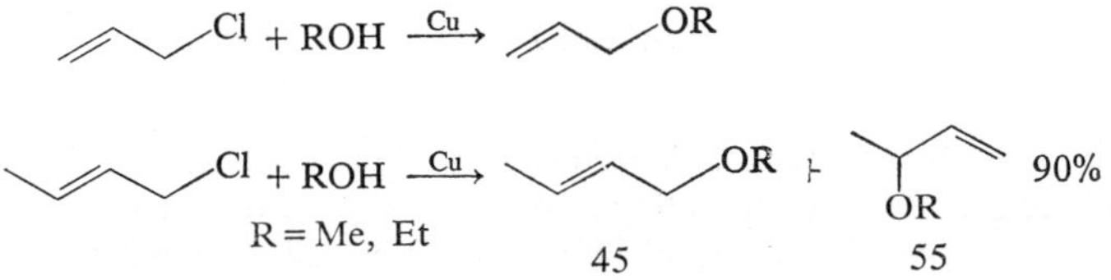

$$\diagdown\!\!\diagup\!\!\diagdown\!\!\text{Cl} + \text{ROH} \xrightarrow{\text{Cu}} \diagdown\!\!\diagup\!\!\diagdown\!\!\text{OR}$$

$$\diagdown\!\!\diagup\!\!\diagdown\!\!\text{Cl} + \text{ROH} \xrightarrow{\text{Cu}} \diagdown\!\!\diagup\!\!\diagdown\!\!\text{OR} + \underset{\text{OR}}{\diagdown\!\!\diagup\!\!\diagdown} \quad 90\%$$

R = Me, Et   45   55

Scheme 206

[634] A. McKillop, J.-C. Fiaud, and R. P. Hug, *Tetrahedron*, 1974, **30**, 1379.
[635] C. A. Brown, D. Barton, and S. Sivaram, *Synthesis*, 1974, 434; *cf.* B. A. Stoochnoff and N. L. Benoiton, *Tetrahedron Letters*, 1973, 21.
[636] R. Fellows, J.-P. Rabine, L. Lizzani-Cuvelier, and R. Luft, *Bull. Soc. chim. France*, 1974, 923.
[637] P. Binger and R. Köster, *Synthesis*, 1974, 350.
[638] B. Fleming and H. I. Bolker, *Canad. J. Chem.*, 1974, **52**, 888.
[639] S. W. Baldwin, R. J. Doll, and S. A. Haut, *J. Org. Chem.*, 1974, **39**, 2470.

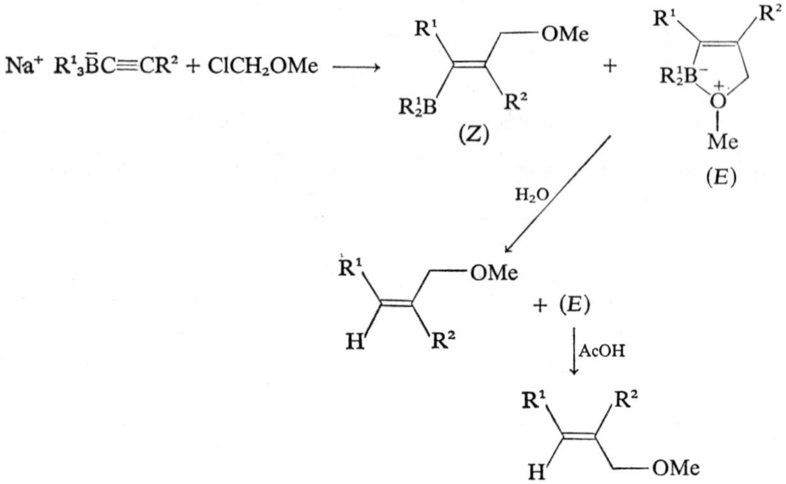

Scheme 207

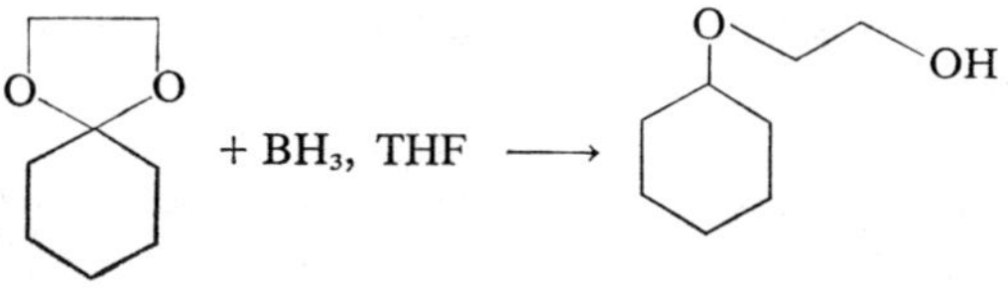

Scheme 208

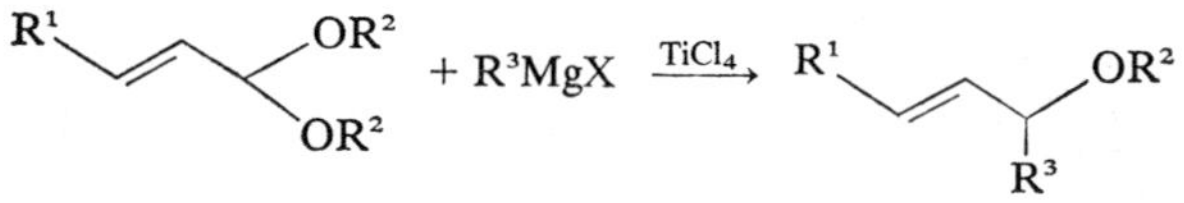

Scheme 209

Allylic acetals react with Grignard reagents in the presence of titanium(IV) chloride to give the corresponding allylic ethers[640] (Scheme 209). The photochemical cleavage of ethers to alcohols was described previously.[543]

13 Sulphur

Sulphoxides.—Co-oxidation of terminal olefins and arene thiols with oxygen produces β-hydroxysulphoxides[641] in good yield (Scheme 210); subsequent Pummerer rearrangement[264] affords protected α-hydroxyaldehydes. The

---
[640] T. Mukaiyama and H. Ishikawa, *Chem. Letters*, 1974, 1077.
[641] S. Iriuchijima, K. Maniwa, T. Sakakibara, and G. Tsuchihashi, *J. Org. Chem.*, 1974, **39**, 1170.

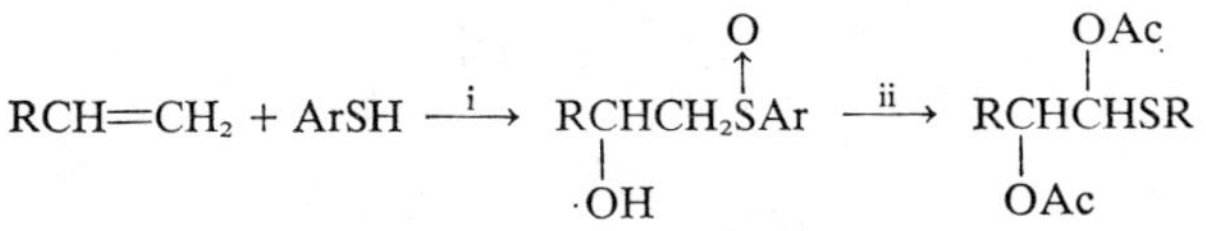

Functional Groups other than Alkanes, Acetylenes, Allenes, and Olefins 193

$$RCH=CH_2 + ArSH \xrightarrow{i} \underset{OH}{RCHCH_2SAr}^{\uparrow O} \xrightarrow{ii} \underset{OAc}{RCHCHSR}^{OAc}$$

Reagents: i, O_2; ii, Ac_2O–NaOAc

Scheme 210

oxidation of sulphides with N-bromocaprolactam in the presence of optically active alcohols affords optically active sulphoxides.[642]

The phosphole (144) efficiently reduces[643] sulphoxides to sulphides under mild conditions (Scheme 211); sulphoxides are also converted into sulphides on heating with reduced iron.[644]

The readily accessible alkoxysulphonium salts (145) are selectively reduced[645] to sulphides by a sodium cyanoborohydride–crown ether system (Scheme 212); this method is tolerant of the presence of other functionalities such as aldehydes or ketones.

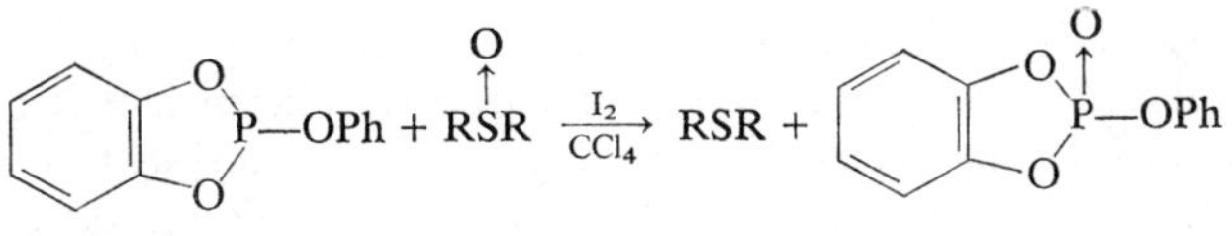

[structure (144) P—OPh] + RŚR $\xrightarrow[CCl_4]{I_2}$ RSR + [structure P—OPh]

(144)

Scheme 211

The first example of asymmetric induction promoted by isotopic dissymmetry has been presented.[646] Halogenation of the chiral sulphoxide (146), followed by oxidation, results in optical activity in the α-halogenosulphone (147).

A further example of the intramolecular transfer of chirality from sulphur to carbon is seen[647] in the Pummerer-type rearrangement of the chiral sulphoxide (148) to the sulphide (149) (Scheme 213).

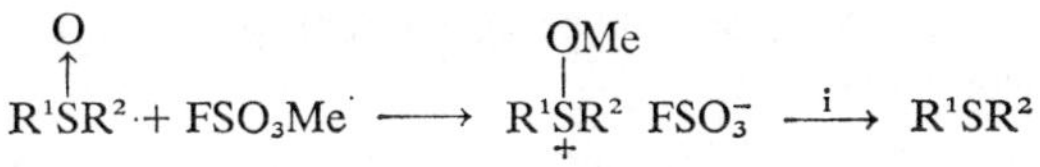

$$R^1\overset{\uparrow O}{S}R^2 + FSO_3Me \longrightarrow R^1\underset{+}{\overset{OMe}{S}}R^2 \; FSO_3^- \xrightarrow{i} R^1SR^2$$

(145)

Reagent: i, $NaCNBH_3$

Scheme 212

[642] M. Kinoshita, Y. Sato, and N. Kunieda, *Chem. Letters*, 1974, 377.
[643] M. Dreux, Y. Leroux, and P. Savignac, *Synthesis*, 1974, 506.
[644] T. Fujisawa, K. Sugimoto, and H. Ohta, *Chem. Letters*, 1973, 1241.
[645] H. D. Durst, J. W. Zubrick, and G. R. Kieczykowski, *Tetrahedron Letters*, 1974, 1777.
[646] M. Cinquini and S. Colonna, *J.C.S. Chem. Comm.*, 1974, 769.
[647] B. Stridsberg and S. Allenmark, *Acta Chem. Scand.*, 1974, **B28**, 591.

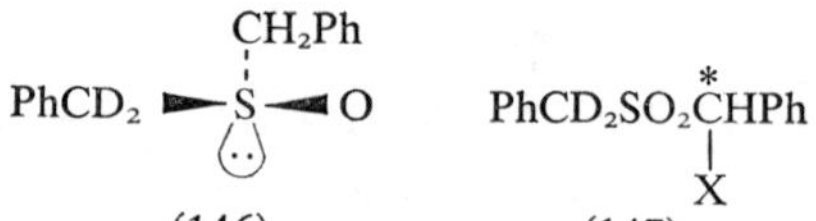

(146) (147)

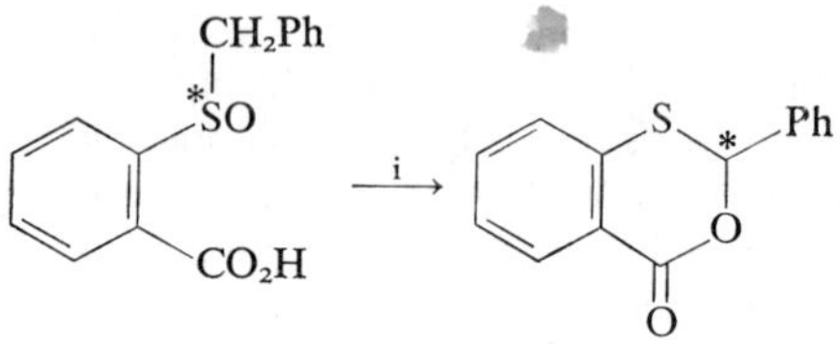

(148) (149)

Reagent: i, dicyclohexylcarbodi-imide

Scheme 213

Optically active sulphoxides can be transformed into optically active sulphoximines,[648] which are useful alkyl transfer reagents.[649] Spontaneous generation of strong acids such as methanesulphonic acid in dimethyl sulphoxide provides a mechanistic rationale[650] linking several diverse reactions of dimethyl sulphoxide.

t-Butyl- and n-butyl-lithium are unsatisfactory bases for the generation of α-sulphinyl carbanions, owing to competitive C—S bond cleavage; the chiral sulphoxide (150) is racemized[651] in a few minutes at −78 °C (Scheme 214). Methyl-lithium or a lithium dialkylamide are much better reagents for such carbanion formation.

Monoalkylcopper(I) reagents add to αβ-acetylenic sulphoxides with almost exclusive *cis* stereochemistry (Scheme 215);[652] lithium dialkylcuprates cause cleavage.

Sulphones.—Büchi has presented a new synthesis (Scheme 216) of bis-allylic sulphones, which act as precursors of poly-olefins by Ramberg–Bäcklund rearrangement.[653]

$$Bu^{t}Li + Bu^{t}\overset{\overset{O}{\uparrow}}{S^{*}}Ar \rightarrow Bu^{t}\overset{\overset{O}{\uparrow}}{S}Ar + Bu^{t}Li$$

(150)

Scheme 214

[648] C. R. Johnson, R. A. Kirchoff, and H. G. Corkins, *J. Org. Chem.*, 1974, **39**, 2458.
[649] C. R. Johnson, *Accounts Chem. Res.*, 1973, **6**, 341.
[650] T. M. Santosusso and D. Swern, *Tetrahedron Letters*, 1974, 4255.
[651] T. Durst, M. J. LeBelle, R. V. den Elzen, and K.-C. Tin, *Canad. J. Chem.*, 1974, **52** 761.
[652] W. E. Truce and M. J. Lusch, *J. Org. Chem.*, 1974, **39**, 3174.
[653] G. Büchi and R. M. Freidinger, *J. Amer. Chem. Soc.*, 1974, **96**, 3332.

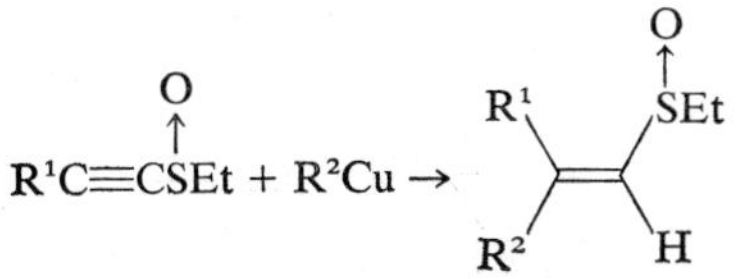

Functional Groups other than Alkanes, Acetylenes, Allenes, and Olefins

$R^1C\equiv CSEt + R^2Cu \rightarrow$ [vinyl sulfide product with R^1, R^2, H, SEt, C=O]

Scheme 215

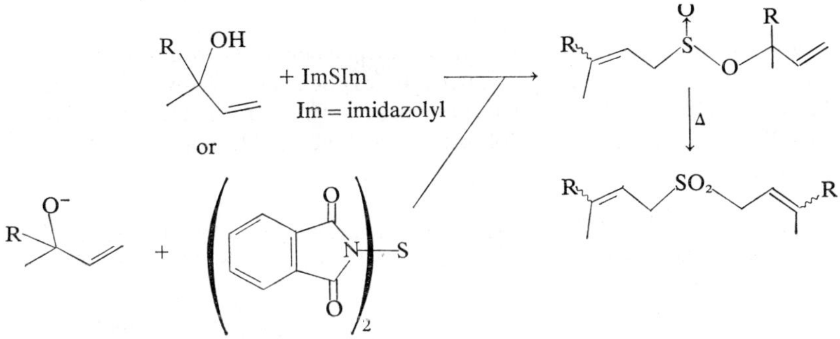

Scheme 216

$$R_3B + BrCH_2SO_2Y \xrightarrow{KOBu^t-Bu^tOH} RCH_2SO_2Y$$

$Y = Ph, Et, OCH_2Bu^t, NEt_2$

Scheme 217

A wide range of α-bromosulphonyl compounds undergo alkylation on treatment with trialkylboranes[654] (Scheme 217)

Hendrickson[655] recommends the trifluoromethylsulphonyl group for methylene activation (Scheme 218). It can also act as a leaving group, and is considerably more powerful than the cyano-group in both of these capacities.

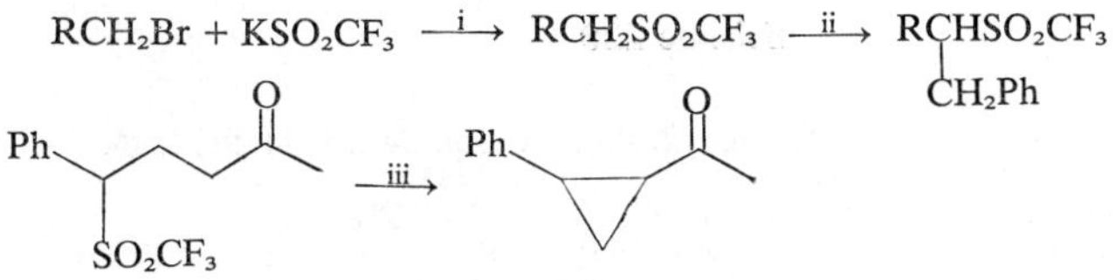

$$RCH_2Br + KSO_2CF_3 \xrightarrow{i} RCH_2SO_2CF_3 \xrightarrow{ii} R\underset{CH_2Ph}{C}HSO_2CF_3$$

Reagents: i, I⁻; ii, K_2CO_3–PhCH₂Br; iii, EtO⁻

Scheme 218

[654] W. E. Truce, L. A. Mura, P. J. Smith, and F. Young, *J. Org. Chem.*, 1974, **39**, 1449.
[655] J. B. Hendrickson, A. Giga, and J. Wareing, *J. Amer. Chem. Soc.*, 1974, **96**, 2275.

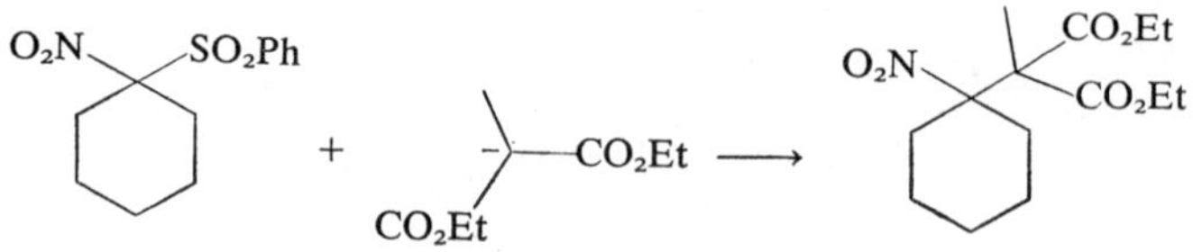

Scheme 219

α-Nitrosulphones undergo a rapid displacement at room temperature when treated with a variety of nucleophiles (Scheme 219) in what seems to be a radical-anion–free-radical chain process.[656] The oxidative addition of metallated phenyl methyl sulphone to isolated double bonds has been reported.[657]

General.—α-Aminonitriles derived from aromatic aldehydes react with hydrogen sulphide under mild conditions[658] to give high yields of thiols (Scheme 220).

The utility of phase transfer catalysis has been further exemplified in a synthesis[659] of dialkyl and aryl alkyl sulphides. Carbenes react with sulphides to give sulphonium ylides in a highly selective manner,[660] whereas the corresponding reaction with nitrenes to give iminosulphuranes is quite unselective.

A comparison[661] has been made of the relative reactivity of different nucleophiles towards sulphenyl *versus* sulphonyl sulphur. Thermally generated sulphenic acids can be advantageously trapped as their trimethylsilyl esters.[662]

New syntheses of sulphinic acid esters include the low-temperature chlorination[663] of disulphides in alcohols, the reaction of t-alkyl Grignard reagents with dialkyl sulphites,[664] and the reaction of sulphinyl chlorides

$$\text{ArCHO} \longrightarrow \underset{\underset{\text{NHR}}{|}}{\text{ArCHCN}} \xrightarrow{\text{i}} \text{ArCH}_2\text{SH}$$

Reagent: i, H$_2$S–py–Et$_3$N

Scheme 220

[656] N. Kornblum, S. D. Boyd, and N. Ono, *J. Amer. Chem. Soc.*, 1974, **96**, 2580.
[657] M. Julia and L. Saussine, *Tetrahedron Letters*, 1974, 3433.
[658] R. Crossley and A. C. W. Curran, *J.C.S. Perkin I*, 1974, 2327.
[659] D. Landini and F. Rolla, *Synthesis*, 1974, 565; see also P. Savignac and P. Coutrot, *ibid.*, p. 818; E. Vowinkel and C. Wolff, *Chem. Ber.*, 1974, **107**, 496.
[660] D. C. Appleton, D. C. Bull, J. McKenna, J. M. McKenna, and A. R. Whalley, *J.C.S. Chem. Comm.*, 1974, 140.
[661] J. L. Kice, T. E. Rogers, and A. C. Warheit, *J. Amer. Chem. Soc.*, 1974, **96**, 8020; see also J. L. Kice and T. E. Rogers, *ibid.*, p. 8009, 8015.
[662] T. S. Chou, *Tetrahedron Letters*, 1974, 725.
[663] I. B. Douglas, *J. Org. Chem.*, 1974, **39**, 563.
[664] M. Mikołajczyk and J. Drabowicz, *Synthesis*, 1974, 124.

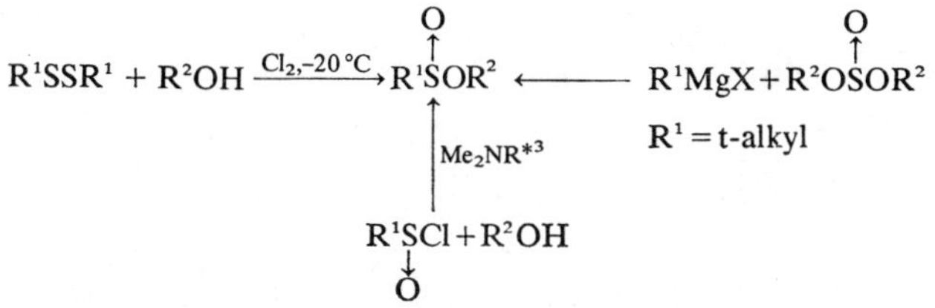

Functional Groups other than Alkanes, Acetylenes, Allenes, and Olefins 197

Scheme 221

$$R^1SSR^1 + R^2OH \xrightarrow{Cl_2, -20\,°C} R^1\overset{\overset{O}{\uparrow}}{S}OR^2 \longleftarrow R^1MgX + R^2O\overset{\overset{O}{\uparrow}}{S}OR^2$$

$$\Big\uparrow Me_2NR^{*3} \quad R^1 = \text{t-alkyl}$$

$$R^1SCl + R^2OH \downarrow O$$

with achiral alcohols[665] in the presence of optically active tertiary amines; optical yields of up to 45% are obtained in the last case (Scheme 221).

Alkanesulphonyl chlorides can be obtained by reaction of alkyllithium reagents with sulphuryl chloride,[666] or *via* pyridinium sulphonates (Scheme 222);[667] alkanesulphonyl fluorides[668] can be prepared by direct oxidation of alkanethiols with nitrogen dioxide in aqueous hydrogen fluoride.

$$RLi + SO_2Cl_2 \xrightarrow{-40\,°C} RSO_2Cl \xleftarrow[SOCl_2]{PCl_5 \text{ or}} RSO_3^-pyH^+$$

Scheme 222

t-Butyl hypochlorite in the presence of methanol converts[669] NN'-di-t-butyl sulphamide into methyl t-butyl sulphamate (Scheme 223).

$$(Bu^tNH)_2SO_2 + Bu^tOCl \longrightarrow [Bu^tN{=}SO_2] \xrightarrow{MeOH} Bu^tNHSO_2OMe$$

Scheme 223

14 Miscellaneous

Diazoalkanes.—Toluene-*p*-sulphonyl azide is a most useful reagent for the direct conversion of active methylene compounds into the corresponding diazoalkanes (Scheme 224); tertiary carbanions experience an efficient azido transfer,[670] rather than diazo transfer. The advantageous use of phase transfer catalysis[671] for such a transformation has been described, as has a simple preparation[672] of a polymer-supported arenesulphonyl azide.

An efficient synthesis of α-diazoesters from α-amino-acid esters proceeds by base-induced scission[673] of the triazine (151) (Scheme 225).

[665] M. Mikołajczyk and J. Drabowicz, *J.C.S. Chem. Comm.*, 1974, 547.
[666] H. Quast and F. Kees, *Synthesis*, 1974, 489.
[667] A. Barco, S. Benetti, G. P. Pollini, and R. Taddia, *Synthesis*, 1974, 877.
[668] C. Comninellis, P. Javet, and E. Plattner, *Synthesis*, 1974, 887.
[669] B. Weinstein and H.-H. Chang, *Tetrahedron Letters*, 1974, 901.
[670] S. J. Weininger, S. Kohen, S. Mataka, G. Koga, and J.-P. Anselme, *J. Org. Chem.*, 1974, **39**, 1591; *cf.* J. O. Reed and W. Lwowski, *ibid.*, 1971, **36**, 2864.
[671] H. Ledon, *Synthesis*, 1974, 347.
[672] W. R. Roush, D. Feitler, and J. Rebek, *Tetrahedron Letters*, 1974, 1391.
[673] J. F. McGarrity, *J.C.S. Chem. Comm.*, 1974, 558.

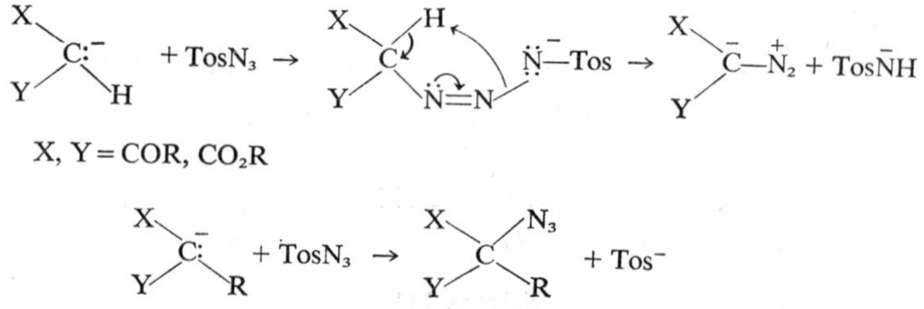

$X, Y = COR, CO_2R$

Scheme 224

Diazomethane can be routinely obtained in good yield by the crown ether-catalysed reaction[674] of hydrazine hydrate with chloroform and potassium hydroxide in dichloromethane. The preparation and properties of disilver diazomethane have been described.[675]

Tetraphenylethylene seems capable[676] of mimicking copper and copper salts in their ability to effect olefin cyclopropanations with diazoalkanes. Copper(II)-catalysed decomposition of the $\beta\gamma$-unsaturated diazomethyl ketone (152) in the presence of methanol leads[677] to the $\gamma\delta$-unsaturated ester (154), possibly via [2,3]sigmatropic rearrangement of the keto-carbene (153); photolysis of (152) gives the expected homologated ester by Wolff

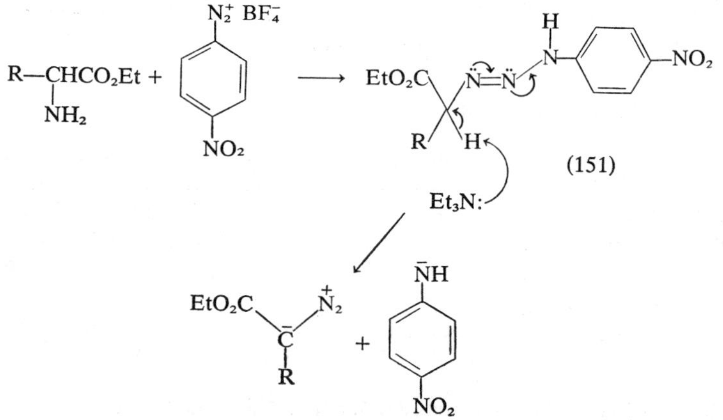

Scheme 225

[674] D. T. Sepp, K. V. Scherer, and W. P. Weber, *Tetrahedron Letters*, 1974, 2983.
[675] E. T. Blues, D. Bryce-Smith, J. G. Irwin, and I. W. Lawston, *J.C.S. Chem. Comm.*, 1974, 466.
[676] C.-T. Ho, R. T. Conlin, and P. P. Gaspar, *J. Amer. Chem. Soc.*, 1974, **96**, 8109.
[677] A. B. Smith, *J.C.S. Chem. Comm.*, 1974, 695.

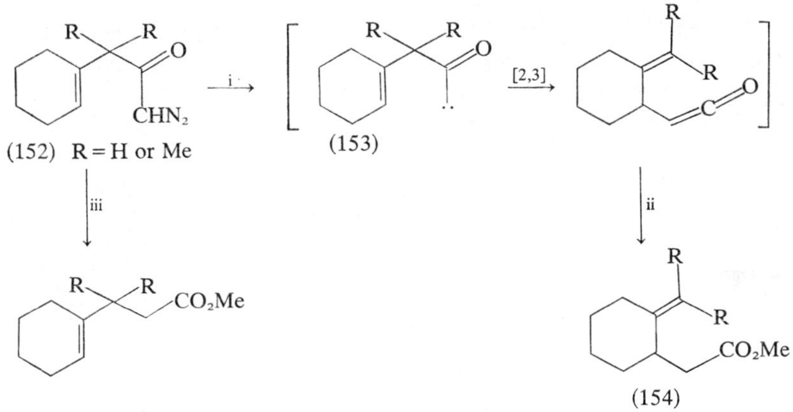

Functional Groups other than Alkanes, Acetylenes, Allenes, and Olefins

Scheme 226

(152) R = H or Me

(153)

(154)

Reagents: i, Cu^{2+}, heat; ii, MeOH; iii, $h\nu$, MeOH

rearrangement (Scheme 226). The physical and chemical properties of alkoxycarbonylcarbenes have been reviewed.[678]

Organometallics and Metal Complexes.—Further evidence has been presented[679] which suggests that the rate-determining step in the formation of alkyl Grignard reagents involves electron transfer from the metal to the alkyl halide. A golden compound, formulated[680] as a salt of the sodium anion, Na^+(cryptate)Na^-, has been isolated by addition of a cryptate to a solution of sodium metal in ethylamine.

A study[681] of the selectivity of organic group transfer in the reactions of mixed lithium diorganocuprates has concluded that the best non-transferable groups are those which form unreactive copper(I) compounds, such as 1-pentynyl. The area of stability of organolithium reagents in hexamethylphosphoramide continues to contract; Eliel[682] has reported that, at low temperatures, the solvent is attacked with formation of N-methyl methyleneimine (155), leading ultimately to secondary amines in good yield (Scheme 227).

Stabilized carbanions, preferably tertiary, are 'phenylated' by π-(chlorobenzene)chromium tricarbonyl[683] (Scheme 228).

---

[678] A. P. Marchand and N. MacBrockway, *Chem. Rev.*, 1974, **74**, 431.
[679] R. J. Rogers, H. L. Mitchell, Y. Fujiwara, and G. M. Whitesides, *J. Org. Chem.* 1974, **39**, 857; cf. H. W. H. J. Bodewitz, C. Blomberg, and F. Bickelhaupt, *Tetrahedron*, 1973, **29**, 719.
[680] J. L. Dye, J. M. Ceraso, M. T. Lok, B. L. Barnett, and F. J. Tehan, *J. Amer. Chem. Soc.*, 1974, **96**, 608.
[681] W. H. Manderville and G. M. Whitesides, *J. Org. Chem.*, 1974, **39**, 400.
[682] A. G. Abatjoglou and E. L. Eliel, *J. Org. Chem.*, 1974, **39**, 3042.
[683] M. F. Semmelhack and H. T. Hall, *J. Amer. Chem. Soc.*, 1974, **96**, 7091, 7092.

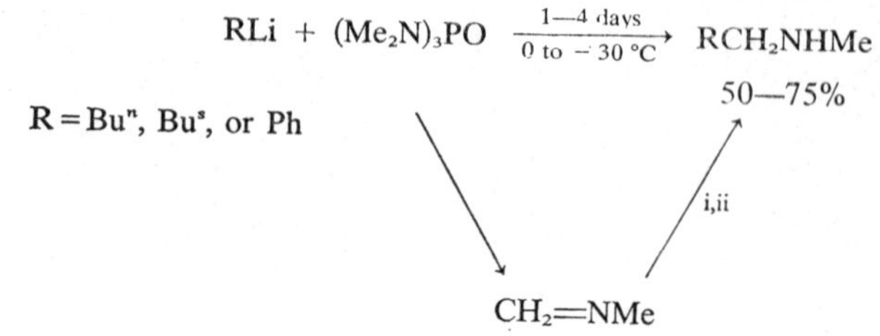

RLi + (Me$_2$N)$_3$PO $\xrightarrow[\text{0 to }-30\,°C]{\text{1—4 days}}$ RCH$_2$NHMe

R = Bu^n, Bu^s, or Ph

50—75%

i, ii

CH$_2$=NMe
(155)

Reagents: i, RLi; ii, H$_2$O

Scheme 227

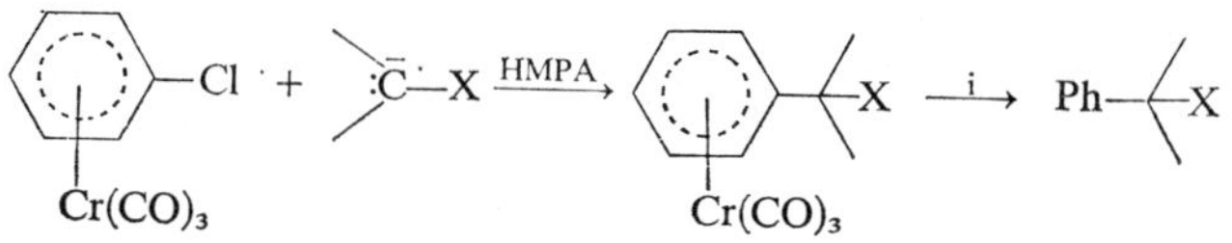

X = CN, CO$_2$Et, etc.

Reagent: i, I$_2$–Et$_2$O, 0 °C

Scheme 228

Titanocene mediates the conversion of a variety of aldehydes, esters, and oxirans into alkanes,[684] with preservation of the original carbon skeleton. The addition of n-butyl-lithium to iron(III) chloride in tetrahydrofuran produces a black solution, which smoothly converts[685] a range of simple oxirans into olefins in excellent yields. The reducing ability of titanium(II) species is described on p. 157.

'Woodward fission' of catechol has been developed[686] into a synthetically useful procedure by means of metal complexes; catechol and molecular oxygen react in the presence of copper(I) chloride in methanol to give the monoester (156) (Scheme 229).

Phase Transfer Catalysts and Crown Ethers.—Phase transfer catalysis has been reviewed.[687] Several applications of phase transfer agents and the not

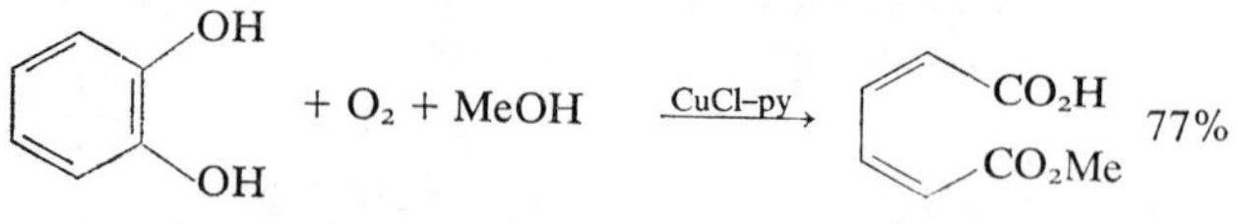

+ O$_2$ + MeOH $\xrightarrow{\text{CuCl-py}}$ 77%

(156)

Scheme 229

[684] E. E. van Tamelen and J. A. Gladysz, *J. Amer. Chem. Soc.*, 1974, **96**, 5290.
[685] T. Fujisawa, K. Sugimoto, and H. Ohta, *Chem. Letters*, 1974, 883.
[686] J. Tsuji and H. Takayanagi, *J. Amer. Chem. Soc.*, 1974, **96**, 7349.
[687] E. V. Dehmlow, *Angew. Chem. Internat. Edn.*, 1974, **13**, 170.

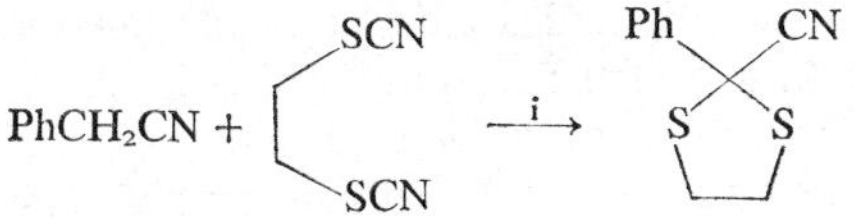

Functional Groups other than Alkanes, Acetylenes, Allenes, and Olefins

PhCH$_2$CN + [SCN / SCN] —i→ [Ph, CN / S, S (ring)]

Reagent: i, NaOH–H$_2$O–PhCH$_2$ÑEt$_3$Cl$^-$

Scheme 230

unrelated (in mode of behaviour) crown ethers as catalysts in anion-promoted two-phase reactions have been described already (see refs. 98–100, 128, 232, 271, 316, 554, 598–601, 605, 634, 645, 659, 671, and 674). Such behavioural equivalence has been further exemplified;[688,689] one example is seen in the use of phase transfer agents to solubilize permanganate ions in benzene: this cheap two-phase oxidizing system is recommended[690] as the method of choice for permanganate oxidations.

Aliphatic and aromatic thiocyanates are effective alkyl- or aryl-thiolating agents[691] for carbanions generated in a two-phase catalytic system (Scheme 230).

A general synthesis[692] of mandelic acids uses phase transfer catalysis for the generation of dichlorocarbene (Scheme 231).

Chiroptical Properties.—Further details[693,694] have been given of the determination of enantiomeric purity by n.m.r. spectroscopy, using chiral lanthanide shift reagents. A kinetic method[695] has been described for the determination of specific rotatory power based on the difference in rate of

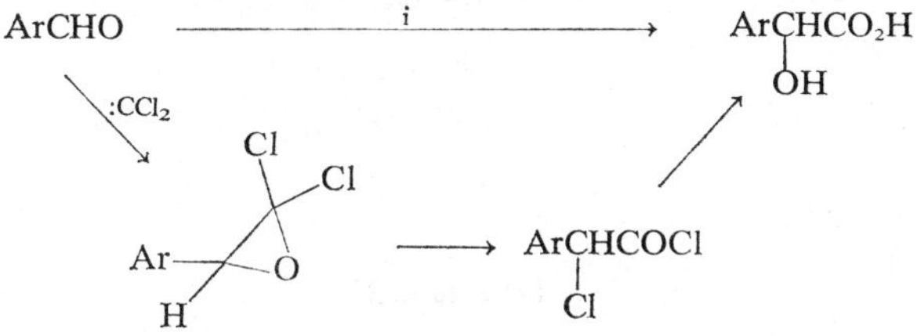

ArCHO —i→ ArCH(OH)CO$_2$H

:CCl$_2$ ↓   ↗

Ar–CH(O)–CCl$_2$ (epoxide) —→ ArCHCOCl / Cl

Reagents: i, NaOH–CHCl$_3$–PhCH$_2$ÑEt$_3$Cl$^-$

Scheme 231

[688] D. Landini, F. Montanari, and F. M. Pirisi, *J.C.S. Chem. Comm.*, 1974, 879.
[689] M. Mąkosza and M. Ludwikow, *Angew. Chem. Internat. Edn.*, 1974, **13**, 665.
[690] A. W. Herriott and D. Picker, *Tetrahedron Letters*, 1974, 1511.
[691] M. Mąkosza and M. Fedoryński, *Synthesis*, 1974, 274.
[692] A. Merz, *Synthesis*, 1974, 724.
[693] M. D. McCrearty, D. W. Lewis, D. L. Wernick, and G. M. Whitesides, *J. Amer. Chem. Soc.*, 1974, **96**, 1038.
[694] H. L. Goering, J. N. Eikenberry, G. S. Koermer, and C. J. Lattimer, *J. Amer. Chem. Soc.*, 1974, **96**, 1493.
[695] J. Y. Conan, F. Guinot, G. Lamaty, and A. Natat, *Tetrahedron Letters*, 1974, 1667.,

decomposition of a racemic and a chiral moiety; this method obviates the necessity for resolution of the substrate, which in the example given is a racemic ketone. Horeau's method for determination of absolute configuration has been applied successfully[696] to some cyclic and acyclic 1,2-diols; a modified method[697] has been used on amines, alcohols, and carboxylic acids.

Computer-designed Synthesis.—Several papers[698–700] have described refinements in the use of computers in synthetic design. Corey[701] has elegantly executed a computer-designed synthesis of a complex alkaloid, porantherine, while Näf and Ohloff, after a short, stereoselective, chemist-designed synthesis[702] of patchouli alcohol, comment 'we would not wish to discourage those who think travelling hopefully with a computer is better than arriving'.

Bases.—Potassium hydride in tetrahydrofuran rapidly metallates hindered alcohols and secondary amines, allowing easy access[703] to the derived bases. Potassium tri-sec-alkylmethoxides function as hindered bases of sufficient strength[704] to convert a range of ketones essentially quantitatively into their respective enolate anions. The various uses of potassium t-butoxide[705] and of sodamide-metal alkoxides[706] have been reviewed.

General.—*Ab initio* calculations[707] on acyl cations have suggested that C—C hyperconjugation is more effective than C—H hyperconjugation.

Phosphorus halides have been incorporated on to a polymeric support, with resulting enhancement[708] of their halogenating utility. Some reactions, such as nucleophilic displacements and additions, are considerably accelerated if allowed to take place inside the pores of high surface area materials

$$Me_2SO + (CF_3CO)_2O \xrightarrow[CH_2Cl_2]{-60\,°C} [Me_2\overset{+}{S}-OCOCF_3]\,\bar{O}COCF_3 \xrightarrow{i,\,ii} Me_2\overset{+}{S}-\bar{N}R$$

Reagents: i, RNH$_2$; ii, NaOH–H$_2$O

Scheme 232

[696] D. P. G. Hamon, R. A. Massy-Westropp, and T. Pipithakul, *Austral. J. Chem.* 1974, **27**, 2205.
[697] H. Brockmann and N. Risch, *Angew. Chem. Internat. Edn.*, 1974, **13**, 664.
[698] W. T. Wipke and T. M. Dyott, *J. Amer. Chem. Soc.*, 1974, **96**, 4825, 4834.
[699] E. J. Corey, W. J. Howe, and D. A. Pensak, *J. Amer. Chem. Soc.*, 1974, **96**, 7724.
[700] H. G. Elernter, N. S. Sridharan, A. J. Hart, S.-C. Yen, F. W. Fowler, and H.-J. Shue, *Fortschr. Chem. Forsch.*, 1973, **41**, 113.
[701] E. J. Corey and R. D. Balanson, *J. Amer. Chem. Soc.*, 1974, **96**, 6516.
[702] F. Näf and G. Ohloff, *Helv. Chim. Acta*, 1974, **57**, 1868.
[703] C. A. Brown, *Synthesis*, 1974, 427.
[704] C. A. Brown, *J.C.S. Chem. Comm.*, 1974, 680.
[705] D. E. Pearson and C. A. Buehler, *Chem. Rev.*, 1974, **74**, 45.
[706] P. Caubere, *Accounts Chem. Res.*, 1974, **7**, 301.
[707] L. Radom, *Austral. J. Chem.*, 1974, **27**, 231.
[708] H. M. Relles and R. W. Schluenz, *J. Amer. Chem. Soc.*, 1974, **96**, 6469.

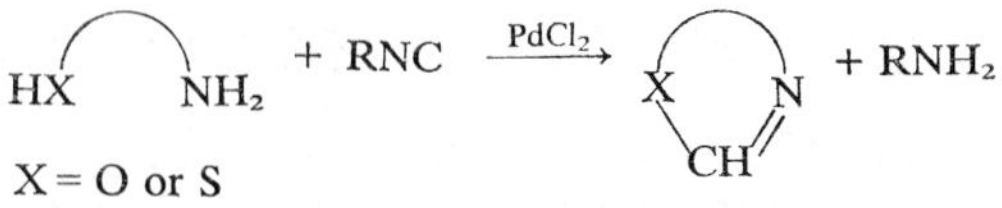

Functional Groups other than Alkanes, Acetylenes, Allenes, and Olefins 203

$$\underset{\substack{HX \quad NH_2 \\ X = O \text{ or } S}}{\bigcap} + RNC \xrightarrow{PdCl_2} \underset{CH}{\overset{X \quad N}{\bigcap}} + RNH_2$$

Scheme 233

such as alumina and silica; examples of such 'capillary techniques' have been given.[709]

Trifluoroacetic anhydride serves as an efficient activating agent for dimethyl sulphoxide in a new synthesis[710] of sulphuranes (Scheme 232).

A facile route[711] to cyclic iminoethers and iminothioethers is illustrated in Scheme 233.

The synthesis has been reported of disparlure (7,8-epoxy-2-methyloctadecane), the sex attractant emitted by the female gypsy moth, *Portheria dispar*. (L.); preliminary results suggest that the natural pheromone is the (+)-*cis* form.[712] The defensive compound in the frontal gland secretion of soldier termites has been identified[713] as 1-nitro-*trans*-pentadec-1-ene.

15 Reviews

Many review articles have been cited in their appropriate subject areas (refs. 30, 123, 240, 286, 278, 327, 340, 347, 441, 457, 465, 470, 508, 595, 596, 620, 649, 678, 687, 705, and 706).

In addition to the excellent review articles in the *Journal of Organometallic Chemistry*, reviews have been published on olefin–carbon monoxide insertion reactions,[714] and on the synthetic utility of π-allyl metal derivatives.[715] Rosenblum[716] has summarized some of the remarkable uses of iron complexes in synthesis, paying considerable attention to the problem of liberating the organic product from the metal. Kochi[717] has given an account of his studies on electron transfer mechanisms in organometallic intermediates.

The reducing abilities of chromium(II) salts[718] have been reviewed, as have rearrangements[719] occurring on lithium aluminium hydride reduction. Some applications of ionic hydrogenation[720] have been given, in which triethylsilane and trifluoroacetic acid provide the required hydride donor–proton

---

[709] M. Hudlicky, *J. Org. Chem.*, 1974, **39**, 3460.
[710] A. K. Sharma and D. Swern, *Tetrahedron Letters*, 1974, 1503.
[711] Y. Ito, I. Ito, T. Hirao, and T. Saegusa, *Synthetic Comm.*, 1974, **4**, 97.
[712] S. Iwaki, S. Marumo, T. Saito, M. Yamada, and K. Katagiri, *J. Amer. Chem. Soc.*, 1974, **96**, 7482.
[713] J. Vrkoč and K. Ubik, *Tetrahedron Letters*, 1974, 1463.
[714] G. P. Chiusoli, *Accounts Chem. Res.*, 1973, **6**, 422.
[715] R. Baker, *Chem. Rev.*, 1973, **73**, 487.
[716] M. Rosenblum, *Accounts Chem. Res.*, 1974, **7**, 122.
[717] J. K. Kochi, *Accounts Chem. Res.*, 1974, **7**, 351.
[718] J. R. Hanson, *Synthesis*, 1974, 1.
[719] S.-C. Chen, *Synthesis*, 1974, 691.
[720] D. N. Kursanov, Z. N. Parnes, and N. M. Loim, *Synthesis*, 1974, 633.

donor pair. The various uses of periodic acid and periodates have been reviewed.[721]

A series of critical reviews on acylation[722] has appeared; aspects discussed include the concept of carbonyl umpolung[347] (dipole inversion), peptide bond formation,[168] Friedel–Crafts acylation,[722b] and enamine acylation;[722c] thioacylating agents[723] are also discussed. Two reviews[724,725] of the uses of polymer supported reagents and substrates have been presented (see also refs. 11, 43, 48, 177, 181, 672, and 708).

Reviews have been published on recent aspects of the chemistry of hydroxamic acids and N-hydroxyimides,[726] alkane diazotates,[727] diketen,[728] halogenovinylene carbonates,[729] and carbon suboxide,[730] C_3O_2, a precursor of, *inter alia*, malonic acid derivatives.

[721] A. J. Fatiadi, *Synthesis*, 1974, 229; see also ref. 447.
[722] (a) D. P. N. Satchell, *Chem. and Ind.*, 1974, 683; (b) P. H. Gore, *ibid.*, p. 727; (c) P. W. Hickmott, *ibid.*, p. 731.
[723] K. M. Doyle and F. Kurzer, *Chem. and Ind.*, 1974, 803.
[724] C. C. Leznoff, *Chem. Soc. Rev.*, 1974, 3, 65.
[725] C. G. Overberger and K. N. Sannes, *Angew. Chem. Internat. Edn.*, 1974, 13, 99.
[726] L. Bauer and O. Exner, *Angew. Chem. Internat. Edn.*, 1974, 13, 376.
[727] R. A. Moss, *Accounts Chem. Res.*, 1974, 7, 421; see also R. A. Moss and J. Banger, *Tetrahedron Letters*, 1974, 3549.
[728] T. Kato, *Accounts Chem. Res.*, 1974, 7, 265.
[729] H.-D. Scharf, *Angew. Chem. Internat. Edn.*, 1974, 13, 520.
[730] T. Kappe and E. Ziegler, *Angew. Chem. Internat. Edn.*, 1974, 13, 491.

3
Naturally Occurring Polyolefinic and Polyacetylenic Compounds

BY G. PATTENDEN

1 Introduction

The layout of this chapter is closely similar to that of last year. Significantly, more and more papers describe the application of ^{13}C n.m.r. spectroscopy to problems of the biosynthesis of polyolefinic compounds, and some refreshing new data in this area have been forthcoming.

2 Naturally Occurring Polyacetylenes

Introduction.—Zeisberg and Bohlmann[1] have examined the ^{13}C n.m.r. spectra of several poly-ynes. These data show that the chemical shifts of all acetylenic ^{13}C atoms differ from one another, and the authors have discussed the relationship between coupling constants and structure in these molecules.

New and Known Polyacetylenes from Nature.—In a re-examination of *Artemesia vulgaris*, triynes (1) and (2) have been characterized from the root extract, and (3)—(5) have been found in the flowers;[2] the triynes (1) and (2) are new compounds to the species. Bohlmann and Zdero[3,4] have reported the presence of the two C_{15}-acetylenes (6) and (7) in *Peyrousia umbellata* and *Thaminophyllum multiflorum*, and also the isolation of (8) and the thiophen-acetylene (9) from roots of *Osmitopsis asteriscoides* (L.) *Cass*. The two diynes (10) and (11) have been isolated from the root of the Japanese *Chrysanthemum boreale* Makino[5] and the polyacetylenes (12a), (12b), and (13) as well as the known alcohol (14) have been detected in

[1] R. Zeisberg and F. Bohlmann, *Chem. Ber.*, 1974, **107**, 3800.
[2] D. Drake and J. Lam, *Phytochemistry*, 1974, **13**, 455.
[3] F. Bohlmann and C. Zdero, *Chem. Ber.*, 1974, **107**, 1044.
[4] F. Bohlmann and C. Zdero, *Chem. Ber.*, 1974, **107**, 1409.
[5] A. Matsuo, Y. Uchio, M. Nakayama, and S. Hayashi, *Tetrahedron Letters*, 1974, 1885.

extracts from *Trachelium caeruleum*.[6] A C_{14}-enediynediol assigned the structure (15a) or (15b) was also detected in *T. caeruleum*.

Jones and co-workers[7] have detected the presence of the (2S)-triynene diol (16), the C_{10}-diynene acids (17) and (18), and the C_9-diynene acid (19) in extracts of the culture fluids of *Fistulina pallida*. Both the glucoside and 3-phenyl-lactate of (16) were also detected, and the *cis*-isomer of (19) was tentatively characterized. The same group of workers has detected the new polyacetylenes (20)—(24) and the known diyne (25) in extracts of the culture fluid of the fungus *Collybia peronata* (Bolt *ex* Fr.) Kummer,[8] and have

---

[6] R. K. Bentley, C. A. Higham, J. K. Jenkins, E. R. H. Jones, and V. Thaller, *J.C.S. Perkin I*, 1974, 1987.
[7] M. Ahmed, G. C. Barley, M. T. W. Hearn, E. R. H. Jones, V. Thaller, and J. A. Yates, *J.C.S. Perkin I*, 1974, 1981.
[8] C. A. Higham, E. R. H. Jones, J. W. Kippling, and V. Thaller, *J.C.S. Perkin I*, 1974, 1991.

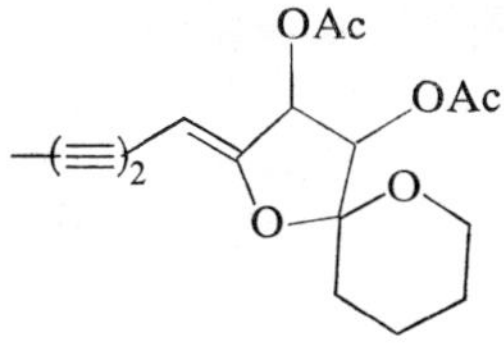

(10)

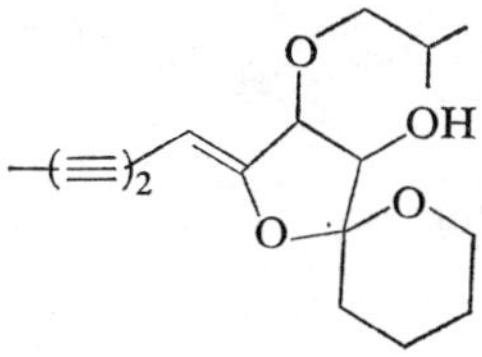

(11)

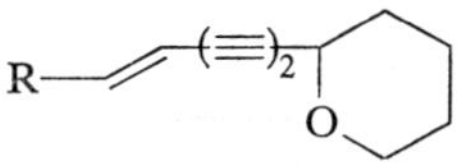

(12) a; R = CH₂OH
     b; R = CHO

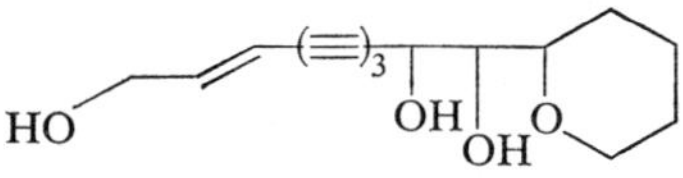

(13)

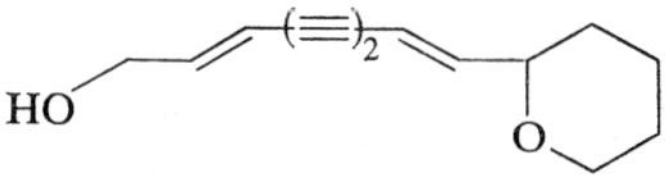

(14)

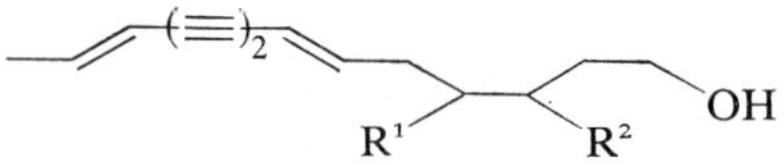

(15) a; R¹ = H, R² = OH
     b; R¹ = OH, R² = H

isolated the new acetylenic nitrile (26) from cultures of the fungus *Lepista glaucocana* (Bres.) Singer.[9]

The diyne-dieneamide (27), accompanied by the thiophen amides (28a—c) and the dieneamide (29), has been isolated from *Otanthus maritimus*,[10] and Vanhaelan-Fastré[11] has reported the presence of several polyacetylenes in *Cricus benedictus L.*

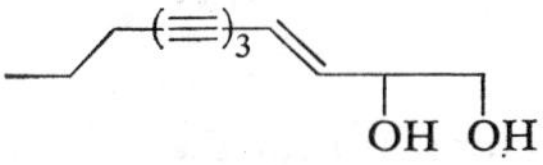

(16)

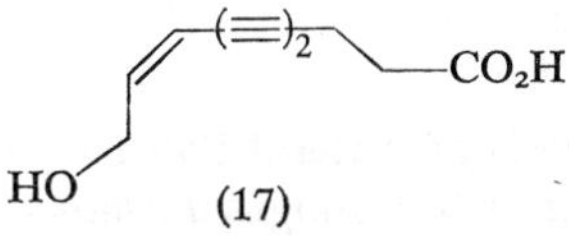

(17)

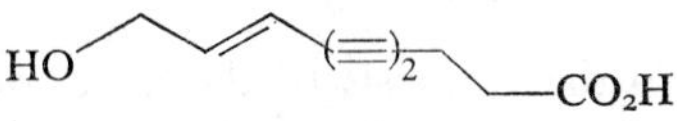

(18)

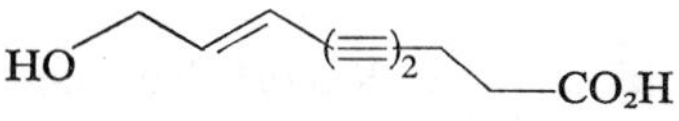

(19)

[9] M. T. W. Hearn, E. R. H. Jones, V. Thaller and J. L. Turner, *J.C.S. Perkin I*, 1974, 2335.
[10] F. Bohlmann, C. Zdero, and A. Suwita, *Chem. Ber.*, 1974, **107**, 1038.
[11] R. Vanhaelen-Fastré, *Planta Med.*, 1974, **25**, 45.

Synthesis of Natural Polyacetylenes.—Structure (16) for the new polyacetylene from the fungus *Fistulina pallida* was confirmed by an unambiguous synthesis (Scheme 1),[7] and a synthesis of ichthyotherol acetate (30), isolated some time ago from *Chrysanthemum serotinum*, has been achieved (Scheme 2).[12]

Bohlmann and Kocur[13] have described the synthesis of the novel diynes (31) and (32) from *Berkheya barbata*, along lines similar to those used in the synthesis of the methyl derivative of (31) published last year (see Vol. 3, p. 262).

[12] F. Bohlmann and D. Vogel, *Chem. Ber.*, 1974, **107**, 654.
[13] F. Bohlmann and J. Kocur, *Chem. Ber.* 1974, **107**, 2115.

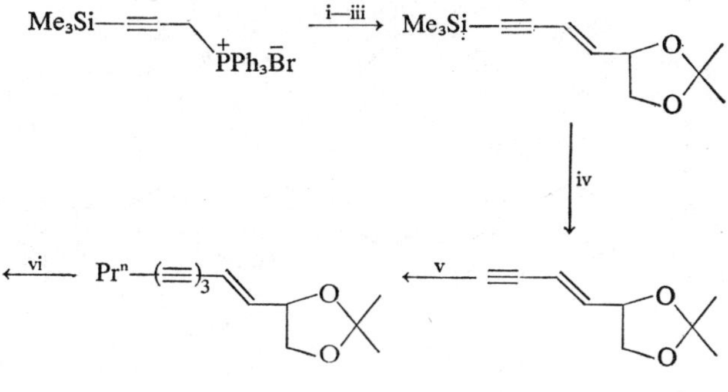

Reagents: i, Bu^{n}Li; ii, CHO-[epoxide]; iii, chromatography; iv, AgNO$_3$–EtOH, KCN; v, CuCl, EtNH$_2$, NH$_2$OH, HCl, Pr-(≡)$_2$Br; vi, HCl–EtOH

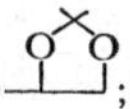

Scheme 1

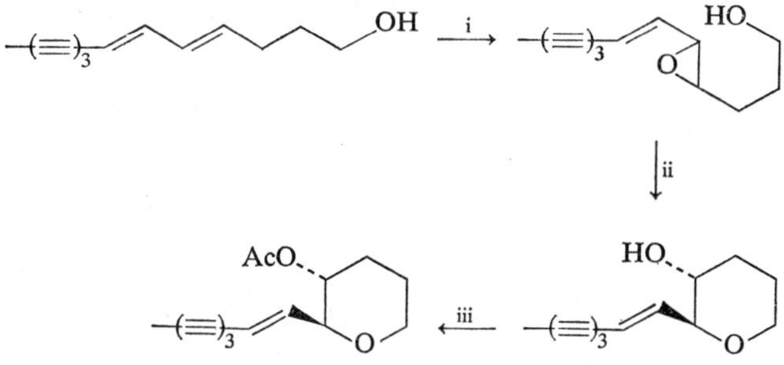

Reagents: i, m-ClC$_6$H$_4$CO$_3$H; ii, BF$_3$; iii, Ac$_2$O

Scheme 2

3 Naturally Occurring Allenes

Four analogues of histrionicotoxin (33) and isodihydrohistrionicotoxin (34) have been isolated from extracts of skins of the arrow poison frog *Dendrobates histrionicus*.[14] The analogues differ from (33) and (34) only in the degree of unsaturation in the olefinic side-chains.

---

[14] T. Tokuyama, K. Llenoyama, G. Brown, J. W. Daly, and B. Witkop, *Helv. Chim. Acta*, **15**, 283.

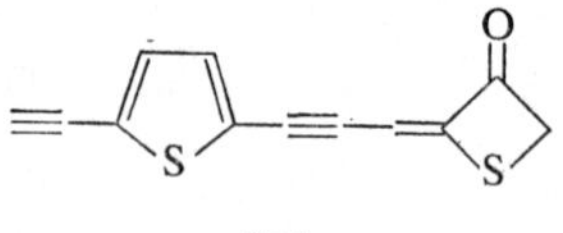

A third synthesis of racemic methyl tetradeca-E-2,4,5-trienoate (35) has been outlined (Scheme 3).[15] The (−)-enantiomer of (35) has been reported to be the sex pheromone of the dried bean beetle.

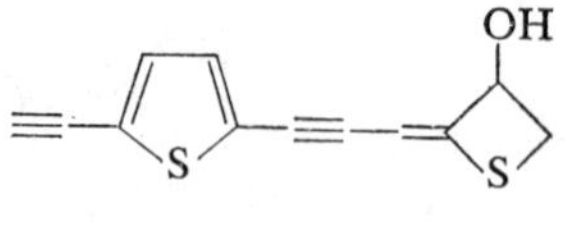

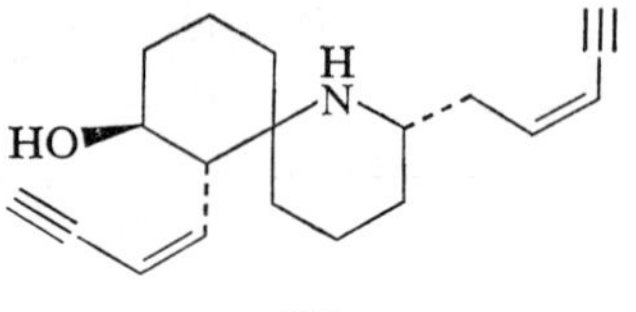

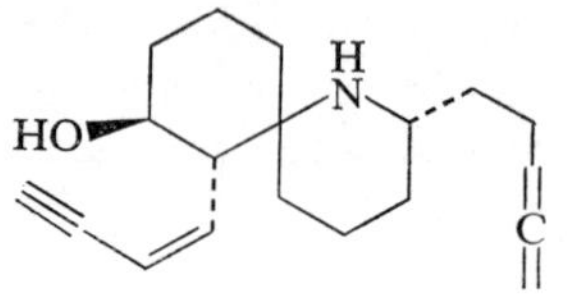

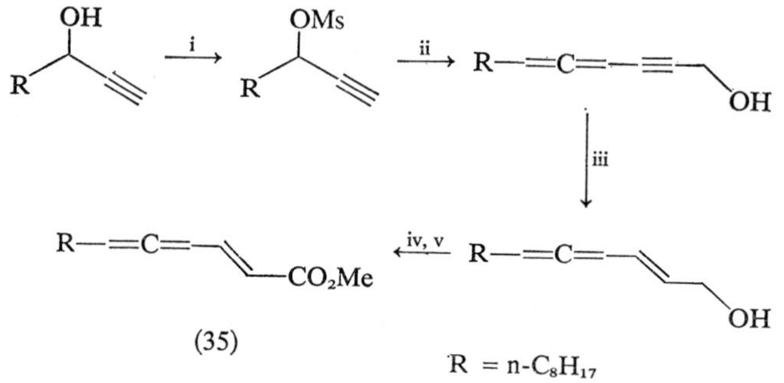

Reagents: i, MeSO$_2$Cl–C$_5$H$_5$N; ii, ≡—CH$_2$OH, CuCl, Bu^{t}NH$_2$; iii, Li(OMe)$_3$AlH; iv, MnO$_2$; v, NaCN–AcOH–MnO$_2$–MeOH

Scheme 3

4 Natural Acetylenes and Olefins from Marine Sources

Introduction.—There has been a resurgence of interest in the search for new compounds from marine sources. The volatile components of the red seaweed (*Asparagopsis taxiformis*) have been analysed, as have the chemicals in a marine annelid (segmented worm) and a marine mollusk. Aspects of the chemistry and biosynthesis of prostaglandins in coral are covered in Chapter 4.

[15] R. Baudoug and J. Gore, *Synthesis*, 1974, 573.

(36) (37)

(38) → [...] → (36)

→ (37)

Polyolefins from *Dictyopteris*.—Moore and co-workers[16] have published full details of their work leading to the structures of the odoriferous C_{11}-hydrocarbons from Hawaiian *Dictyopteris*; most of this work has been published earlier in preliminary form (see Vols. 2 and 3). The structures of multifidene and aucantene, two C_{11}-hydrocarbons found in the essential oil from the gynogametes (eggs) of the brown algae *Cutleria multifida* have been determined as (36) and (37), respectively.[17] The two compounds are obviously biogenetically related to the C_{11}-hydrocarbons from *Dictyopteris*, and it has been suggested that the trienol (38) could serve as a common intermediate.

Billups and co-workers[18] have described a new synthesis of Dictyopterin A (41) starting from (39). Treatment of the dichlorocyclopropane (39) with KOBu^t in DMSO led first to (40) as a mixture of *syn*- and *anti*-isomers. Prolonged treatment of (40) with base then resulted in formation of (41) and its other geometrical isomers. Thermolysis of the mixture of isomers of (41) gave Dictyopterin C (42).

Bromo-olefins from *Verongia* Sponges.—Borders *et al.*[19] have reported the isolation of a new compound from the sponge *Verongia lacunosa*, collected off the coast of Puerto Rico. The compound has been designated LL-PAA216 and assigned structure (43), which contains oxazolidin-2-one rings.

[16] R. E. Moore, J. A. Pettus, jun., and J. Mistysyn, *J. Org. Chem.*, 1974, **39**, 2201.
[17] L. Jaenicke, D. G. Muller, and R. E. Moore, *J. Amer. Chem. Soc.*, 1974, **96**, 3324.
[18] W. E. Billups, W. Y. Chow, and J. H. Cross, *J.C.S. Chem. Comm.*, 1974, 252.
[19] D. B. Borders, G. O. Morton, and E. R. Wetzel, *Tetrahedron Letters*, 1974, 2709.

Enyne Cyclic Bromo-ethers from *Laurencia* (Family Rhodomelaceae).

A new bromoenyne, designated rhodophytin, has been isolated from the marine algae *Chondria oppositiclada*,[20] and structure (44), containing a novel vinyl peroxide moiety, has been suggested. The structure of chondriol, from the same *Laurencia* source, has been re-assigned as (45) as a result of X-ray studies.[21]

A number of new sesquiterpenes containing bromine and chlorine have been isolated from *Laurencia* species.[22-25]

Halogen Compounds from the Sea Hare, *Aplysia*.

The sesquiterpene prepacifenol epoxide (46)[26] and dactyloxene-B (47)[27] have been isolated from the digestive gland of the sea hare.

[20] W. Fenical, *J. Amer. Chem. Soc.*, 1974, **96**, 5580.
[21] W. Fenical, K. B. Gifkens, and J. Clardy, *Tetrahedron Letters*, 1974, 1507.
[22] J. A. McMillan, I. C. Paul, R. H. White, and L. P. Hager, *Tetrahedron Letters*, 1974, 2039.
[23] S. M. Waraszkiewicz and K. L. Erickson, *Tetrahedron Letters*, 1974, 2003.
[24] M. Suzuki, E. Kurosawa, and T. Irie, *Tetrahedron Letters*, 1974, 821, 1807.
[25] J. J. Sims, G. H. Y. Lin, and R. M. Wing, *Tetrahedron Letters*, 1974, 3487.
[26] D. J. Faulkner, M. O. Stallard, and C. Ireland, *Tetrahedron Letters*, 1974, 3571.
[27] F. J. Schmitz and F. J. McDonald, *Tetrahedron Letters*, 1974, 2541.

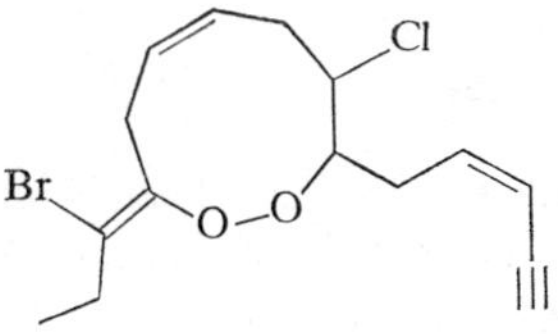

(44) 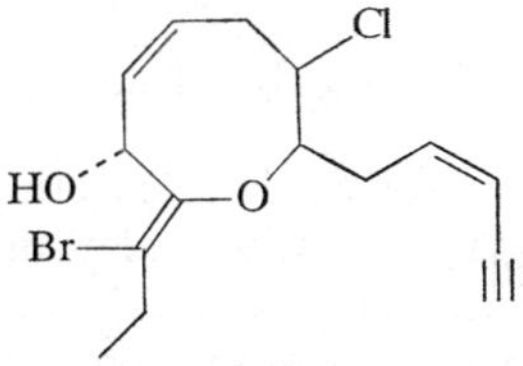   (45)

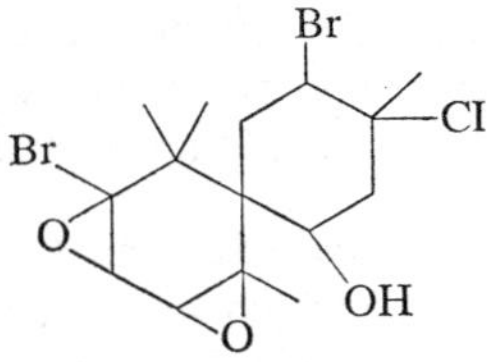

(46) 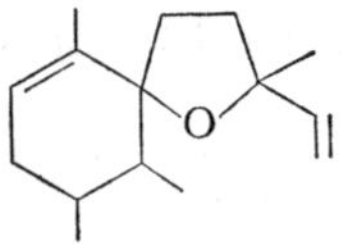   (47)

Miscellaneous Compounds from Marine Sources.—Fenical[28] has identified 1,3-dibromoacetone, 1,1,3-tribromoacetone, 1,1,3,3-tetrabromoacetone, 1-bromo-3-chloroacetone, 1-chloro-3,3-dibromoacetone, 1-chloro-1,3-dibromoacetone, 1-chloro-1,3,3-tribromoacetone, tribromobut-3-en-2-one, chlorodibromobut-3-en-2-one, chlorotribromobut-3-en-2-one, and tetrabromobut-3-en-2-one as volatile components of the red seaweed (*Asparagopsis taxiformis* (Bonnemaisoniaceae). The halogeno-compounds violacene (48)[29] and cartilagineal (49)[30] have been isolated from the red algae *Plocamium violaceum* and *P. cartilagineum*, respectively, and several brominated metabolites have been found in the marine annelid (segmented worm) *Thelepus setosus*.[31]

Two constituents of the marine mollusk *Stylocheilus longicauda*, designated aplysiatoxin and debromoaplysiatoxin, have been separated and characterized,[32] and two new cembrene derivatives have been isolated from a Pacific soft coral, *Nephthea* sp.[33]

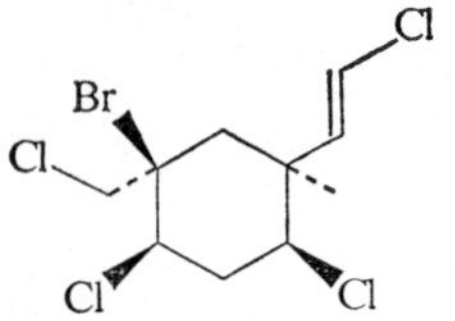

(48) 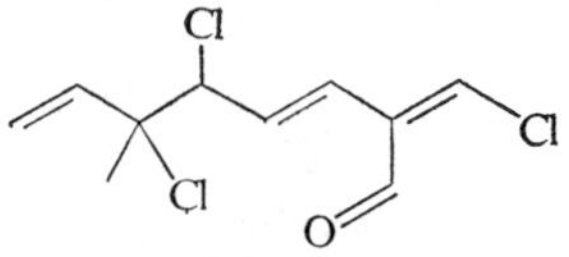   (49)

[28] W. Fenical, *Tetrahedron Letters*, 1974, 4463.
[29] J. S. Mynderse and D. J. Faulkner, *J. Amer. Chem. Soc.*, 1974, **96**, 6771.
[30] P. Crews and E. Kho, *J. Org. Chem.*, 1974, **39**, 3303.
[31] T. Higa and P. J. Schener, *J. Amer. Chem. Soc.*, 1974, **96**, 2246.
[32] Y. Kato and P. J. Scheuer, *J. Amer. Chem. Soc.*, 1974, **96**, 2245.
[33] F. J. Schmitz, D. J. Vanderah, and L. S. Ciereszko, *J.C.S. Chem. Comm.*, 1974, 407.

5 Polyolefinic Microbial Metabolites

A new sub-section, covering natural butenolides, has been introduced this year; otherwise the layout of this section is identical with that of previous Reports. Tamm has written a useful review of the chemistry of the antibiotic complex of the verrucarins and roridins.[34]

Polyolefinic Macrolides.—Rinehart and co-workers[35] have shown that field desorption mass spectrometry is the method of choice for the determination of the molecular weights of non-volatile or thermally unstable antibiotics.

Primycin. In a series of papers,[36-38] de Mayo and MacLean and their respective co-workers have presented evidence which supports structure (50) for primycin, the antibiotic produced by *Streptomyces primycini*.

Amphotericin B. A synthesis of ^{14}C-labelled amphotericin B has been reported.[39]

Leucomycins. The absolute configuration at C-9 in leucomycin A_3 (51) has been revised to R;[40] this observation has suggested that the same centres in the antibiotics spiramycin and maridomycin II are also R. Structures for the chemically modified 16-membered lactone aglycones of leucomycin A_3

(50)

[34] C. Tamm, *Fortschr. Chem. org. Naturstoffe*, 1974, **31**, 63.
[35] K. L. Rinehart, jun., J. C. Cook, K. H. Maurer, and U. Rapp, *J. Antibiotics*, 1974, **27**, 1.
[36] J. Aberhart, R. C. Jain, T. Fehr, P. de Mayo, and I. Szilagyi, *J.C.S. Perkin I*, 1974, 816.
[37] D. E. F. Gracy, L. Baczynskyj, T. I. Martin, and D. B. MacLean, *J.C.S. Perkin I*, 1974, 827.
[38] T. Fehr, R. C. Jain, P. de Mayo, O. Motl, I. Szilagyi, L. Baczynskyj, D. E. F. Gracy, H. L. Holland, and D. B. MacLean, *J.C.S. Perkin I*, 1974, 836.
[39] H. A. B. Linke, W. Mechlinski, and C. P. Schaffner, *J. Antibiotics*, 1974, **27**, 155.
[40] L. A. Freiberg, R. S. Egan, and W. H. Washburn, *J. Org. Chem.*, 1974, **39**, 2474.

Naturally Occurring Polyolefinic and Polyacetylenic Compounds 215

(51)

(52) R = H or Ac

have been presented,[41–43] and structure–activity relationships in 16-membered macrolide antibiotics generally have been considered.[44]

Lankacidins. The aglycone of the lankacidin group of antibiotics from *Streptomyces* sp. 6642 GC_1 has been shown to have constitution (52).[45] Mass spectral data of the antibiotics have been recorded and fragmentation patterns discussed.[46]

Antibiotic XL-704. The aglycone of the major components of the antibiotic XL-704 complex produced by *Streptomyces platensis* subsp. *malvims* MCRL 0388 has been found to have structure (53); minor components of the complex have structures containing the modified aglycones (54)—(56).[47]

Vermiculine. Vermiculine, an antibiotic produced by *Penicillium vermiculatum* and which had previously been assigned structure (57), has now been shown by X-ray analysis to have the constitution (58).[48] Cytotoxic effects of the antibiotic have been presented.[49]

Other Polyolefinic Macrolides. The isolation of two hexaene macrolides, designated candihexin I and candihexin II and produced by the mutant

---

[41] S. Omura, A. Nakagawa, K. Suzuki, T. Hata, A. Jakubowski, and M. Tishler, *J. Antibiotics*, 1974, **27**, 147.
[42] A. Nakagawa, K. Suzuki, K. Iwasaki, T. Hata, and S. Omura, *Chem. and Pharm. Bull. (Japan)*, 1974, **22**, 1426.
[43] S. Omura, A. Nakagawa, K. Suzuki, and T. Hata, *J. Antibiotics*, 1974, **27**, 370.
[44] S. Rakhit and K. Singh, *J. Antibiotics*, 1974, **27**, 221.
[45] S. Harada and T. Kishi, *Chem. and Pharm. Bull. (Japan)*, 1974, **22**, 99.
[46] M. Uramoto and N. Otake, *Agric. and Biol. Chem. (Japan)*, 1974, **38**, 855.
[47] A. Kinumaki, I. Takamori, Y. Sugawara, M. Suzuki, and T. Okuda, *J. Antibiotics*, 1974, **27**, 107, 117.
[48] R. K. Boeckmann, J. Fayos, and J. Clardy, *J. Amer. Chem. Soc.*, 1974, **96**, 5954.
[49] J. Fuska, L. Ivanitskaya, K. Horakova, and I. Kuhr, *J. Antibiotics*, 1974, **27**, 141.

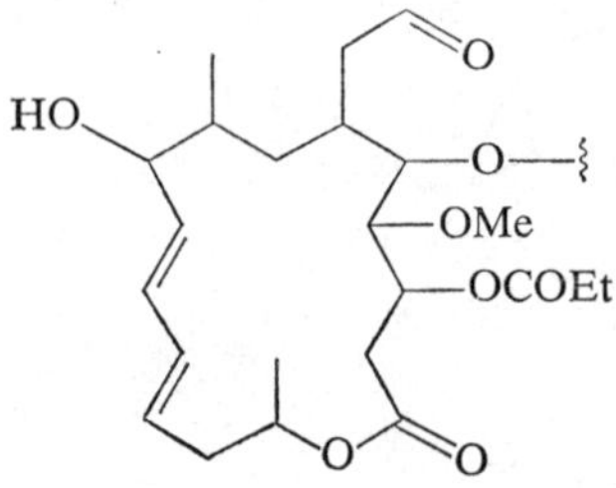

(53)    (54) 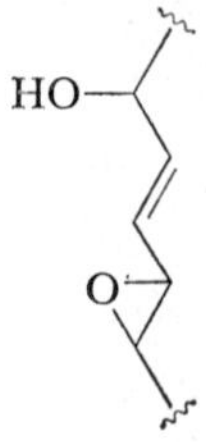   (55) 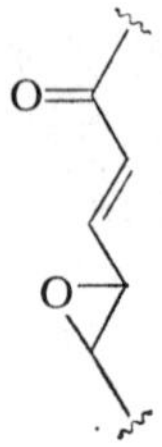   (56)

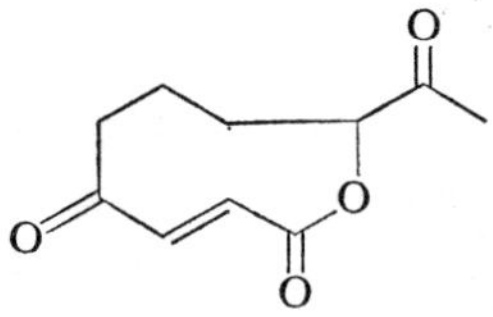

(57)

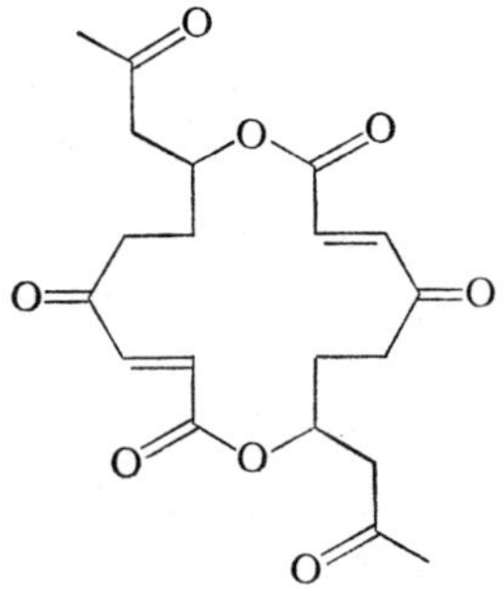

(58)

18A2 of *Streptomyces viridoflavus* IMRU 3685, has been reported,[50] and further studies of aspects of the biosynthesis of the verrucarins have been published.[51]

Polyolefinic Macrocycles (Non-macrolide).—*The Cytochalasans.* The fungal pathogen of koda millet, *Phomopsis paspalli*, has been found to produce the new cytochalasans (59) and (60);[51] the compounds have been named kodo-cytochalasin I and II, respectively. Biosynthetic experiments in cultures of *Zygosporium masonii*, with $^{14}C,^{3}H$-[52] and ^{13}C-labelled[53] precursors, have demonstrated that the cytochalasans are built up from phenylalanine, methionine, and a C_{18} or C_{16} polyketide part (see Scheme 4).

X-Ray studies have established structure (61) for cytochalasan G, isolated from an unidentified *Nigrosabulum* sp.[54]

[50] J. F. Martin and L. E. McDaniel, *J. Antibiotics*, 1974, **27**, 726.
[51] R. Achini, B. Muller, and C. Tamm, *Helv. Chim. Acta*, 1974, **57**, 1442.
[52] C. R. Lebet and C. Tamm, *Helv. Chim. Acta*, 1974, **57**, 1785.
[53] W. Graf, J. L. Robert, J. C. Vederas, C. Tamm, P. H. Solomon, I. Miura, and K. Nakanishi, *Helv. Chim. Acta*, 1974, **57**, 1801.
[54] A. F. Cameron, A. A. Freer, B. Hesp, and C. J. Strawson, *J.C.S. Perkin II*, 1974, 1741.

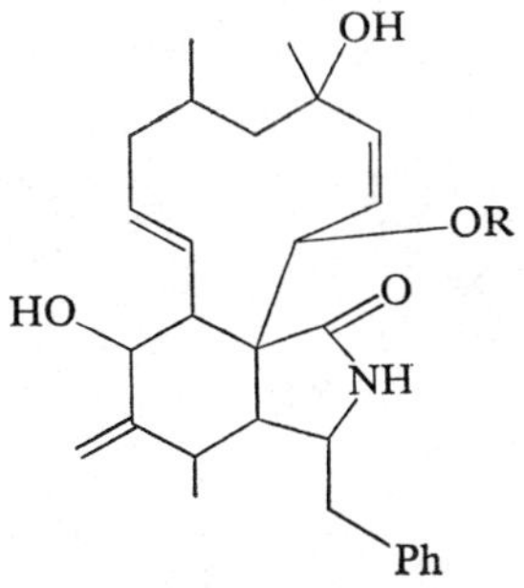

(59) R = Ac
(60) R = H

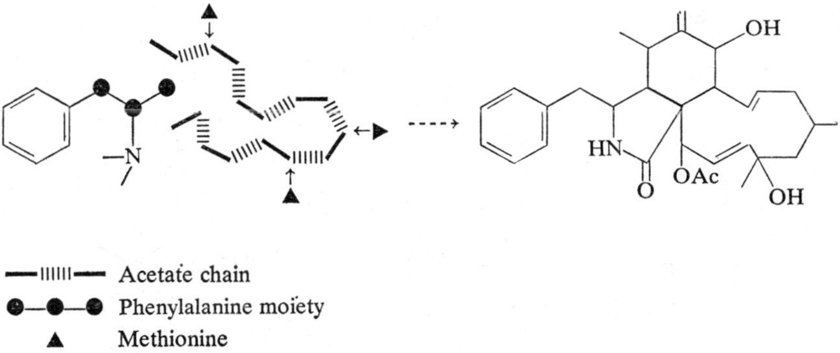

──||||||── Acetate chain
●─●─● Phenylalanine moiety
▲ Methionine

Scheme 4

Ansamycin Antibiotics.—The 'halomycins', a new group of antibiotics produced by *Micromonospora halophytica*, have now been added to the list of ansa macrolides.

Maytansines (Maytansinoides). Four new maytansinoides, maytanvaline (62), maysine (63), normaysine (64), and mayseine (65), have been found in the alcoholic extract of wood and bark of *Maytenus buchananii* (Loes) R. Wilczek.[55] Meyers and Shaw[56] have outlined an efficient route to the cyclic carbinolamide moiety (66) found in the maytansinoides.

Rifamycins. The structure of rifamycin W, isolated from a mutant strain of *Nocardia mediterranei*, has been shown to be (67).[57] Biosynthetic studies

---

[55] S. M. Kupchan, Y. Komoda, A. R. Branfman, R. G. Dailey, and V. A. Zimmerby, *J. Amer. Chem. Soc.*, 1974, **96**, 3706.
[56] A. I. Meyers and C. C. Shaw, *Tetrahedron Letters*, 1974, 717.
[57] E. Martinelli, G. G. Gallo, P. Antonini, and R. J. White, *Tetrahedron*, 1974, **30**, 3087.

(61)

(62)

(63) R = Me
(64) R = H

(65)

(66)

(67)

Naturally Occurring Polyolefinic and Polyacetylenic Compounds 219

with ^{13}C-labelled acetates and propionates have established that the alignment of acetate and propionate units in rifamycin W corresponds to that previously found in rifamycin S.[58] Rifamycin W is suggested to be an intermediate in the biosynthesis of other rifamycins.

Discrepancies in the literature have led Beynon and co-workers[59] to re-examine the [13]C n.m.r. assignments given previously to rifamycin S, and all ambiguities now seem to have been ironed out. The interesting ^{13}C data recorded for Ca- and Na-complexed rifamycin S have been discussed,[60] and some chemistry of derivatives of rifamycin has been reported.[61] White and collaborators[62] have discussed further aspects of the biosynthetic origin of the ansa chain of the antibiotics.

Halomycins. The halomycins are a new group of antibiotics produced by *Micromonospora halophytica* and are highly active against gram positive bacteria. From chemical and spectral data, Ganguly and co-workers[63] have proposed structure (68) for halomycin B. Halomycin B yields rifamycin S on treatment with nitrous acid.

Geldanamycin. Studies of the biosynthesis of geldanamycin in *Streptomyces hygroscopicus* var. *geldanus* var. *nova* have established that the antibiotic is

(68)

[58] R. J. White, E. Martinelli, and G. Lancini, *Proc. Nat. Acad. Sci. U.S.A.*, 1974, **71**, 3260.
[59] E. Martinelli, R. J. White, G. G. Gallo, and P. J. Beynon, *Tetrahedron Letters*, 1974, 1367.
[60] D. Liebfritz, *Tetrahedron Letters*, 1974, 4125.
[61] R. Crichio, G. Lancini, G. Tamborini, and P. Sensi, *J. Medicin. Chem.*, 1974, **17**, 396.
[62] A. Karlsson, G. Sartori, and R. J. White, *European J. Biochem.*, 1974, **47**, 251.
[63] A. K. Ganguly, S. Szmulewicz, O. Z. Sarre, D. Greeves, J. Morton, and J. McGlotten, *J.C.S. Chem. Comm.*, 1974, 395.

formed in a closely similar manner to that found for streptovaricin and rifamycin (see Vol. 3, pp. 280—282 and Scheme 5.)[64]

Pyran-pyranoid Compounds.—*Aurovertin B.* Structure (69) has been proposed for aurovertin B, a metabolite from the fungus *Calcarisporium arbuscula* (*Preuss*).[65] The structure is closely similar to that determined for citreoviridin.

Aureothin. Cardillo *et al.*[66] have presented evidence which shows that the nitro-aromatic portion of aureothin (70) is derived from *p*-aminophenylalanine; the biosynthetic operation takes place through β-hydroxylation to *p*-aminoserine. In a separate study, it was established that oxidation of the *p*-NH_2 to *p*-NO_2 takes place very late in the overall sequence leading to (70).

Pestalotin (LL-P880α). Pestalotin (71), a gibberellin synergist isolated from *Pestalotia cryptomeriaecola* has been shown to be identical with the metabolite LL-P880α found in an unidentified *Penicillium*.[67] Two groups[68,69] have

(69)

---

[64] R. D. Johnson, A. Haber, and K. L. Rinehart, jun., *J. Amer. Chem. Soc.*, 1974, **96**, 3316.
[65] L. J. Mulheirn, R. B. Beechey, D. P. Leworthy, and M. D. Osselton, *J.C.S. Chem. Comm.*, 1974, 874.
[66] R. Cardillo, C. Fuganti, D. Ghiringhelli, D. Giangrasso, P. Grasselli, and A. Santopietro-Amisano, *Tetrahedron*, 1974, **30**, 459.
[67] G. M. Strunz, C. J. Heissner, M. Kakushima, and M. A. Stillwell, *Canad. J. Chem.*, 1974, **52**, 825.
[68] Y. Kimura and S. Tamura, *Agric. and Biol. Chem. (Japan)*, 1974, **38**, 875.
[69] D. Seebach and H. Meyer, *Angew. Chem. Internat. Edn.*, 1974, **13**, 77.

(70), (71), (72), (73)

independently outlined a convenient synthesis of (71) and of some of its analogues. Pestalotin and the related pyrone (72) have also been found in another unidentified fungus.

Phacidin. Structure (73) has been determined for phacidin, isolated from the liquid culture of the canker fungus, *Potebniamyces balsamicola* Smerlis var *boycei* Funk.[70] The metabolite exhibits strong inhibitory effects on the growth of a variety of fungi.

[70] G. A. Poulton, M. E. Williams, and E. E. McMullan, *Tetrahedron Letters*, 1974, 2611.

(74)   (75)

Metabolite LL-Z 1220. Borders and Lancaster[71] have reported full details of the structural elucidation of antibiotic LL-Z 1220, the first natural product containing a benzene dioxide moiety, from an unidentified fungus (see Vol. 2, p. 245).

Other Pyran-pyranoid Compounds. Structure (74) has been proposed for ochrephilone from *Penicillium multicolar* NRRL 2060,[72] and the pyrone structure (75) has been determined for a metabolite designated LL-N 313 from *Sporormia affinis*.[73] Dean and co-workers[74] have developed a useful synthesis of di-*O*-methylcitromycin (76), and Vleggaar *et al.*[75] have identified the fungal metabolite austdiol, from *Aspergillus ustus* as (77).

(76)   (77)

N-Heterocyclic Polyolefinic Compounds.—*Antibiotic X-5108 (Goldinodox) and Mocimycin (Delvomycin).* The Hoffman-La Roche group have examined several stereochemical features of the antibiotics X-5108 and mocimycin.[76,77] X-Ray studies on a derivative of a degradation product of X-5108, which contained the tetrahydrofuryl moiety intact, have established that the full absolute stereochemistry of X-5108 and mocimycin is as shown in formula (78).[78]

Tenellin and Bassiamin. Structures (79a) and (79b) have been determined for

[71] D. B. Borders and J. E. Lancaster, *J. Org. Chem.*, 1974, **39**, 435.
[72] H. Seto and M. Tanabe, *Tetrahedron Letters*, 1974, 651.
[73] W. J. McGahren, G. A. Ellestad, J. E. Lancaster, G. O. Morton, and M. P. Kunstmann, *J. Amer. Chem. Soc.*, 1974, **96**, 1616.
[74] F. M. Dean, S. Murray, and W. Taylor, *J.C.S. Chem. Comm.*, 1974, 440.
[75] R. Vleggaar, P. S. Steyn, and D. W. Nagel, *J.C.S. Perkin I*, 1974, 45.
[76] H. Maehr, J. F. Blount, M. Leach, and A. Stempel, *Helv. Chim. Acta*, 1974, **57**, 936.
[77] H. Maehr, T. H. Williams, M. Leach, and A. Stempel, *Helv. Chim. Acta*, 1974, **57**, 212.
[78] H. Maehr, M. Leach, J. F. Blount, and A. Stempel, *J. Amer. Chem. Soc.*, 1974, **96**, 4034.

(78) R = Me or H

(79) a; $n = 1$
     b; $n = 2$

tenellin and bassiamin, respectively, two pigments isolated from the insect pathogenic fungi *Beauveria tenella* (Delacroix) Siem, and *B. bassiana* (Bals) Vuill.[79] Studies of the biosynthesis[80] of tenellin in cultures of *B. bassiana*, using singly and doubly ^{13}C-labelled precursors, have established that the molecule is derived from phenylalanine, methionine, and acetate.

Butenolide Metabolites.—*Penicillic Acid.* Two groups have independently investigated the biosynthesis of penicillic acid (81) in *Penicillium cyclopium*. In the first application of ^{3}H n.m.r. spectroscopy to a biosynthetic investigation, Thomas and co-workers[81] examined the incorporation of [^{3}H]acetate by *P. cyclopium*. These studies established that the intermediate (80) undergoes 4,5-cleavage in producing penicillic acid, and the mechanism shown in Scheme 6 was proposed. Similar conclusions were arrived at independently by Seto *et al.*[82] based on consideration of ^{13}C–^{13}C couplings in (81) formed from [^{13}C]acetates.

Multicolic and Multicolosic Acids. Structures (82) and (83) have been determined for two new tetronic acids, designated multicolic acid and multicolosic acid respectively, isolated from *Penicillium multicolar*.[83] Biosynthetic studies, with singly and doubly labelled [^{13}C]acetate, have suggested

---
[79] A. G. McInnes, D. G. Smith, C. K. Wat, L. C. Vining, and J. L. C. Wright, *J.C.S. Chem. Comm.*, 1974, 281.
[80] A. G. McInnes, D. G. Smith, J. A. Walter, L. C. Vining, and J. L. C. Wright, *J.C.S. Chem. Comm.*, 1974, 282.
[81] J. M. A. Al-Rawi, J. A. Elvidge, D. K. Jaiswal, J. R. Jones, and R. Thomas, *J.C.S. Chem. Comm.*, 1974, 220.
[82] H. Seto, L. W. Cary, and M. Tanabe, *J. Antibiotics*, 1974, **27**, 558.
[83] J. A. Gudgeon, J. S. E. Holker, and T. J. Simpson, *J.C.S. Chem. Comm.*, 1974, 636.

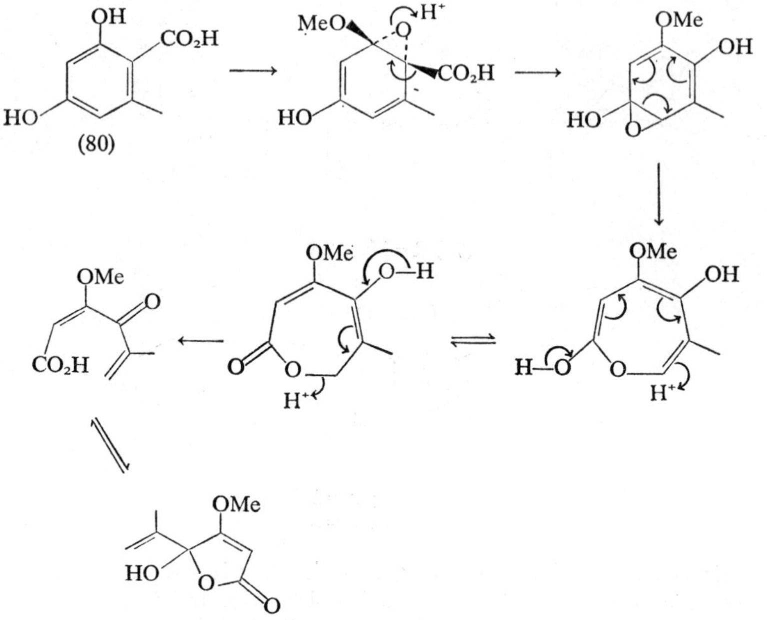

(80) → ... → (81)

Scheme 6

that the two acids are derived from acetate *via* specific 4,5-oxidative fission of a preformed aromatic precursor (Scheme 7).

Patulin. Further experiments by Scott on the biosynthesis of patulin (84) in *Penicillium patulum* have supported the view that a dioxygenase mechanism applies for its biosynthesis from gentisaldehyde (Scheme 8).[84]

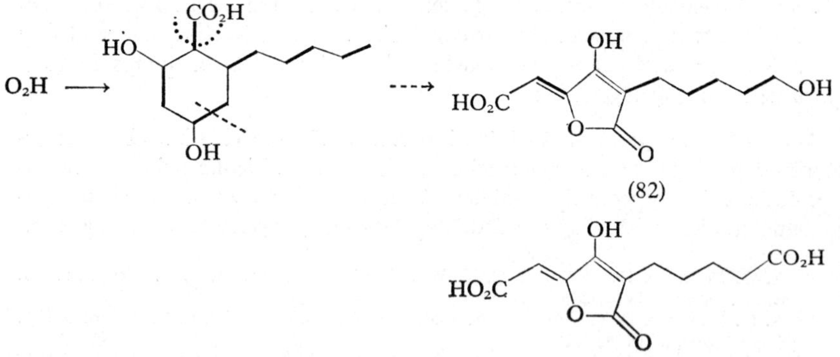

(82)

(83)

Scheme 7

[84] A. I. Scott and L. Beadling, *Bioorganic Chemistry*, 1974, **3**, 281.

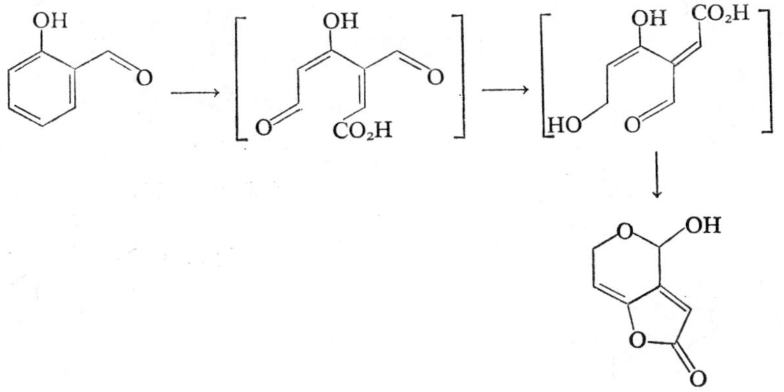

Scheme 8

Independent studies by Lynen and co-workers[85] have shown that hydroxymethylation of m-cresol to m-hydroxybenzyl alcohol probably represents an important stage in the sequence leading to patulin biosynthesis.

Miscellaneous Polyolefinic Microbial Metabolites.—Sasaki *et al.*[86] have determined structure (85) for a new metabolite from the fermented broth of the fungus *Ascochyta viciae* and the new metabolites (86) and (87) have been isolated from *Colletotrichum nicotianae*.[87]

A new antibiotic, named magnesidin, has been isolated from *Pseudomonas magnesiorubia*.[88] The antibiotic has been shown to be a 1:1 mixture of the magnesium salts of the two tetramic acids (88a) and (88b).

The novel cyclobutane structure (89) has been established for moniliformin from *Fusarium moniliforme*,[89] and the diene (90) has been isolated from culture broth of actinomyces strain MD736-C6.[90]

Arison and Omura[91] have presented new evidence to support the revised structure (91) for cerulenin, an antifungal antibiotic from *Cephalosporium caerulens*.

The absolute configuration of oryzoxymycin (92) from *Streptomyces*

---

[85] G. Murphy, G. Vogel, G. Krippahl, and F. Lynen, *European J. Biochem.*, 1974, **49**, 443.
[86] H. Sasaki, T. Hosokawa, Y. Nawata, and K. Ando, *Agric. and Biol. Chem. (Japan)*, 1974, **38**, 1463.
[87] Y. Kosuge, A. Suzuki, and S. Tamura, *Agric. and Biol. Chem. (Japan)*, 1974, **38**, 1265, 1553.
[88] H. Kohl, S. V. Bhat, J. R. Patell, N. M. Gandhi, J. Nazareth, P. V. Divekar, N. J. de Souza, H. G. Berscheid, and H. W. Fehlhaber, *Tetrahedron Letters*, 1974, 983.
[89] J. P. Springer, J. Clardy, R. J. Cole, J. W. Kirksey, R. K. Hill, R. M. Carlson, and J. L. Ididor, *J. Amer. Chem. Soc.*, 1974, **96**, 2267.
[90] Y. Kumada, H. Naganawa, M. Hamada, T. Takeuchi, and H. Umezawa, *J. Antibiotics*, 1974, **27**, 726.
[91] B. H. Arison and S. Omura, *J. Antibiotics*, 1974, **27**, 28.

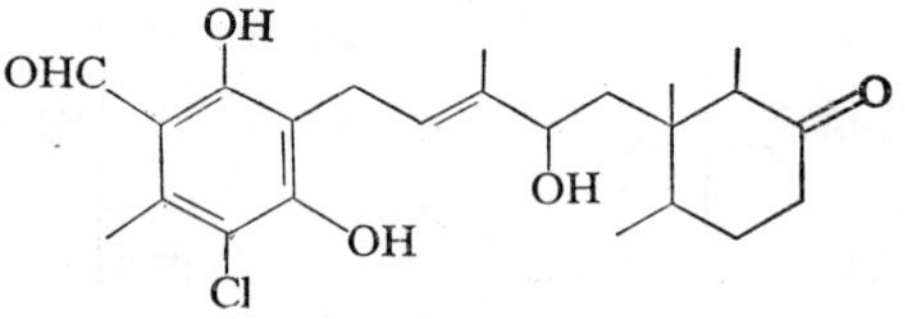

(85)

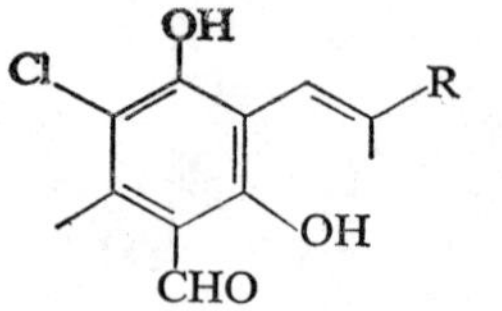

(86) a; R = Me
b; R = [structure]

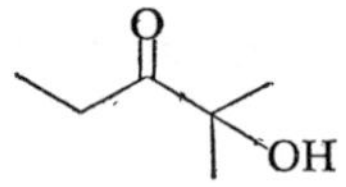

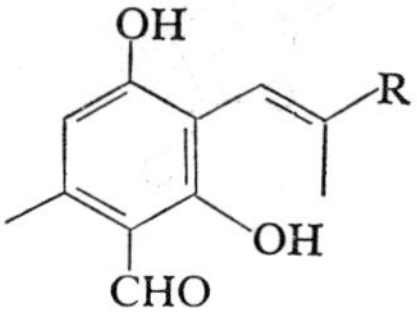

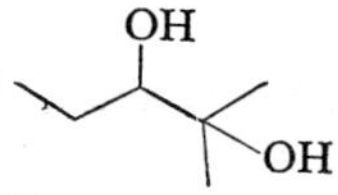

(87) a; R = [structure]
b; R = [structure]

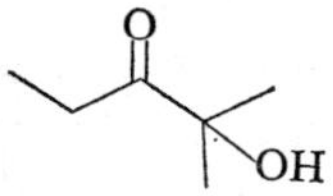

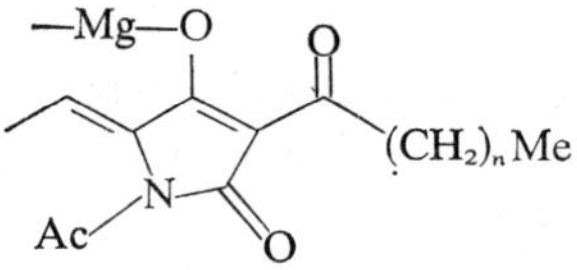

(88) a; n = 4
b; n = 6

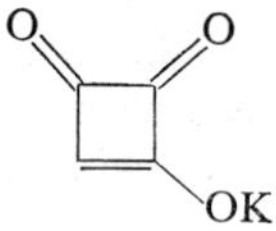

(89)

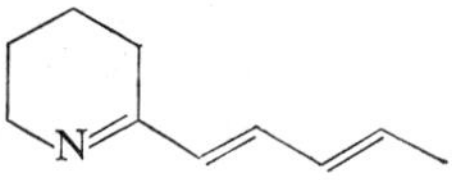

(90)

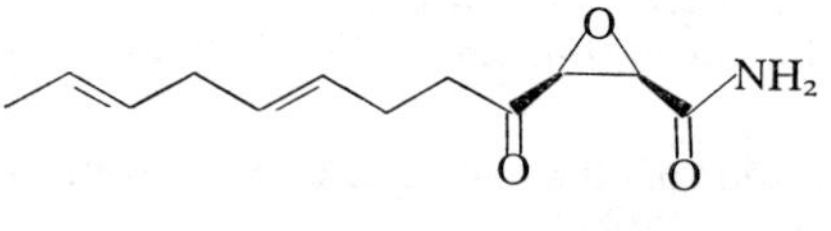

(91)

venezuelae var. *oryzoxymyceticus* has been shown to be $2'R[5S,6S]$,[92] and structures (93) and (94) have been assigned to methylenomycins A and B, respectively, two antibiotics produced by *Streptomyces violaceoruber* No. 2416.[93] Steglich *et al.*[94] have outlined the application of n.m.r. to the determination of the structures of some aryl-hydroxylated pulvinic acids. The

[92] T. Hashimoto, S. Kondo, H. Naganawa, T. Takita, K. Maeda, and H. Umezawa' *J. Antibiotics*, 1974, **27**, 86.
[93] T. Haneishi, A. Terahara, M. Arai, T. Hata, and C. Tamura, *J. Antibiotics*, 1974, **27**, 393.
[94] W. Steglich, H. Besl, and K. Zipfel, *Z. Naturforsch.*, 1974, **29b**, 1.

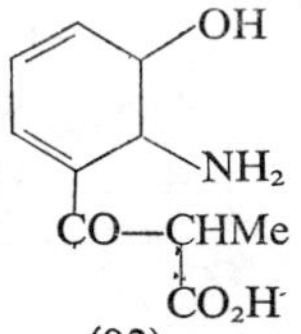

(92)

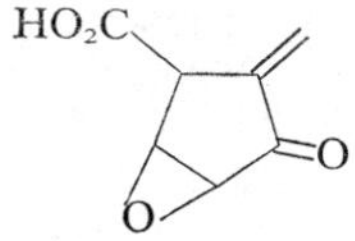

(93)

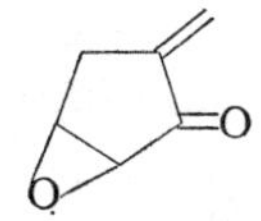

(94)

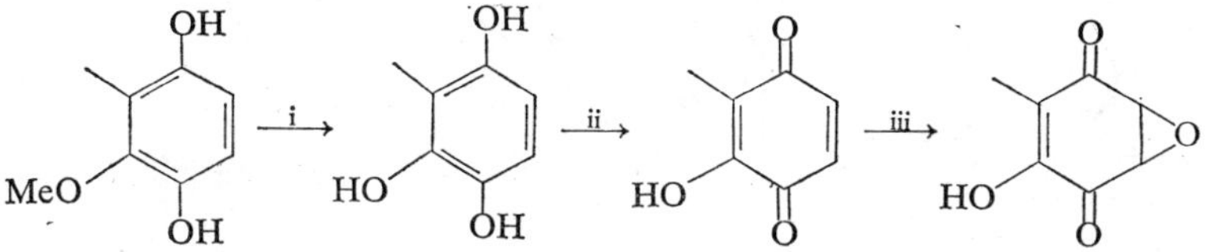

(95)

Reagents: i, BBr_3; ii, Ag_2O; iii, $NaBO_4$

Scheme 9

first synthesis of terreic acid (95), an antibiotic produced by *Aspergillus terreus*, has been reported (Scheme 9).[95]

6 Cyclopentenone Polyolefinic Compounds

Two new sub-sections, covering the chemistry of terrein and cryptosporiopsin, have been introduced this year.

Jasmone.—Ho[96] has summarized, in flow-chart format, the various approaches which have been employed in the synthesis of jasmone; almost fifty approaches are covered. Four new synthetic approaches to jasmone were published during 1974, and several papers outlined new routes to dihydrojasmone.

Wakamatsu and co-workers[97] have described a novel five-stage synthesis of the 1,4-diketone precursor (99) of jasmone (100), which starts from methyl levulinate (Scheme 10). Acyloin condensation of methyl levulinate ethylene acetal in the presence of trimethylchlorosilane first gave (96), which on addition of base and *Z*-1-bromohex-3-ene led to (97). Reduction of (97) with sodium borohydride, followed by oxidation of the resulting diol with lead tetra-acetate, produced (98), which gave the dione (99) on hydrolysis.

In a new approach to jasmone, outlined by McCurry and Abe,[98] use was

[95] J. C. Sheehan and Y. S. Lo, *J. Medicin. Chem.*, 1974, **17**, 371.
[96] T. L. Ho, *Synth. Comm.*, 1974, **4**, 265.
[97] T. Wakamatsu, K. Akasaka, and Y. Ban, *Tetrahedron Letters*, 1974, 3883.
[98] P. M. McCurry, jun. and K. Abe, *Tetrahedron Letters*, 1974, 1387.

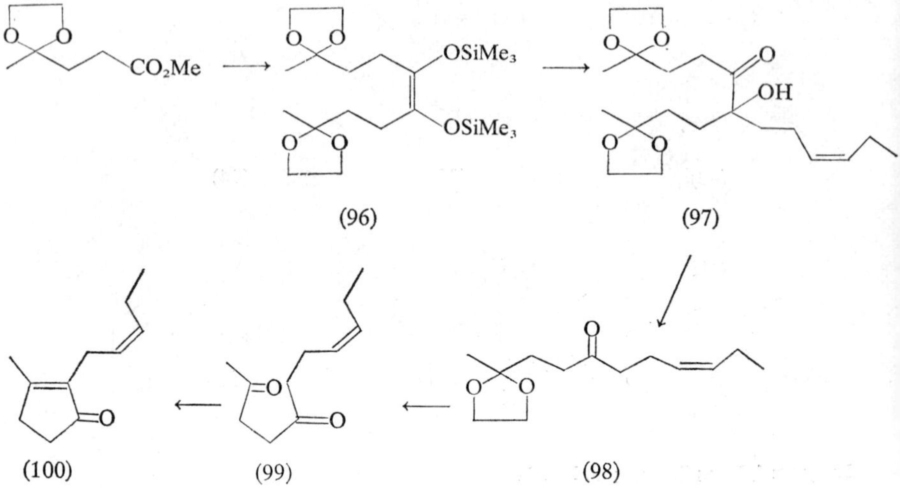

(96) (97)

(100) (99) (98)

Scheme 10

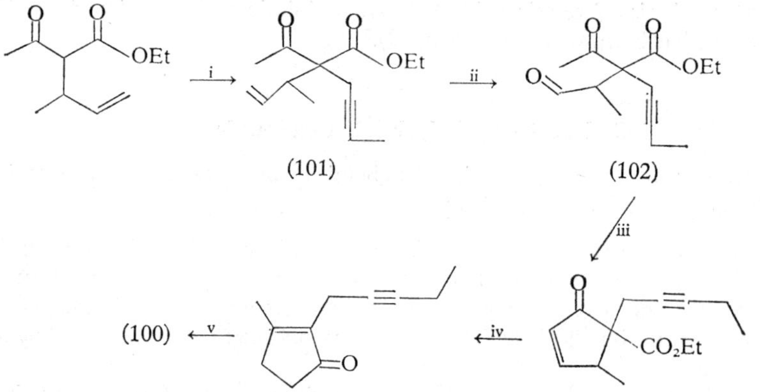

(101) (102)

(100)

Reagents: i, NaH–BrCH$_2$–≡–Et; ii, O$_3$, –78°C, Zn–HOAc; iii, piperidinium acetate; iv, DMSO–NaCl; v, H$_2$–Lindlar catalyst

Scheme 11

made of the selective ozonolysis of an enyne intermediate, *viz*.(101), to generate the desired functionality in the acyclic precursor (102) prior to cyclopentenone ring-formation (Scheme 11).

In separate studies of jasmone synthesis, McCurry and Singh[99] have shown that cyclization of 1,4-diones, *e.g.* (104), in aqueous ethanolic base is a

[99] P. M. McCurry, jun. and R. K. Singh, *J. Org. Chem.*, 1974, **39**, 2317.

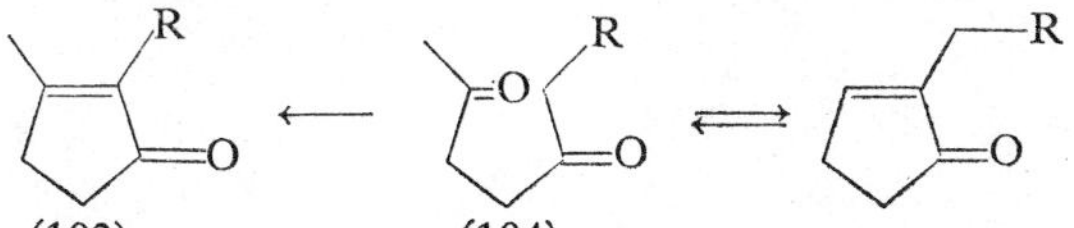

Scheme 12

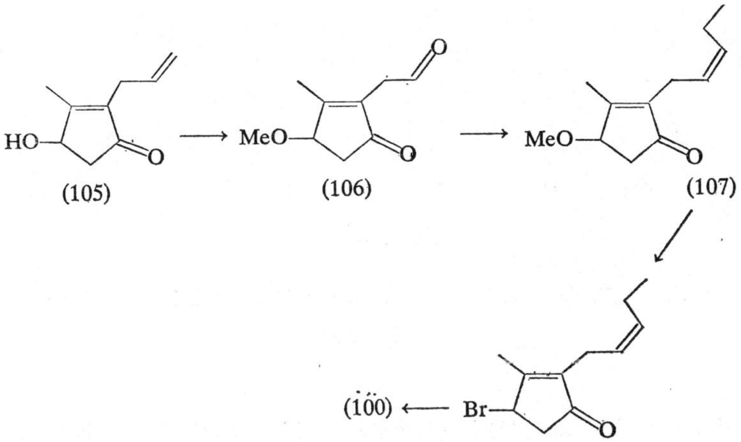

kinetically controlled reaction, giving the cyclopentenone (103) as the major product.

Pattenden and Storer[100] have described the application of the readily available alcohol (105) to a new synthesis of Z-jasmone (Scheme 12). Conversion of the methyl ether derived from (105) into the aldehyde (106), followed by Wittig reaction, produced the Z-alkene (107) almost exclusively. Treatment of (107) with BBr$_3$ then led to bromide (108) which on reduction gave (100). In another new approach (Scheme 13) Bahurel[101] and co-workers have applied the interesting *cis*-vinylspirannic ketone (109) → *cis*-dienone (110) conversion to obtain the desired cyclopentenone functionality.

Several groups of workers have outlined new approaches to dihydro-jasmone (115) during the period covered by this Report.[102-107] A particular

[100] G. Pattenden and R. Storer, *J.C.S. Perkin I*, 1974, 1603.
[101] Y. Bahurel, L. Cottier, and G. Descotes, *Synthesis*, 1974, 118.
[102] A. Barco, S. Benetti, and G. P. Pollini, *Synthesis*, 1974, 33.
[103] T. Cuvigny, M. Larchevêque, and H. Normant, *Tetrahedron Letters*, 1974, 1237.
[104] U. Ravid and R. Ikan, *J. Org. Chem.*, 1974, **39**, 2637.
[105] P. A. Grieco and C. S. Pogonowski, *J. Org. Chem.*, 1974, **39**, 732.
[106] I. Kawamoto, S. Muramatsu, and Y. Yura, *Tetrahedron Letters*, 1974, 4223.
[107] T. Hiyama, M. Tsukanaka, and H. Nozaki, *J. Amer. Chem. Soc.*, 1974, **96**, 3713.

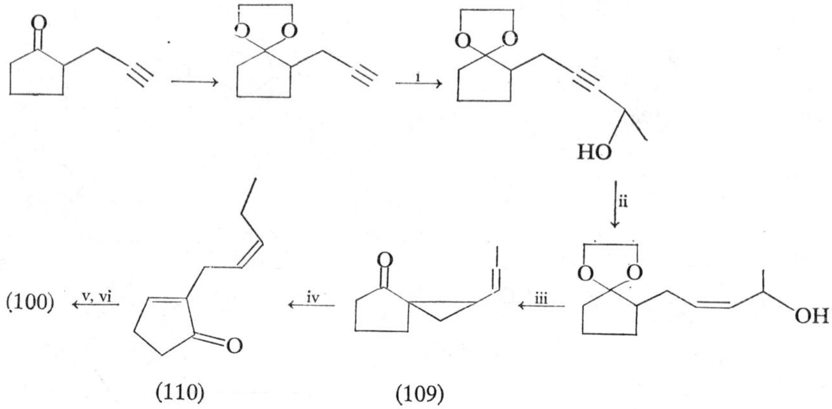

Reagents: i, RMgX, MeCHO; ii, H$_2$-Lindlar catalyst; iii, DCC, Δ; iv, Δ; v, LiMe; vi, CrO$_3$

Scheme 13

novel approach[107] to (115) involves the addition of dichlorocarbene to the unsaturated alcohol (111) followed by treatment of the resulting dichlorocyclopropane with 47% HCl at 100 °C. The ring-opening–ring-formation sequence from (112) to (115) is thought to occur *via* conjugate dehydration [to (113)], ionization [to (114)], and thermal conrotatory ring-closure to a cyclopentenyl cation. Deprotonation and hydrolysis then lead to (115) (Scheme 14). The route, unfortunately, is probably not applicable to

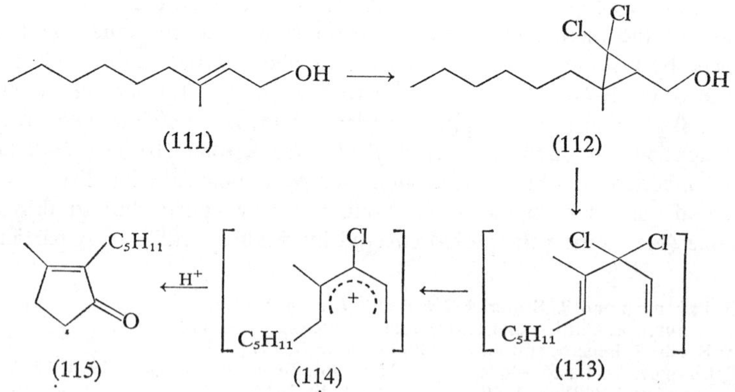

Scheme 14

Naturally Occurring Polyolefinic and Polyacetylenic Compounds 231

(116)   (117)   (118)

jasmone synthesis because of isomerization of the side-chain double bond during the acid-treatment step.

Three new jasmone analogues (116)—(118) have been found in trace amounts in jasmin absolute prepared from a concrète of Italian origin.[108]

Terrein.—A synthesis of racemic terrein (120), a metabolite of *Aspergillus terreus*, has been achieved starting from the epoxycyclopentane (119) (Scheme 15).[109]

Cryptosporiopsin.—Strunz and co-workers[110] have isolated the new metabolite (121) together with cryptosporiopsin (122) from cultures of a *Cryptosporiopsis* sp.

Rethrolones.—Romanet and Schlessinger[111] have outlined a highly efficient synthesis (*ca.* 75% overall yield) of cinerolone (126a) which proceeds according to Scheme 16. Thiomethylacetone, keten thioacetal monoxide (123) and Z-iodobut-2-ene first reacted together in the presence of a base to give (124). Hydrolysis of (124) produced the keto-aldehyde (125) which was converted into Z-cinerolone (126a) on treatment with KOBu^t.

The alcohol (105) has been used as starting material in another route to (126a) and to jasmololone (126b).[100] Conversion of (105) into the bromide (108), as outlined in Scheme 12, followed by hydrolysis, produces jasmololone (126b); application of ethylphosphorylide instead of the propyl ylide in stage (106) → (107) leads by a similar sequence to cinerolone (126a).

Pattenden and Storer[112] have published full details of the novel acid-catalysed rearrangements of rethrolones outlined last year (Vol. 3, p. 297).

Hop Constituents.—Structures (128)—(132) have been assigned to seven new oxidation products of colupulone (127).[113-115] In addition, oxidation of

---

[108] R. Kaiser and D. Lamparsky, *Tetrahedron Letters*, 1974, 3413.
[109] J. Auerbach and S. M. Weinreb, *J.C.S. Chem. Comm.*, 1974, 298.
[110] G. M. Strunz, P. I. Kazinoti, and M. A. Stillwell, *Canad. J. Chem.*, 1974, **52**, 3623.
[111] R. F. Romanet and R. H. Schlessinger, *J. Amer. Chem. Soc.*, 1974, **96**, 3701.
[112] G. Pattenden and R. Storer, *J.C.S. Perkin I*, 1974, 1606.
[113] J. Coates, D. R. J. Laws, and J. A. Elvidge, *J.C.S. Perkin I*, 1974, 36.
[114] R. van den Bossche, M. Anteunis, and M. Verzele, *Bull. Soc. chim. belges*, 1974, **83**, 77, 277.
[115] R. van den Bossche, M. Anteunis, F. Borremans, and M. Verzele, *Bull. Soc. chim. belges*, 1974, **83**, 271.

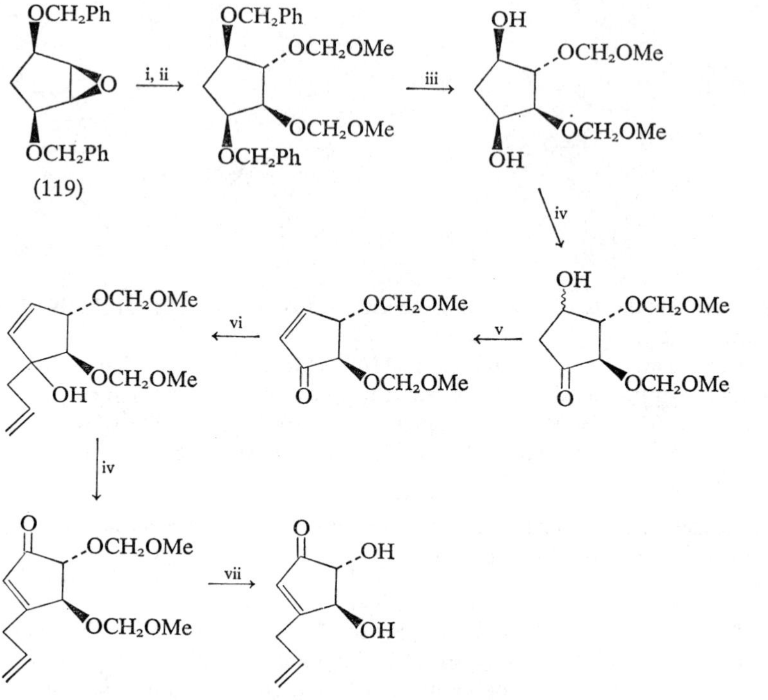

(119)

(120)

Reagents: i, H^+; ii, $ClCH_2OMe$; iii, Na–NH_3; iv, Jones oxidation; v, Ac_2O–C_5H_5N; vi, CH_2=$CHCH_2MgX$; vii, NaOH

Scheme 15

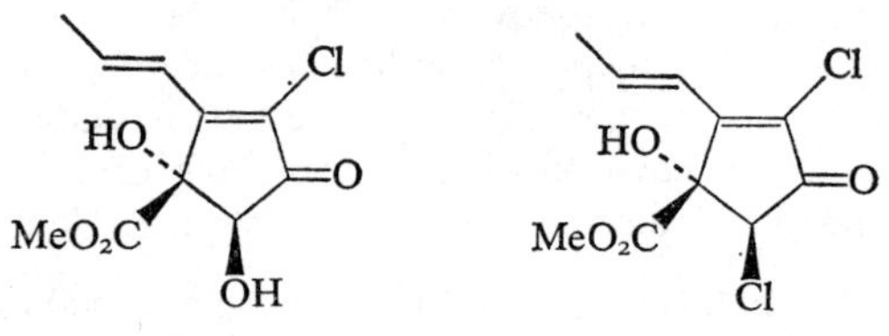

(121) (122)

humulone (133) in acid at 0 °C with an excess of lead tetra-acetate, has been found to produce the novel tricyclic structure (134);[116] interestingly this compound has been detected in samples of stored hops (*ca.* 0.3%) and also in beers (*ca.* 4 p.p.m.).

Biosynthetic studies have established that acetic acid is incorporated into

[116] J. A. Elvidge, D. R. J. Laws, J. D. McGuinness, and P. V. R. Shannon, *Chem. and Ind.*, 1974, 573.

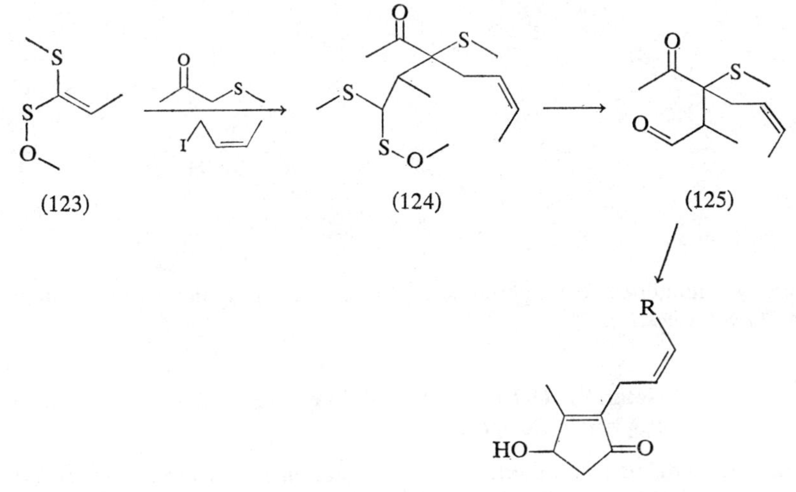

(123)   (124)   (125)

(126) a; R = Me
      b; R = Et

Scheme 16

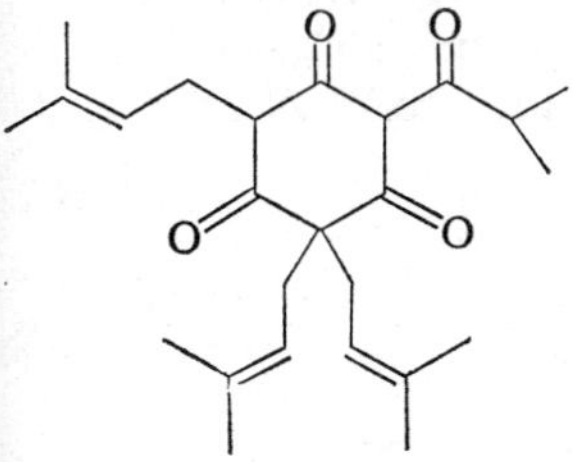

(127) 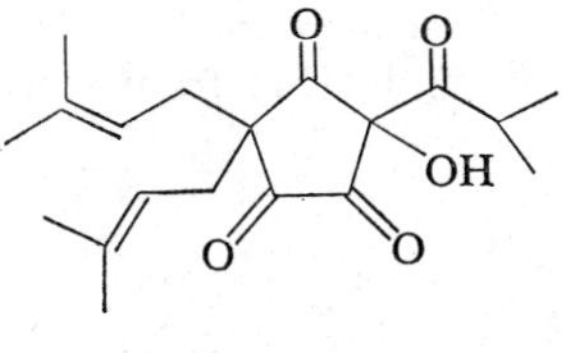   (128) 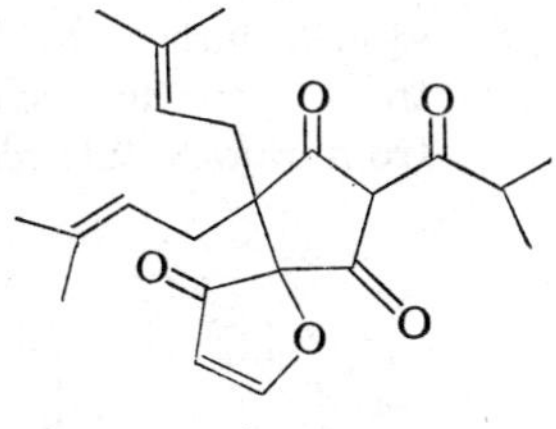   (129)

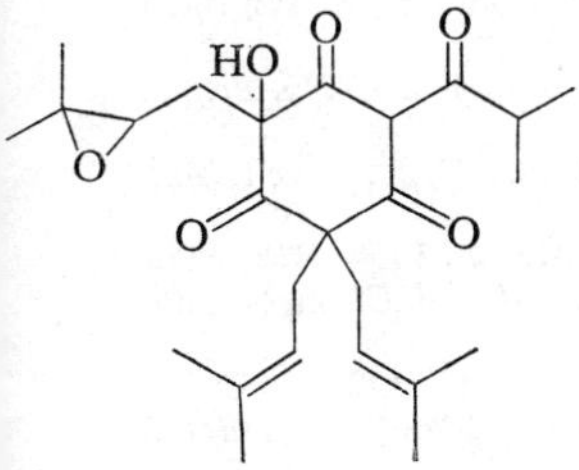

(130) 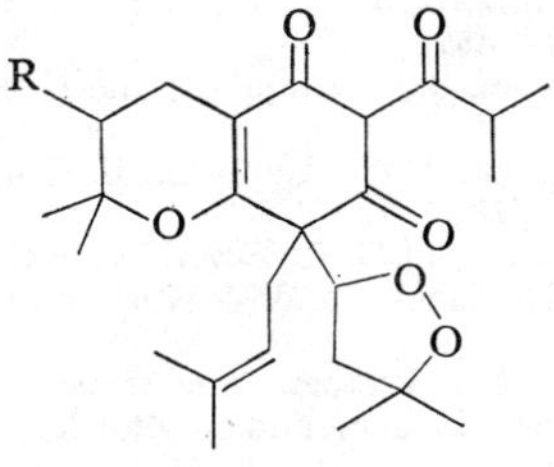   (131) a; R = OH
              b; R = OOH
        (132) a; R = OH 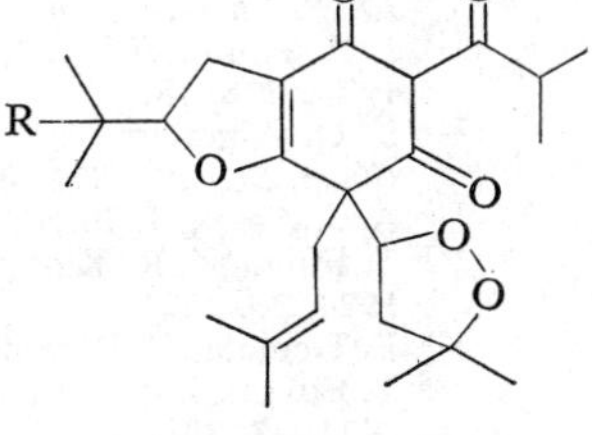
              b; R = OOH

(133)   (134)

desoxycohumulone, desoxyhumulone, humulone, colupulone, and lupulone in *Humulus lupulus*.[116a]

7 Naturally Occurring Polyolefinic Degraded and/or Modified Isoprenoid Compounds

Edulans.—Edulan I and edulan II, trace components in the flavour of the purple passion fruit *Passiflora edulis* Sims, have now been shown to be the isomeric tetrahydrobenzpyrans (135) and (136), respectively;[117] the structures followed from total syntheses, and the configurations from the results of microhydrogenation data.

Constituents of Tobacco.—Several new compounds have been found in Greek, Burley, and Turkish tobaccos. Compounds (137)—(141) were isolated from Greek tobacco, *Nicotiana tabacum* L,[118-120] compounds (142)—(150) from Burley tobacco,[121-123] and (151) from Turkish tobacco.[124] A novel

(135)   (136)

[116a] F. Drawert and J. Beier, *Phytochemistry*, 1974, **13**, 2749.
[117] D. R. Adams, S. P. Bhatnagar, R. C. Cookson, G. Stanley, and F. B. Whitfield, *J.C.S. Chem. Comm.*, 1974, 469.
[118] J. R. Hlubucek, A. J. Aasen, S. O. Almqvist, and C. R. Enzell, *Acta Chem. Scand.*, 1974, **B28**, 18.
[119] S. O. Almqvist, A. J. Aasen, J. R. Hlubucek, B. Kimland, and C. R. Enzell, *Acta Chem. Scand.*, 1974, **B28**, 528.
[120] A. J. Aasen, J. B. Hlubucek, and C. R. Enzell, *Acta Chem. Scand.*, 1974, **B28**, 285.
[121] T. Fujimori, R. Kasuga, H. Kaneko, and M. Noguchi, *Agric. and Biol. Chem. (Japan)*, 1974, **38**, 2293.
[122] E. Demole, C. Demole, and D. Berthet, *Helv. Chim. Acta*, 1974, **57**, 192.
[123] T. Fujimori, R. Kasuga, M. Noguchi, and H. Kaneko, *Agric. and Biol. Chem. (Japan)*, 1974, **38**, 891.
[124] T. Chuman, H. Kaneko, T. Fukuzumi, and M. Noguchi, *Agric. and Biol. Chem. (Japan)*, 1974, **38**, 2295.

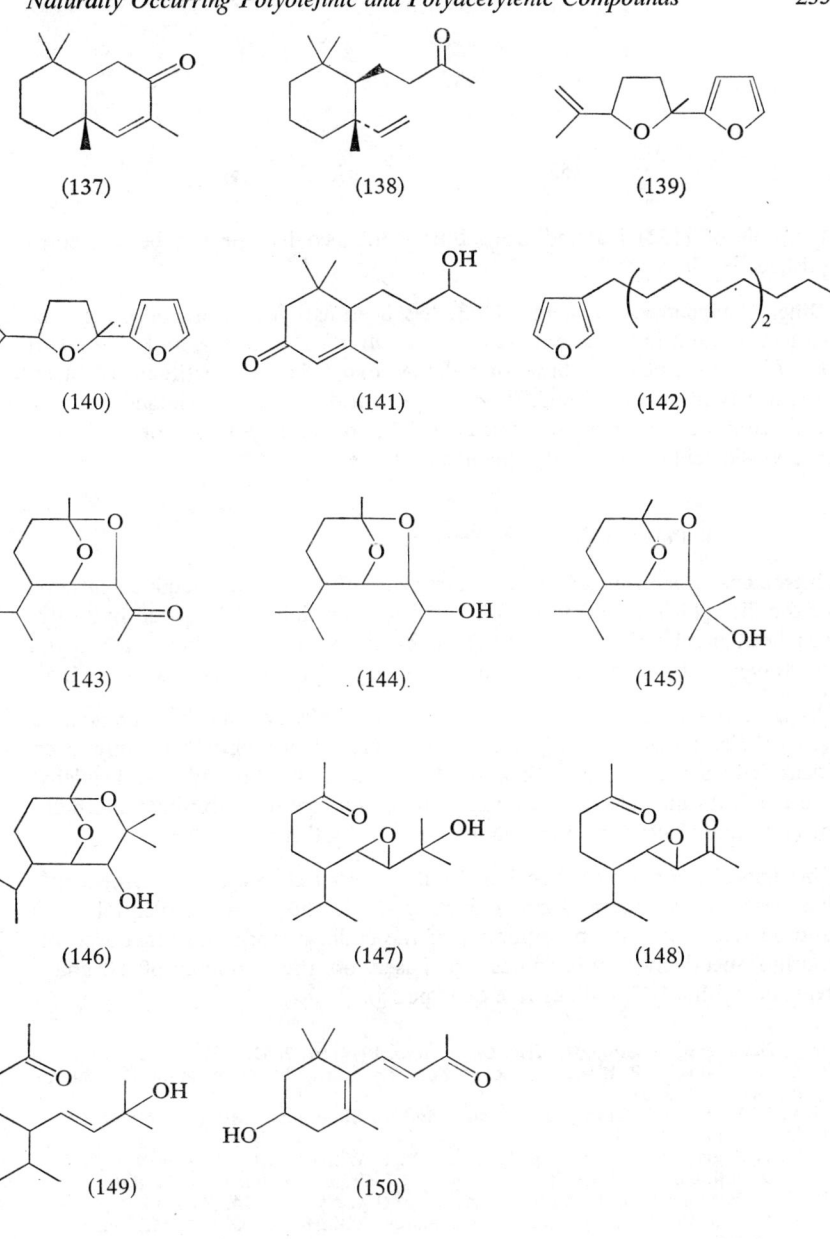

(153)   (154)

synthesis of (152) isolated from Burley tobacco flavour has been accomplished.[125]

Other Compounds.—Structure (153) has been assigned to an odorous constituent present in the essential oils of *Cedrus atlantica Manet* and *C. deodara Lond*,[126] and a new synthesis of α-damascone (154), a constituent of black tea, has been accomplished.[127] Mass spectral data for abscisic acid and related compounds have been discussed,[128] and more syntheses of analogues of abscisic acid have been forthcoming.[129]

8 Polyolefinic Insect Pheromones

Pheromonal Secretion of Queen Butterfly.—A new stereoselective synthesis of the diol (156), found in the pheromonal secretion of the queen butterfly, has been described.[130] The key stage in this synthesis (Scheme 17) is the rearrangement of the intermediate allyl siloxyvinyl ether derived from (155).

Propylure.—The true pheromone of the pink bollworm moth, *Pectinophora gossypiella*, previously assigned structure (157) ('propylure') has now been identified as a mixture of Z,Z- and Z,E-isomers of hexadeca-7,11-dienyl acetate (158) and (159).[131] Two groups of workers have outlined syntheses of (158) and (159) during the period covered by this Report.[132,133]

Pheromonal Secretions of Codling Moth and Red Bollworm Moth.—Mori[134] has described simple and convenient synthetic routes to the alcohol (160) and acetate (161), sex pheromones of the codling moth and red bollworm moth respectively; both routes are based on the coupling of Grignard reagents with allylic halides (see Scheme 18).

[125] E. Demole and P. Enggist, *Helv. Chim. Acta*, 1974, **57**, 2087.
[126] D. R. Adams, S. P. Bhatnagar, R. C. Cookson, and R. M. Tuddenham, *Tetrahedron Letters*, 1974, 3903.
[127] Y. Nakatani, K. Kubota, R. Tahara, and Y. Shigematsu, *Agric. and Biol. Chem. (Japan)*, 1974, **38**, 1351.
[128] R. T. Gray, R. Mallaby, G. Ryback, and V. P. Williams, *J.C.S. Perkin II*, 1974, 919.
[129] T. Oritani and K. Yamashita, *Agric. and Biol. Chem. (Japan)*, 1974, **38**, 801.
[130] J. A. Katzenellenbogen and K. J. Christy, *J. Org. Chem.*, 1974, **39**, 3315.
[131] H. E. Hummel, L. D. Gaston, H. H. Shorey, R. S. Kaae, K. J. Byrne, and R. M. Silverstein, *Science*, 1973, **181**, 873.
[132] P. E. Sonnet, *J. Org. Chem.*, 1974, **39**, 3793.
[133] K. Mori, M. Tominaga, and M. Matsui, *Agric. and Biol. Chem. (Japan)*, 1974, **38**, 1551.
[134] K. Mori, *Tetrahedron*, 1974, **30**, 3807.

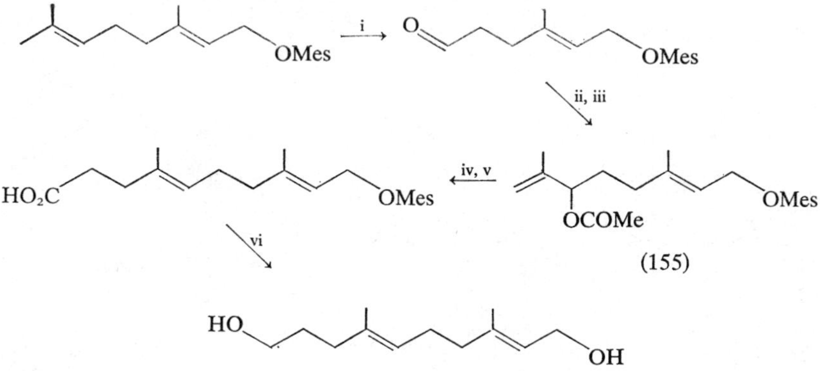

[Scheme 17 with structures 155, 156 and reagents:]

Reagents: i, O₃; ii, ⟋MgBr; iii, Ac₂O–C₅H₅N; iv, Bu^tMe₂SiCl, then Claisen rearrangement; v, H⁺; vi, LiAlH₄

Scheme 17

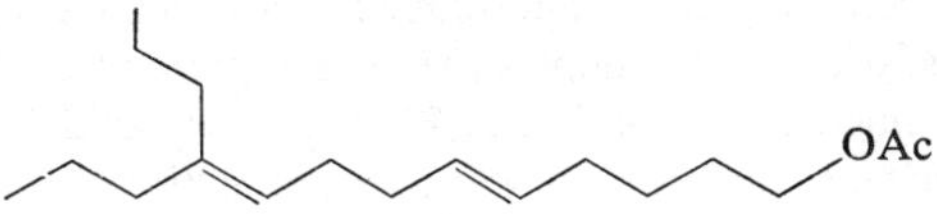

(157)

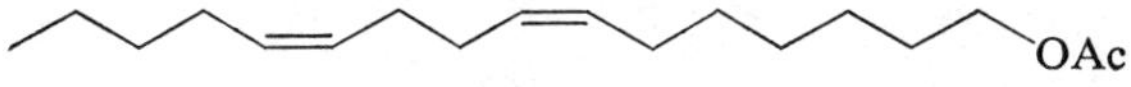

(158)

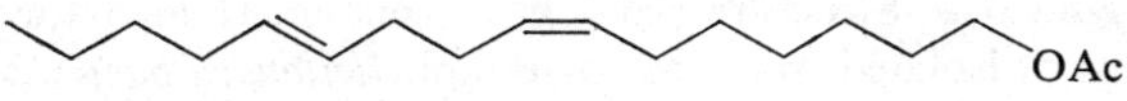

(159)

Pheromone of the Lesser Peachtree Borer.—The pheromone of the lesser peachtree borer *Synanthedon pictipus* has been identified as a mixture of *E,Z*- and *Z,Z*-isomers of octadecadien-1-ol acetate.[135]

Pheromone of the Angoumois Grain Moth.—The pheromone of the female Angoumois grain moth, *Sitotroga cerealella* has been identified as the diene-acetate (159).[136]

[135] J. H. Tumlinson, C. E. Yonce, R. R. Doolittle, R. R. Heath, C. R. Gentry, and E. R. Mitchell, *Science*, 1974, **185**, 614.
[136] K. W. Vick, H. C. F. Su, L. L. Sower, P. G. Mahany, and P. C. Drummond, *Experientia*, 1974, **30**, 17.

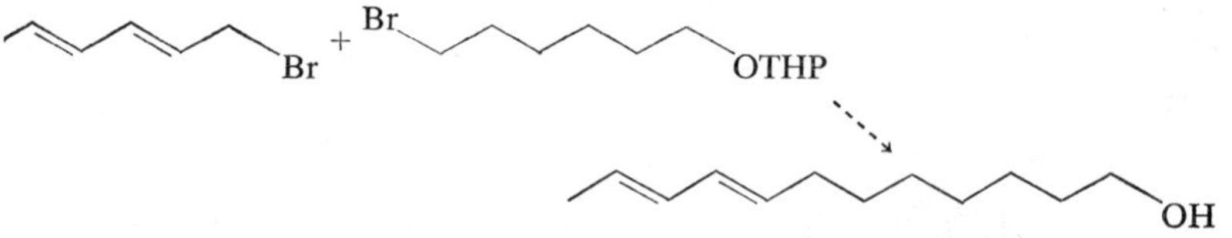

Scheme 18

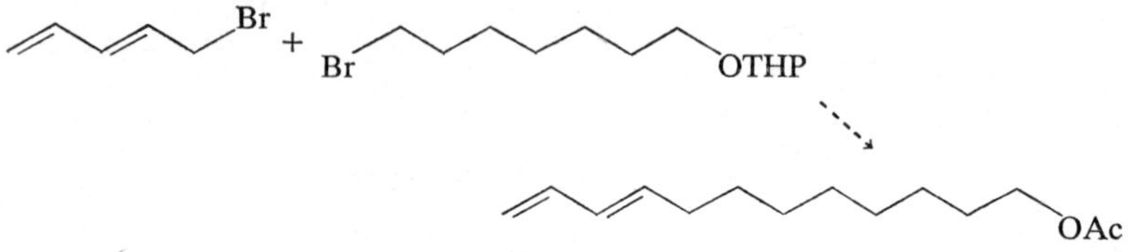

Other Pheromones.—Silverstein and co-workers have outlined an improved synthesis of the alcohol (162), a pheromone of the bark beetle *ips paraconfusus* Lanier,[137] and the pheromone of the furniture carpet beetle *Anthrenus flavipes* LeConte has been identified as Z-dec-3-enoic acid.[138]

Goto *et al.*[139] have examined the structures of the photo-oxidation products of the sex pheromone Z,E-9,12-tetradecadienyl-1-acetate.

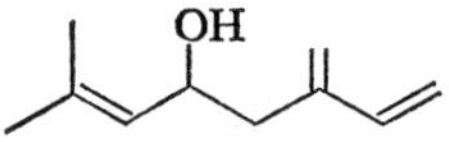

9 Miscellaneous Natural Polyolefins

Flexirubin Pigments.—Structure (163) has been determined for flexirubin, the main pigment isolated from the bacterium *Gleitenden bacterien*.[140]

Polyene Amides.—Several new polyene amides have been isolated from species of the genus *Piper*. Amides (164) and (165) have been identified in *Piper guineense* Schum. and Thonn,[141] (166) in *P. chaba*,[142] (167) in *P.*

---

[137] R. G. Riley and R. M. Silverstein, *J. Org. Chem.*, 1974, **39**, 1957.
[138] H. Fukui, F. Matsumura, M. C. Ma, and W. E. Burkholder, *Tetrahedron*, 1974, 3567.
[139] G. Goto, Y. Masuoka, and K. Hiraga, *Chem. Letters*, 1974, 1275.
[140] H. Achenbach, W. Kohl, H. Reichenbach, and H. Kleinig, *Tetrahedron Letters*, 1974, 255.
[141] J. I. Okogun and D. E. U. Ekong, *J.C.S. Perkin I*, 1974, 2195.
[142] A. Patra and A. Ghosch, *Phytochemistry*, 1974, **13**, 2889.

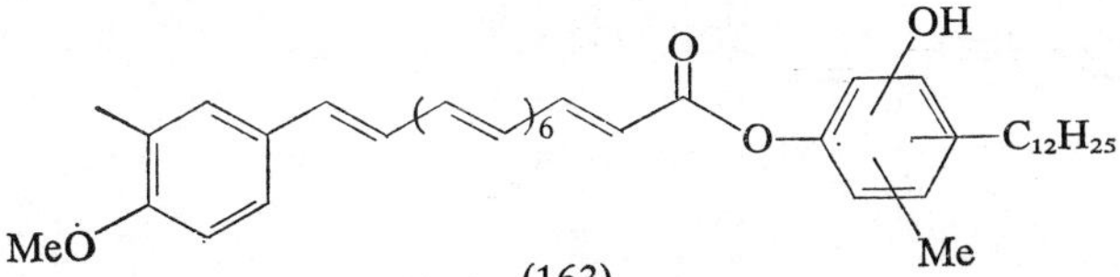

(163)

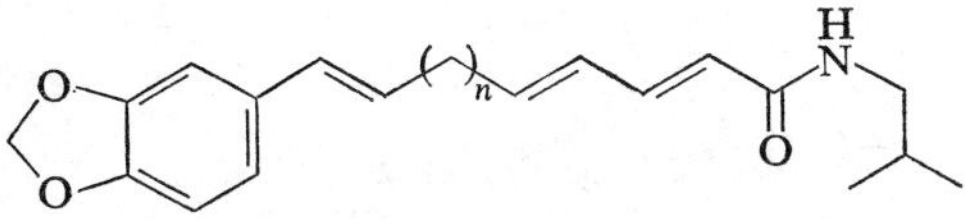

(164) n = 4 or 6

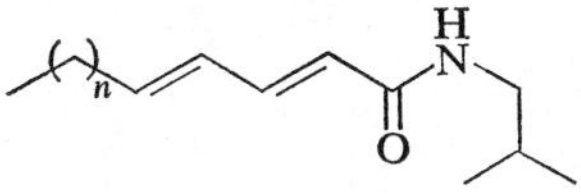

(165) n = 10, 12, or 14

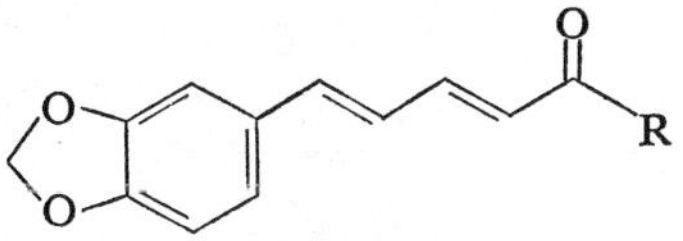

(166) a; R = piperidine b; R = isobutylamine

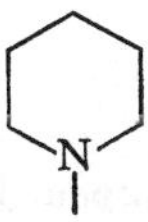

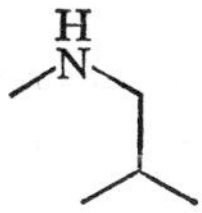

c; R =

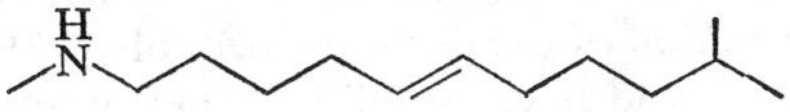

trichostachyon,[143] and (168) in *P. sylvaticum* Roxb.[144] The closely related amide (169) has been found in *Achillea millefolium*.[10] In studies of the biosynthesis of the polyene amides (172) and (173) in *Echinacea purpurea*, Bohlmann and Dallwitz[145] have established that the C_{18}-ester (170) and anacyclin (171) are efficient precursors.

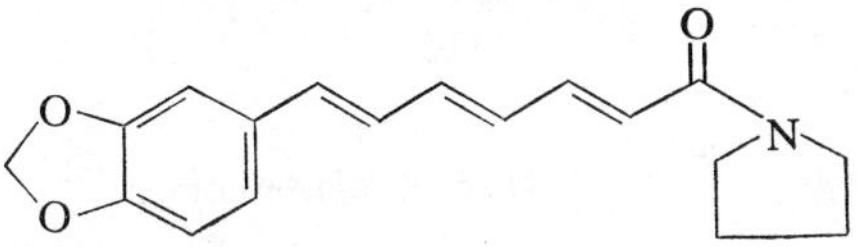

(167)

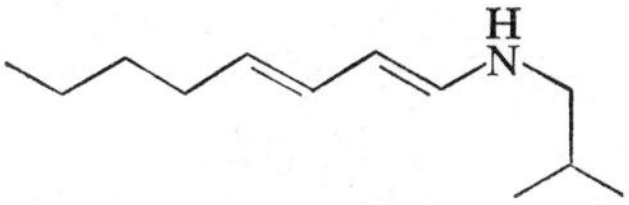

(168)

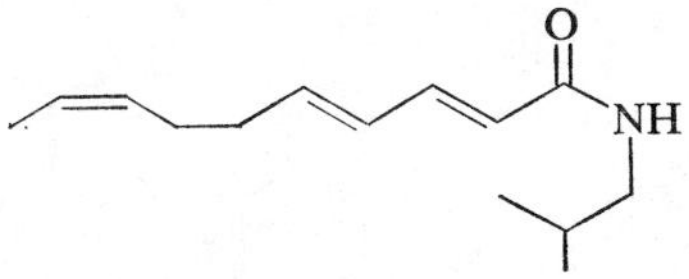

(169)

[143] J. Singh, M. A. Potdar, C. K. Atal, and K. L. Dhar, *Phytochemistry*, 1974, **13**, 677.
[144] A. Banerji, R. N. Rej, and P. G. Ghosch, *Experientia*, 1974, **30**, 223.
[145] F. Bohlmann and E. Dallwitz, *Chem. Ber.*, 1974, **107**, 2120.

(170) [structure: CH≡C chain with (CH$_2$)$_7$CO$_2$Me]

(171) [structure with isobutyl amide]

(172) [structure with isobutyl amide]

(173) [structure with terminal diyne and isobutyl amide]

Kawa Lactones and Related Compounds.—A synthesis of dihydrokawain-5-ol (176) involving oxidation of kawain (174) with selenium dioxide and hydrogenation of the resulting alcohol (175) has been described.[146] Achenbach and Theobald have established that natural (175) has a 6S- and natural (177) a 6R-configuration.[147]

(174) →[SeO$_2$] (175) + diastereomer

(177) (176)

[146] H. Achenbach and H. Hutch, *Tetrahedron Letters*, 1974, 119.
[147] H. Achenbach and N. Theobald, *Chem. Ber.*, 1974, **107**, 735.

Oxidation of (178) with m-chloroperbenzoic acid has been shown to lead to (179) and not the expected product (180).[148] The butenolide (180) co-occurs with the methyl ether of (175) in *Piper sanctum*.

Other Compounds.—The dienol (181) has been isolated from watermelon, *Citrullus vulgaris*, and *Cucumis melo*,[149] and the piscidal constituent (182) has been found in the twig and bark of *Excoecaria agallocha*.[150]

All-*cis*-6,9,12,15-heneicosatetraene and all-*cis*-3,6,9,12,15-heneicosapentaene have been identified in spores of the moss *Polytrichum commune*

[148] R. Hansel, J. Schulz, and A. Leuschke, *Chem. Ber.*, 1974, **107**, 3337.
[149] T. R. Kemp, D. E. Knavel, L. P. Stoltz, and R. E. Lundin, *Phytochemistry*, 1974, **13**, 1167.
[150] H. Ohigashi, H. Katsumata, K. Kawazu, K. Kohimizu, and T. Mitsui, *Agric. and Biol. Chem. (Japan)*, 1974, **38**, 1093.

(183) R = Me, CH_2OH, CH_2OAc, or CO_2H

(184) R = Me or CH_2OH

Hedw,[151] and the mitorubrin derivatives (183) have been found in the fruiting bodies of *Hypoxylon fragiforme*.[152]

Masamune and co-workers[153] have identified 9,10,13-trihydroxy-11,15-octadecadienoic acid in the roots of kidney bean, *Phaseolus vulgaris*. L., and the esters (184) and (185) have been isolated from the soft resin fraction of *Palas* seedlac.[154]

Syntheses of the germination stimulant strigol (186) have been outlined,[155] and the biosynthesis of securimine (187) in *Securinega suffruticosa* has been examined.[156]

[151] P. Karunen, *Phytochemistry*, 1974, **13**, 2209.
[152] W. Steglich, M. Klaar, and W. Furtner, *Phytochemistry*, 1974, **13**, 2874.
[153] M. Takasugi, M. Anetai, and T. Masamune, *Chem. Letters*, 1974, 947.
[154] A. N. Singh, A. B. Upadhye, V. V. Mhaskar, and S. Dev, *Tetrahedron*, 1974, **30**, 867.
[155] G. A. MacAlpine, R. A. Raphael, A. Shaw, A. W. Taylor, and H. K. Wild, *J.C.S. Chem. Comm.*, 1974, 834.
[156] U. Sankawa, K. Yamasaki, and Y. Ebizuka, *Tetrahedron Letters*, 1974, 1867.

4
Chemistry of the Prostaglandins

BY G. PATTENDEN

1 Introduction

This year has witnessed an enormous drop in the number of 'original' publications dealing with the synthesis of PG's and their analogues. Less than thirty papers on synthesis were published during 1974, and many of these contained material published previously in preliminary form. Sadly, only a handful of papers describe new or original contributions to the plethora of synthetic methodology now available in PG chemistry. Some eminent chemists have expressed the view that the synthetic chemist has made his contribution to PG chemistry and that it is now up to the biologist to find the best therapeutic uses for the compounds;[1] the drop in the number of synthesis papers in 1974 perhaps inadvertently echoes these sentiments.

Several reviews of PG chemistry have been published,[2-5] and a book, 'A Guide to the PG Literature 1906—1970', has appeared.[6] A summary of the history, development, and potential of PG's has also been written.[7]

2 Nomenclature

Nelson has written a useful review of PG nomenclature,[8] and discusses the problems associated with describing stereochemical configurations, homologues, nor-compounds, and analogues, *etc.* of PG's.

3 Synthesis of the Primary Prostaglandins

The layout of this section is the same as last year. Few significant developments in PG synthesis were recorded during 1974 (see comments above).

[1] R. L. Rawls, *Chem. Eng. News*, 1974, **52**, 18.
[2] E. W. Horton, in 'Biochemistry of Lipids', ed. T. W. Goodwin, MTP International Review of Science, Biochemistry Series One. Vol. 4, Butterworths, London, 1973, p. 237.
[3] J. E. Pike, *Prostaglandines, Semin. INSERM* 1973.
[4] J. S. Bindra and R. Bindra, *Fortschr. Arzneitmittelforsch.*, 1973, **17**, 410.
[5] E. W. Horton, *Sci. Basis Med.*, 1973, 58.
[6] 'Prostaglandins Abstract. A Guide to the Literature, 1906—1970', Vol. 1., ed. R. M. Sparks, Plenum, New York, 1974.
[7] J. E. Pike, *Stud. Trop. Oceanog.*, 1974, **12**, 9.
[8] N. A. Nelson, *J. Medicin. Chem.*, 1974, **17**, 911.

Two reviews of PG synthesis have been published,[9,10] one of which[10] gives particular attention to the synthesis of PG analogues.

Corey's Bicyclo[2,2,1]heptane Route.—*Refinements.* Ranganathan and co-workers[11] have outlined an improved route to the intermediate (2), which has as a key stage the oxidation of (1) in sodium methoxide with $TiCl_3$–NH_4OAc. The nitroethylene adduct (1) is readily available from 5-methoxymethyl-cyclopentadiene and nitroethylene.

Bowler et al.[12] have found that aluminium isopropoxide is an extremely useful alternative to borohydride reducing agents for the crucial reduction step (3) → (4) in PG synthesis.

New Routes to Key Intermediates. An I.C.I. group[13] has outlined a highly attractive synthesis of the Corey aldehyde (11) starting from 6-acetoxyfulvene (5). Acid-catalysed Diels–Alder reaction between (5) and 2-chloroacrylonitrile led first to the adduct (6), obtained as a mixture of C-5 epimers. Hydrolysis produced the *anti*-aldehyde (7), which isomerized in hot acid to the *syn*-aldehyde (8). Protection of the aldehyde as the dimethyl acetal

---

[9] M. Vandewalle, *Chem. Weekblad*, 1974, **70**, 14.
[10] M. K. Bigami, *Pharmacos.*, 1973, **18**, 59.
[11] S. Ranganathan, D. Ranganathan, and A. M. Mehrotra, *J. Amer. Chem. Soc.*, 1974, **96**, 5261.
[12] J. Bowler, K. B. Mallion, and R. A. Raphael, *Synth. Comm.*, 1974, 211.
[13] E. D. Brown, R. Clarkson, T. J. Leeney, and G. E. Robinson, *J.C.S. Chem. Comm.*, 1974, 642.

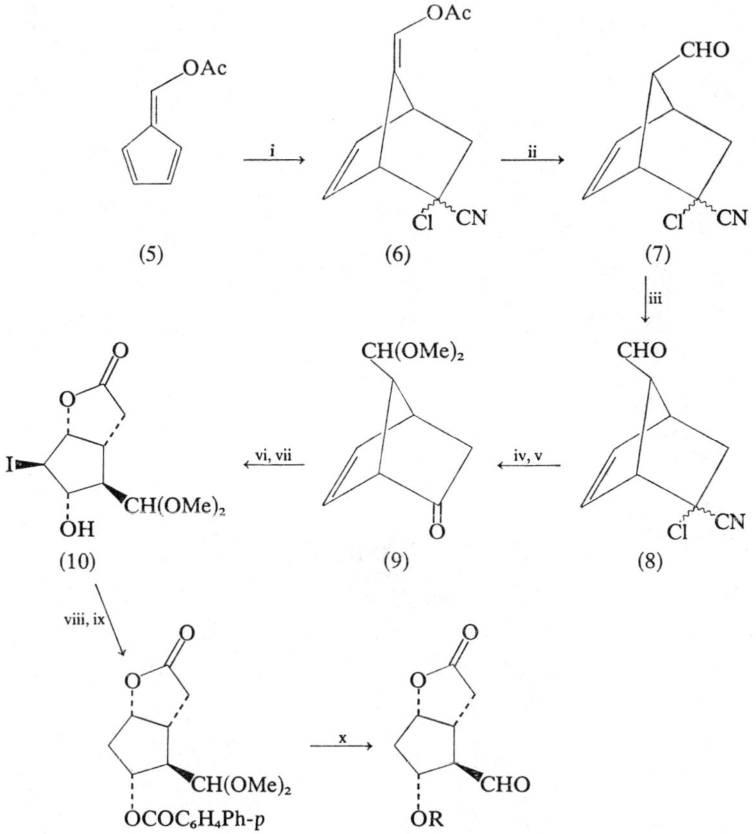

Scheme 1

Reagents: i, $CH_2=C(Cl)CN$; ii, HCl; iii, HCl, 84 °C; iv, $HC(OMe)_3-H^+$; v, KOH; vi, H_2O_2-alkali; vii, $KI-I_2$; viii, $p-PhC_6H_4COCl$; ix, Bu_3SnH; x, H^+

followed by hydrolysis of the chloronitrile moiety then gave the ketone (9). Baeyer–Villiger oxidation of (9) and iodolactonization of the intermediate lactone gave the iodohydrin (10), which was converted into (11) as outlined in Scheme 1. This route to (11) *via* construction of a bicycloheptene offers several advantages over Corey's original route involving the use of thallium cyclopentadienide as starting material.

In a second new approach to (11) published during 1974, Vandewalle and co-workers[14] have exploited the use of the cyclopentene-1,3-dione (12) as key starting material. Treatment of (12) with $HC(OEt)_3$–EtOH in acid led

[14] J. van Hooland, P. De Clercq, and M. Vandewalle, *Tetrahedron Letters*, 1974, 4343.

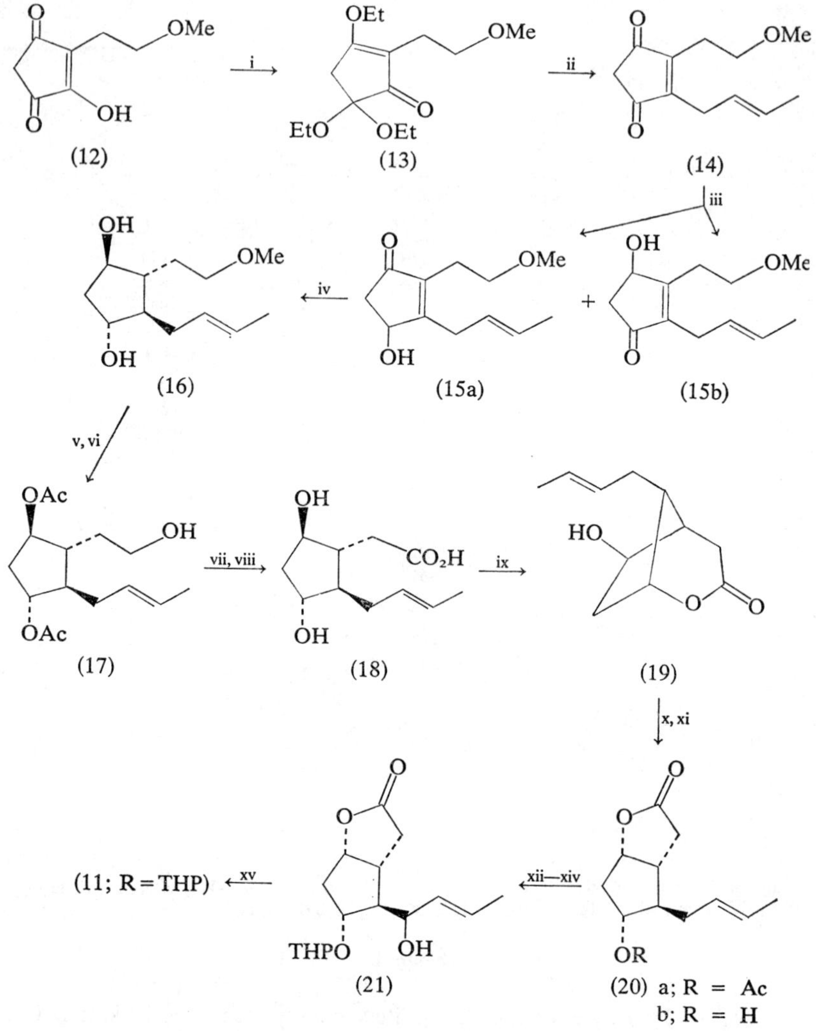

Reagents: i, HC(OEt)$_3$–H$^+$; ii, ⋀⋀Br–Zn; then H$^+$; iii, Zn–HOAc; iv, Li–NH$_3$–PhOH; v, Ac$_2$O–C$_5$H$_5$N; vi, BBr$_3$; vii, Jones oxidation; viii, HCl; ix, p-TsOH; x, Et$_4$$\overset{+}{\text{N}}OAc^-$; xi, HCl; xii, dihydropyran–H$^+$; xiii, NBS; xiv, AgCO$_3$–Celite; xv, OsO$_4$–NaIO$_4$

Scheme 2

first to (13), which in a Grignard reaction with (*E*)-4-bromobut-2-ene gave (14). Reduction of (14) with Zn–HOAc produced a mixture of hydroxycyclopentenones (15a) and (15b), which on further reduction with Li–NH$_3$ led to a mixture of *cis*- and *trans*-cyclopentane-1,4-diols. The *trans*-diol (16) was separated by chromatography and converted into the carboxylic acid (18) *via* the carbinol (17). After lactonization, the tosylate derived from (19) was treated with tetraethylammonium acetate; this effected inversion of the hydroxy-function in (19) and led to the acetate (20a). Treatment of (20a) with acid gave (20b), which was converted into (21) following protection of the hydroxy-function as the tetrahydropyranyl ether, allylic bromination, and hydrolysis. Conversion of (21) into (11; R = THP) was then completed by treatment with OsO$_4$–NaIO$_4$.

Full details have been published of a route to the PG synthons (22a) and (22b) based on oxidative cleavage of appropriately substituted norbornenes;[15] this approach has been discussed previously (see Vol. 2, p. 262).

The synthesis of the keto-acid (23) *via* a Prins reaction of norbornadiene and paraformaldehyde was discovered contemporaneously by Corey and by Sutherland. Corey's route from (23) to (28) was described last year (Vol. 3, p. 315). Sutherland converted (23) into (28) *via* the bromide (24) and the lactone (25) according to Scheme 3.[16] An advantage of Sutherland's route to (26) is that the conversion (24) → (26) can be effected in a 'one-pot process'; alternatively (26) can be obtained from (24) *via* (27). In a closely similar approach, Roberts[17] has outlined a synthesis of the potentially useful PG precursor (31) starting from bicyclo[3,2,0]hept-2-en-6-one (29). Treatment of (29) with *N*-bromoacetamide in methanol, for example, gave (30), which, on treatment with a catalytic amount of sodium methoxide in methanol saturated with KCN, rearranged to the 5-*endo*-methoxy-7-*anti*-cyanobicyclo-[2,2,1]heptan-2-one (31; R = Me); this structure is closely similar to (24), and oxidative ring expansion could provide a new and useful approach to PG synthons such as (26) *etc*.

Corey[18] has shown that reaction of the epoxide (32) with lithium divinylcuprate to give (33) is 81% regiospecific. The alcohol (33) can be smoothly

(22) a; R^1 = Me
b; R^1 = CPh$_3$

---

[15] G. Jones, R. A. Raphael, and S. Wright, *J.C.S. Perkin I*, 1974, 1676.
[16] R. Peel and J. K. Sutherland, *J.C.S. Chem. Comm.*, 1974, 151.
[17] S. M. Roberts, *J.C.S. Chem. Comm.*, 1974, 948.
[18] E. J. Corey, K. C. Nicolaou, and D. J. Beames, *Tetrahedron Letters*, 1974, 2439.

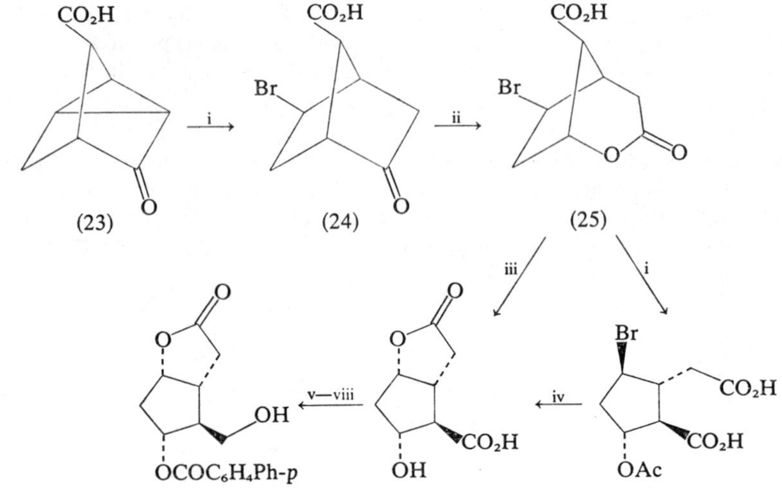

(23) (24) (25)

(28) (26) (27)

Reagents: i, HBr–HOAc; ii, $MeCO_3H$; iii, NH_2OH; iv, aq. $NaHCO_3$; v, *p*-$BrC_6H_4COCH_2X$; vi, *p*-PhC_6H_4COCl; vii, Zn–HOAc; viii, B_2H_6

Scheme 3

converted into the PG intermediate (34) as shown. The regiospecificity observed by Corey is closely similar to that found by Fried and Sih in a similar reaction between (32) and ethynylalane reagents (see Vol. 3, p. 320). The lactone (35) employed in one of Corey's new approaches to PG's (Vol. 3, p. 316; the lactone is wrongly formulated on p. 314 of Vol. 3) has now been resolved,[19] and it was established that the *laevo*-form possesses the absolute configuration shown in (35), which corresponds to the natural PG skeleton.

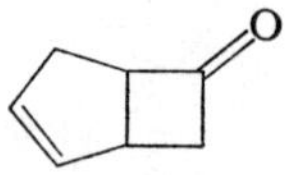

(29)

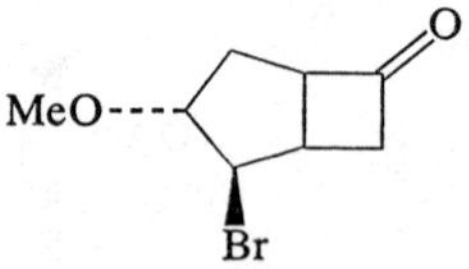

(30)

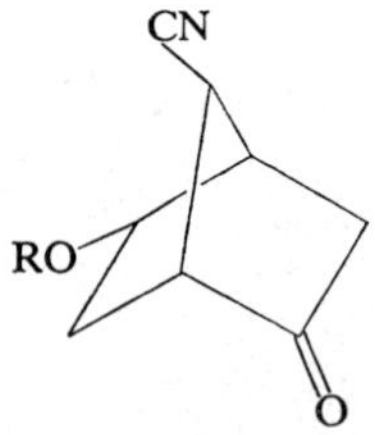

(31)

[19] E. J. Corey and B. B. Snider, *J. Org. Chem.*, 1974, **39**, 256.

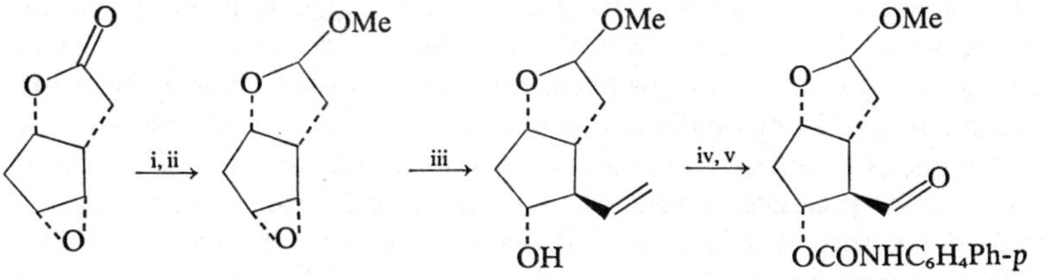

Chemistry of the Prostaglandins

(32) → (33) → (34)

Reagents: i, Bu$^i_2$AlH; ii, MeOH–BF$_3$; iii, (CH$_2$=CH)$_2$CuLi; iv, p-PhC$_6$H$_4$N=C=O; v, OsO$_4$–NaIO$_4$

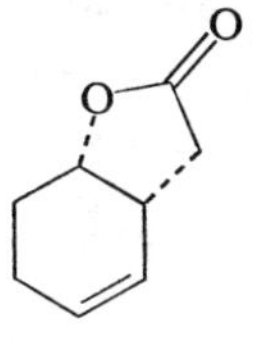

(35)

Routes by the Merck Group.—The Merck group has published full details of its synthesis of PGE$_1$ which employs the Diels–Alder adduct of *trans*-piperylene and maleic anhydride as starting material.[20] This route was outlined in preliminary form in 1971 (see Volume 2, p. 269).

Routes involving Cyclopentane Ring Synthesis.—Clercq and Vandewalle[21] have described a synthesis of (+)-PGF$_{1\alpha}$ which proceeds in 25 steps from the cyclopentene-1,3-dione (36). Reduction of (36) with Zn–HOAc first produced a 1:1 mixture of the hydroxycyclopentenones (37a) and (37b),

(36) → (37a)

+

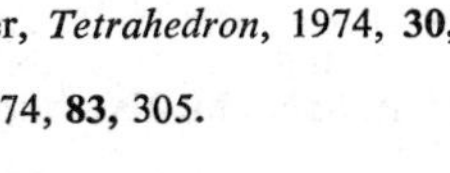

(37b)

[20] H. L. Slates, Z. S. Zelawski, D. Taub, and N. L. Wendler, *Tetrahedron*, 1974, **30**, 819.
[21] P. De Clercq and M. Vandewalle, *Bull. Soc. chim. belges*, 1974, **83**, 305.

which on further reduction with Na–NH$_3$ led to a complex mixture of the three diols (38)—(40) and their fully saturated analogues. The diol mixture was converted into a mixture of diacetates which was oxidized with KMnO$_4$–NaIO$_4$ to a mixture of isomers of the acid (41). Hydrolysis of (41) gave a mixture of isomeric diols, one of which was immediately separated as the α-lactone (42). When the remaining two diols were heated under reflux in benzene in the presence of toluene-p-sulphonic acid, one of them was converted into a second γ-lactone (43). In this manner, the most abundant isomer (38) from the Li–NH$_3$ reduction of (37a + b) could be separated. [A discussion of the configurational assignments of diols (38)—(40) based on ^{1}H n.m.r. was presented in accompanying papers.[22,23]] The γ-lactone (43) was converted into (44), which after tosylation and reaction with tetraethylammonium acetate gave the inverted acetate (45). The latter compound was converted into racemic PGF$_{1\alpha}$ via the route shown in Scheme 4.

As part of a biogenetically patterned investigation of PG synthesis via epoxy polyunsaturated fatty acids, Sih and co-workers[24] have described a synthesis of ($\pm$)-eicosa-cis-14,15-epoxy-cis,cis-8,11-dienoic acid (46) (Scheme 5). When (46) was treated with bovine seminal vesicle microsomes, however, no significant quantity of PGE$_1$ was detected, and instead the diol (47) was obtained.

Routes incorporating Conjugate Addition of Vinylcopper Reagents to a Functionalized Cyclopentenone.—The lactone (48), with the configuration

---
[22] P. De Clercq, P. Van Haver, D. Tavernier, and M. Vandewalle, *Tetrahedron*, 1974, **30**, 55.
[23] D. Van Haver, D. Tavernier, M. Anteunis, and M. Vandewalle, *Tetrahedron*, 1974, **30**, 105.
[24] R. Sood, M. Nagasawa, and C. J. Sih, *Tetrahedron Letters*, 1974, 423.

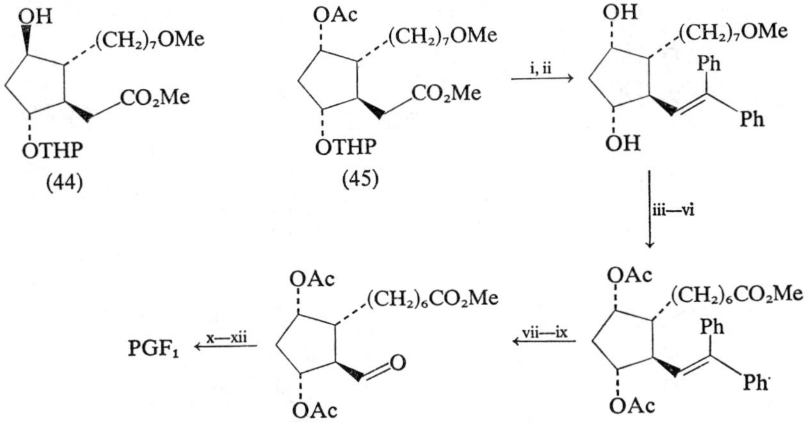

Chemistry of the Prostaglandins

Reagents: i, PhMgX; ii, H⁺; iii, Ac₂O–C₅H₅N; iv, BBr₃; v, CrO₃; vi, CH₂N₂; vii, RuO₄–NaIO₄; viii, B₂H₆; ix, Collins oxidation, x, (EtO)₂POCHCOC₅H₁₁; xi, ZnBH₄; xii, hydrolysis

Scheme 4

shown, has not previously found use in PG synthesis, since it does not lead to the natural PG skeleton. Gruber and co-workers[25] have now shown, however, that this compound can be used for the synthesis of the optically active hydroxy-cyclopentenone (49) (Scheme 6), an intermediate employed extensively in earlier syntheses of PGE₂ and modified PG's (see Vol. 2,

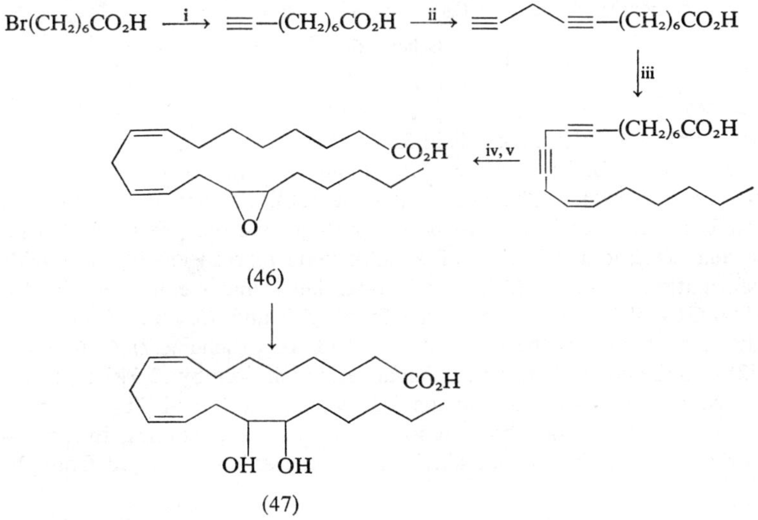

Reagents: i, LiC≡CH; ii; EtMgBr then ≡—CH₂Br; iii, Br⌒⌒⌒⌒; iv, MeCO₃H; v Lindlar catalyst–H₂

Scheme 5

[25] L. Gruber, I. Tömösközi, E. Major, and G. Kovács, *Tetrahedron Letters*, 1974, 3729.

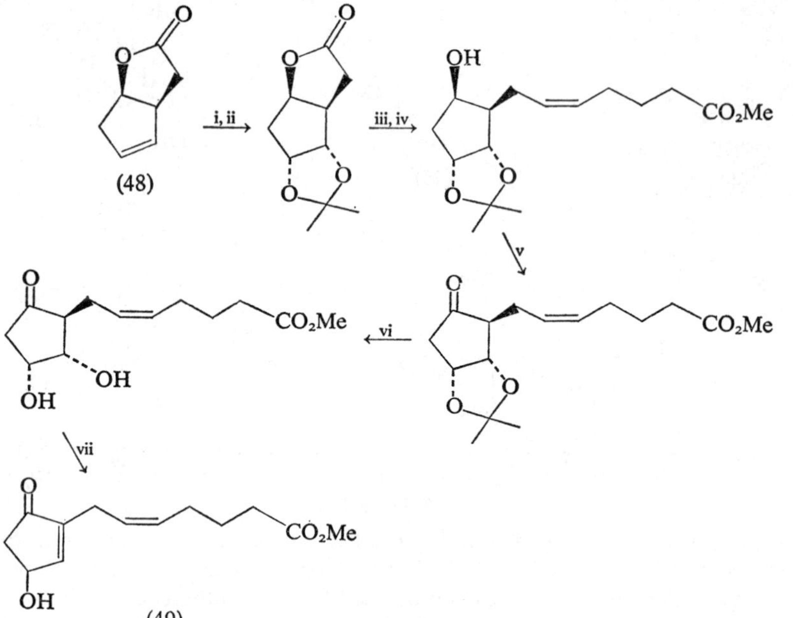

(48)

(49)

Reagents: i, OsO$_4$; ii, MeC(OMe)$_2$Me–H$^+$; iii, Bu$^i_2$AlH; iv, MeO$_2$C(CH$_2$)$_3$C̄HP̄Ph$_3$; v, Collins oxidation; vi, H$_2$SO$_4$; vii, –H$_2$O

Scheme 6

p. 274; Vol. 3, p. 327). Floyd[26] has also outlined a closely similar approach to (49) from (48) which proceeds in 35—40% overall yield.

As a result of further studies of the conjugate addition of the *cis*-vinylcopper reagent (51) to the cyclopentenone (50), the 'stereoselectivity' of the addition has been shown to be lower than reported previously.[27] In a somewhat elegant application of a sulphenate rearrangement–sulphoxide transformation sequence Miller and co-workers[27] have employed the 13-*cis*-15β-PGE$_1$ (52) product prepared from (50) and (51) in an efficient, nearly stereospecific synthesis of natural (±)-PGE$_{1α}$ (Scheme 7). Conversion of (52) into the corresponding sulphenate ester followed by [2,3]sigmatropic rearrangement led to the sulphoxide (53), which gave (±)-PGE$_{1α}$ on treatment with trimethyl phosphite. In some closely parallel studies, Evans and co-workers[28] have shown that alkylation of the anion produced from the

[26] M. B. Floyd, *Synth. Comm.*, 1974, 317.
[27] J. G. Miller, W. Kurz, K. G. Untch, and G. Stork, *J. Amer. Chem. Soc.*, 1974, **96**, 6775.
[28] D. A. Evans, T. C. Crawford, T. T. Fujimoto, and R. C. Thomas, *J. Org. Chem.*, 1974, **39**, 3176.

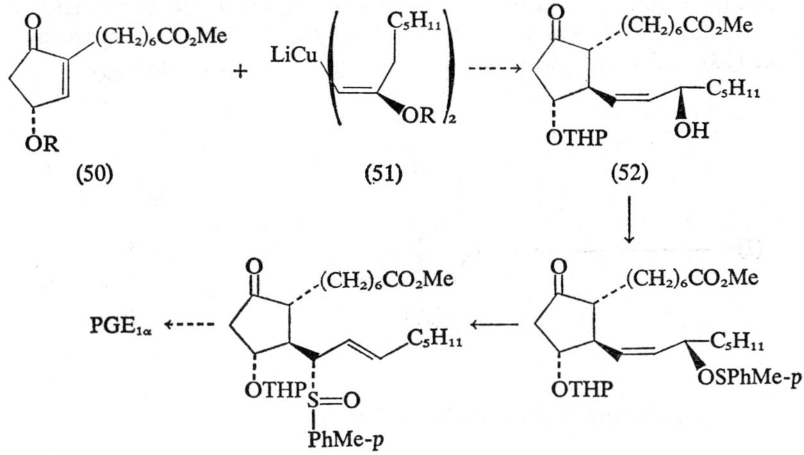

Chemistry of the Prostaglandins 253

Scheme 7

sulphoxide (54) followed by rearrangement of the resulting sulphoxide (55) and cleavage provides an efficient route to the potentially useful PG intermediate (56).

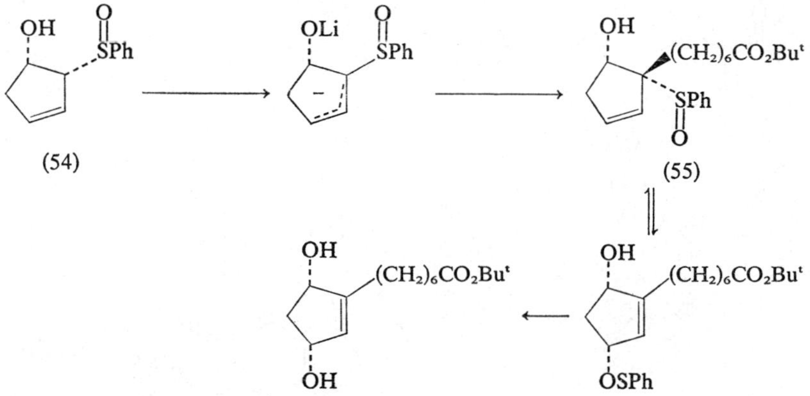

4 Synthesis of A-Prostaglandins

There have been no major developments in new approaches to PGA synthesis during the period covered by this Report. Ranganathan and co-workers[11] have, however, outlined a preparation of the PGA synthon (57) which involves treatment of the sodium salt of (1) with HCl; conversion of (1) into (57) in this way was accompanied by formation of the hydroxamic ester (58), which on treatment with HNO_2 gave (57) quantitatively.

5 Synthesis of C-Prostaglandins

Corey and Cyr[29] have described a convenient method for the conversion of PGA_2 (59) into PGC_2 (61). The method is based on the formation of the extended enolate anion (60) by abstraction of the C-12 hydrogen atom in (59) with $KOBu^t$, followed by α-protonation at C-10. Kelly and co-workers[30] have described a closely similar synthetic approach to PGC_2 to that outlined last year by Corey (Volume 3, p. 333).

[29] E. J. Corey and C. R. Cyr, *Tetrahedron Letters*, 1974, 1761.
[30] R. C. Kelly, I. Schletter, and R. L. Jones, *Prostaglandins*, 1974, **4**, 653.

6 Synthesis of Modified Prostaglandins

A review of the synthesis of modified PG's has been published.[10] Considerable attention has been given to the synthesis and biological evaluation of members of the oxa- and thia-PG's during the period covered by this Report.

11-Desoxyprostaglandins.—Abraham[31] has developed his synthesis of 11-desoxy-PGE$_2$, which was described last year (Vol. 3, p. 335), to produce optically active PGE$_2$, and the May and Baker group[32] has modified its earlier approach to 11-desoxy-PGF$_1$ [starting from (62)] to provide a useful synthesis of 11-desoxy-PGE$_1$ (63) (cf. Vol. 2, p. 284).

(62) ----→ (63)

From studies of the alkylation of enolate anions generated from conjugate addition of vinylcuprates to cyclopentenone Patterson and Fried[33] have developed new and useful synthetic routes to 11-desoxy-PGE$_2$ (64), 5,6-dihydro-11-desoxy-PGE$_2$ (65), and 11,15-desoxy-PGE$_2$ (66). In each case conjugate addition of the appropriate vinylcuprate led first to an enolate anion, which was then converted into a silyl enol ether. Alkylation of the latter, followed by hydrolysis, then gave the desoxy-PG's; the general method is illustrated in Scheme 8 for 11-desoxy-PGE$_2$. Similar studies of the alkylation of enolate anions generated *in situ via* conjugate addition reactions were reported contemporaneously by Posner *et al.*[34]

Trost and Kurozumi[35] have outlined the application of the oxaspiropentane intermediate (67) and the vinylcyclopropane rearrangement (68) → (69) to the synthesis of (70), an intermediate used extensively in 11-desoxy-PG synthesis (Scheme 9). Novak and Szantay[36] have also outlined a new route to (70).

A novel synthesis of cyclopentenones, which could be useful in PG synthesis, has been reported by Hirano *et al.*,[37] based on acid-catalysed cyclization of a cross-conjugated dienone moiety (Scheme 10). Bartmann and

---

[31] N. A. Abraham, *Tetrahedron Letters*, 1974, 1393.
[32] M. P. L. Caton, E. C. J. Coffee, and G. L. Watkins, *Tetrahedron Letters*, 1974, 585.
[33] J. W. Patterson and J. H. Fried, *J. Org. Chem.*, 1974, **39**, 2506.
[34] G. H. Posner, C. E. Whitten, J. J. Stirling, and D. J. Brunelle, *Tetrahedron Letters*, 1974, 2591.
[35] B. M. Trost and S. Kurozumi, *Tetrahedron Letters*, 1974, 1929.
[36] L. Novak and C. Szantay, *Synthesis*, 1974, 353.
[37] S. Hirano, T. Higama, and N. Nozaki, *Tetrahedron Letters*, 1974, 1429.

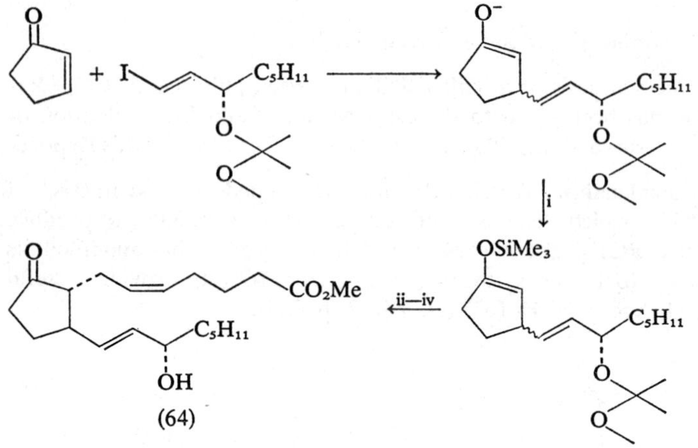

256   *Aliphatic Chemistry*

Reagents: i, Me$_3$SiCl; ii, LiNH$_2$; iii, Br⁀⁀⁀CO$_2$Me; iv, H$^+$

Scheme 8

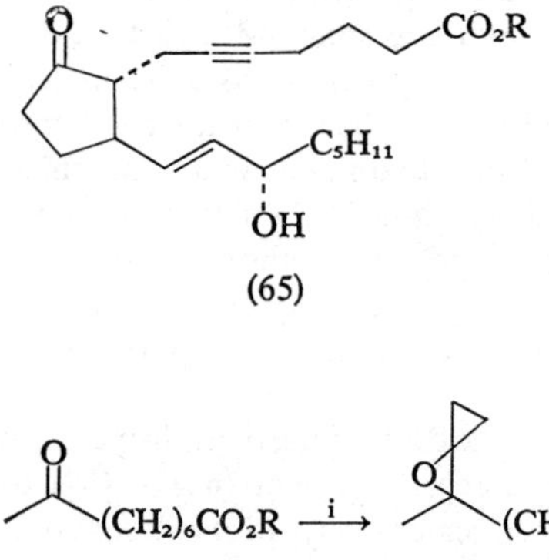

(65) 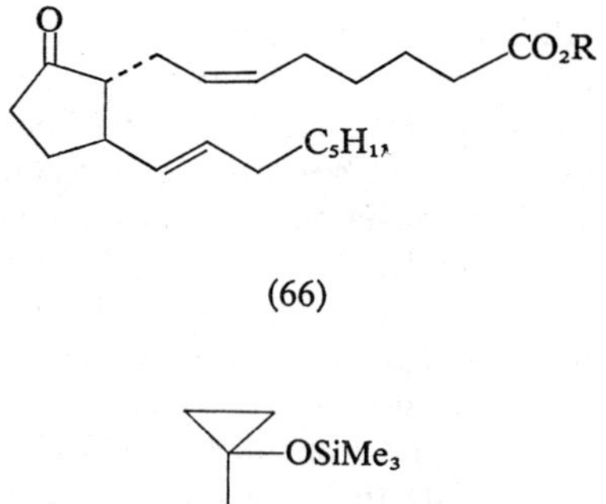   (66)

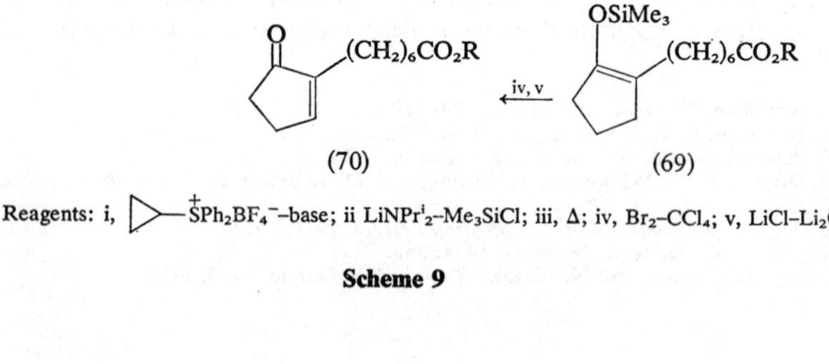

(67)   (68)

(70)   (69)

Reagents: i, ▷—$\overset{+}{S}$Ph$_2$BF$_4^-$–base; ii LiNPr$^i_2$–Me$_3$SiCl; iii, Δ; iv, Br$_2$–CCl$_4$; v, LiCl–Li$_2$CO$_3$

Scheme 9

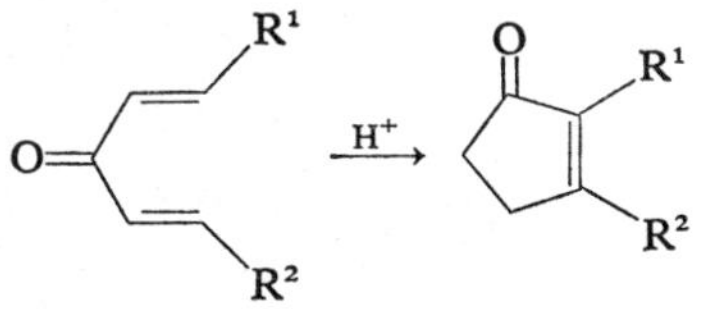

Scheme 10

co-workers[38] have outlined a new synthesis of 11-desoxy-PGE$_2$ which starts from the hemi-acetal (71) and proceeds as outlined in Scheme 11.

A synthetic route to 9-desoxy-$\Delta^{9,10}$-PG's has been devised,[39] and a May and Baker group[40] has outlined a simple route to 2-alkylcyclopent-2-enones based on isomerization of the product resulting from reaction of cyclopentanone enamines with aldehydes.

Oxaprostaglandins.—A few years ago it was shown that 7-oxa-derivatives of the PG's could function as either PG antagonists or agonists. This observation has led to a great deal of interest in these molecules, and during the period covered by this Report the first syntheses of members of the 9-, 10-, and 11-oxa-PG's and of a bis-oxa-PG have been accomplished.

9-*Oxaprostaglandins*. A synthesis of 9-desoxy-9-oxa-PGE$_1$ (74) has been described which proceeds according to Scheme 12.[41] The tetrahydrofuranone unit in (73) was constructed from the $\alpha\beta$-unsaturated ester (72) by reaction with ethyl sodioglycolate; the usual chain-extension reactions employed in PG synthesis then led to (74).

10-*Oxaprostaglandins*. Hauser and Huffman[42] have outlined a useful synthesis of 10-oxa-11-desoxy-PGE$_1$ (76) starting from diethyl (3-cyclo-octenyl) malonate (75) (Scheme 13).

11-*Oxaprostaglandins*. A Canadian group has described an interesting approach to the synthesis of 11-oxa-PG's which utilizes 1,4-anydro-D-glucitol (77) as starting material[43] (Scheme 14).

In closely similar approaches, two groups have described the syntheses of 11-desoxy-11-oxa-PGE$_1$ (78) and 11-desoxy-11-oxa-PGE$_2$ (79); in each case the five-membered ring was constructed from an $\alpha\beta$-unsaturated ester and the sodium salt of ethyl glycolate.[44,45]

---

[38] W. Bartmann, G. Beck, and U. Lerch, *Tetrahedron Letters*, 1974, 2441.
[39] C. Gandolfi and G. Doria, *Farmaco, Ed. sci.*, 1974, **29**, 405.
[40] M. P. L. Caton, E. C. J. Coffee, T. Parker, and G. L. Watkins, *Synth. Comm.*, 1974, 303.
[41] I. Vlattas and L. DellaVecchia, *Tetrahedron Letters*, 1974, 4455.
[42] F. M. Hauser and R. G. Huffman, *Tetrahedron Letters*, 1974, 905.
[43] S. Hanessian, P. Dextraze, A. Fougerousse, and Y. Guindon, *Tetrahedron Letters*, 1974, 3983.
[44] I. T. Harrison, V. R. Fletcher, and J. H. Fried, *Tetrahedron Letters*, 1974, 2733.
[45] I. Vlattas and A. O. Lee, *Tetrahedron Letters*, 1974, 4451.

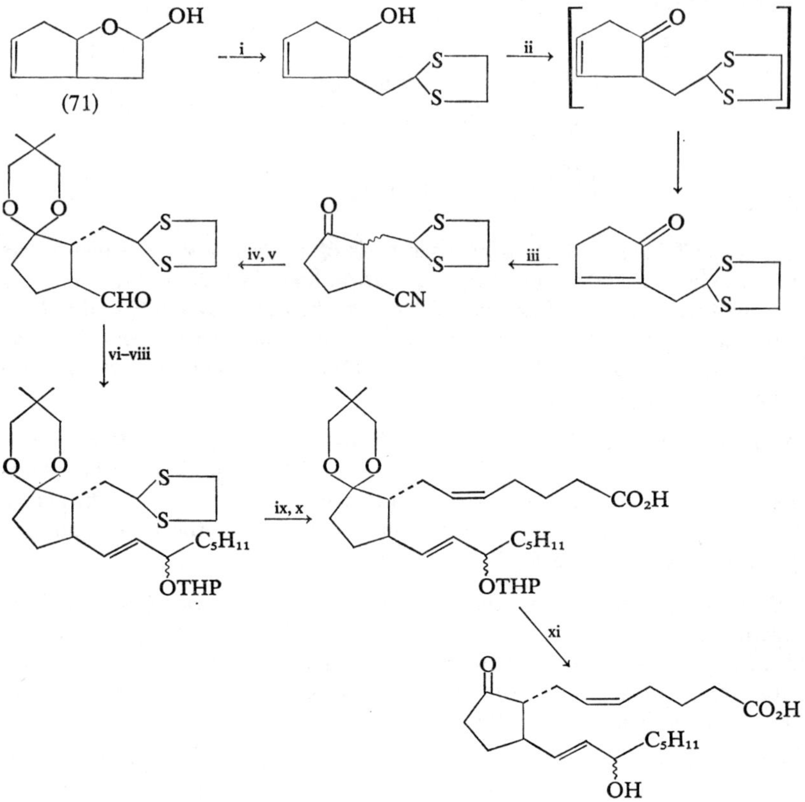

Reagents: i, (HSCH$_2$)$_2$; ii, DMSO–DCC; iii, HCN; iv, HOCH$_2$CMe$_2$CH$_2$OH; v, Bu$^i_2$AlH; vi, C$_5$H$_{11}$COCHPO(OEt)$_2$; vii, NaBH$_4$; viii, dihydropyran–H$^+$; ix, DMF–MeI–CaCO$_3$; x, CO$_2$(CH$_2$)$_3$CHPPh$_3$; xi, H$^+$

Scheme 11

A Syntex group has described synthetic routes to the interesting bis-oxa-PG's (82) and (84).[46] Treatment of the diol (80) with paraformaldehyde in the presence of acid led to the cyclic acetal (81), which was then elaborated to (82) by standard methods (Scheme 15). The cyclic carbonate (83), prepared from (80) and phosgene, was elaborated similarly to produce (84).

Thiaprostaglandins.—The interesting biological properties shown by members of the oxa-PG's has inevitably led to the synthesis and biological evaluation of thia-PG's. Synthetic routes to members of the 7- and 9-thia-PG's were published this year.

[46] I. T. Harrison and V. R. Fletcher, *Tetrahedron Letters*, 1974, 2729.

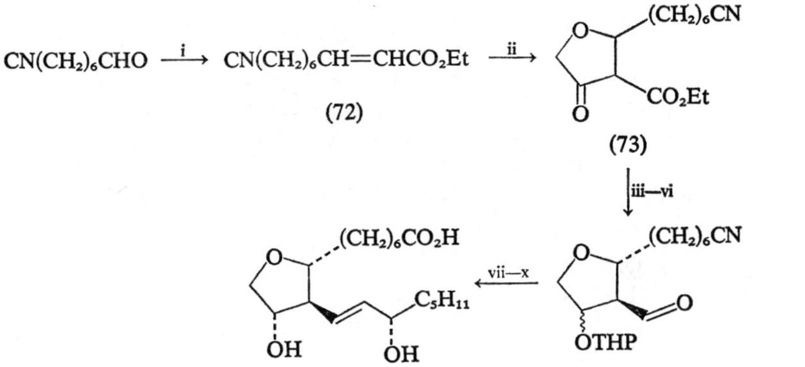

Chemistry of the Prostaglandins

$CN(CH_2)_6CHO \xrightarrow{i} CN(CH_2)_6CH=CHCO_2Et \xrightarrow{ii}$ (73) [structure with $(CH_2)_6CN$ and CO_2Et]

(72)

$\downarrow$ iii—vi

[structures (74) with $(CH_2)_6CO_2H$, C_5H_{11}, OH, OH] ←vii—x— [structure with $(CH_2)_6CN$, OTHP]

(74)

Reagents: i, $EtO_2CCH_2PO(OEt)_2$–NaH; ii, EtO_2CCH_2ONa; iii, $NaBH_4$; iv, dihydropyran–H^+; v, $LiAlH_4$; vi, Collins oxidation; vii, $C_5H_{11}CO\bar{C}HPO(OEt)_2$; viii, $NaBH_4$; ix, H^+; x, KOH

Scheme 12

9-Thiaprostaglandins. Vlattas and DellaVecchia[47] have described a synthesis of 9-thia-PG (90) by a route closely similar to that used by Corey in his synthesis of the primary PG's published in 1971. Michael addition of the anion from mercaptoacetaldehyde diethylacetal to (85) led to (86), which was converted into (87) by a Wittig reaction. Acetalization of (87) gave (88), which on treatment with acid gave a mixture of diastereoisomers of (89). The diastereoisomer (89) was separated by chromatography and elaborated

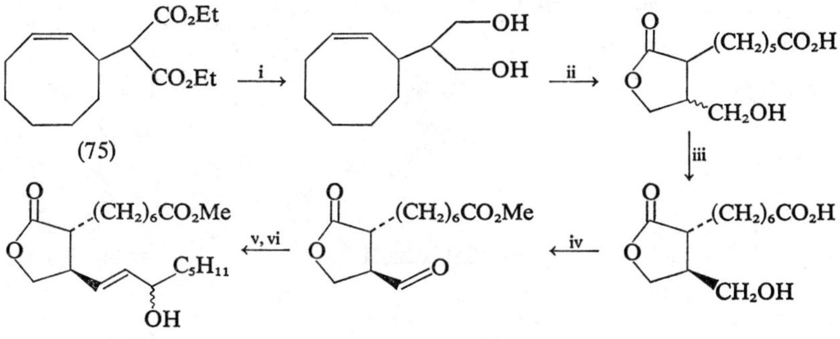

[structures (75), (76), and intermediates with CO_2Et, OH, $(CH_2)_5CO_2H$, CH_2OH, $(CH_2)_6CO_2Me$, $(CH_2)_6CO_2H$, C_5H_{11}]

(75) →i→ →ii→ →iii→

(76) ←v,vi— ←iv—

Reagents: i, $LiAlH_4$; ii, O_3; iii, chain homologation; iv, Collins oxidation; v, $C_5H_{11}CO\bar{C}HPO(OEt)_2$; vi, $NaBH_4$

Scheme 13

[47] I. Vlattas and L. DellaVecchia, *Tetrahedron Letters*, 1974, 4459.

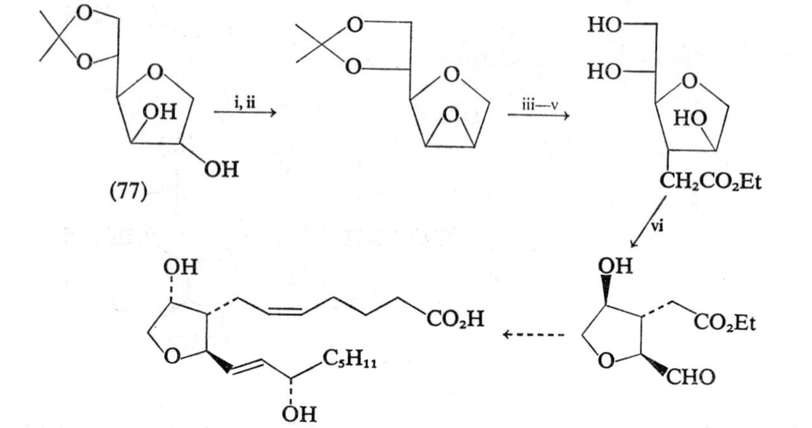

Aliphatic Chemistry

Reagents: i, MeSO$_2$Cl; ii, NaOMe; iii, CH$_2$(CO$_2$Et)$_2$–Na–EtOH; iv, chromatographic separation of isomers; v, NaOH; vi, NaIO$_4$

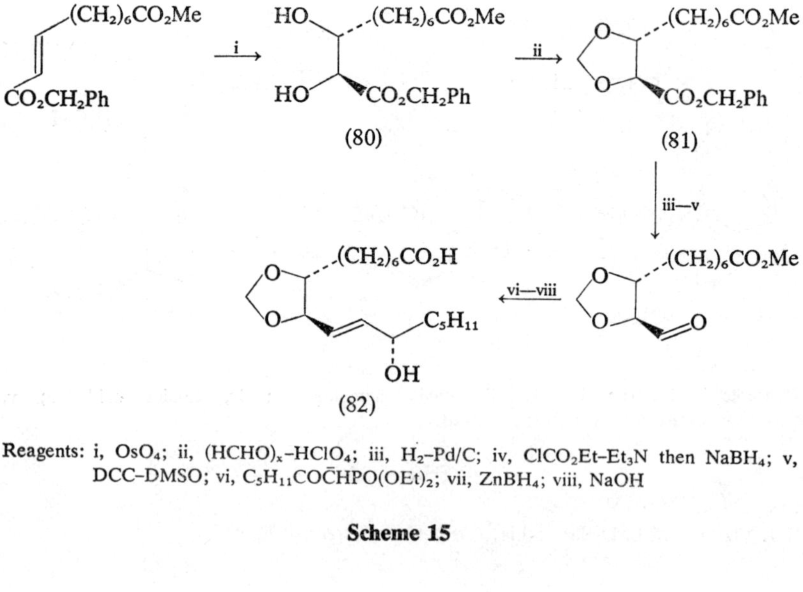

Scheme 14

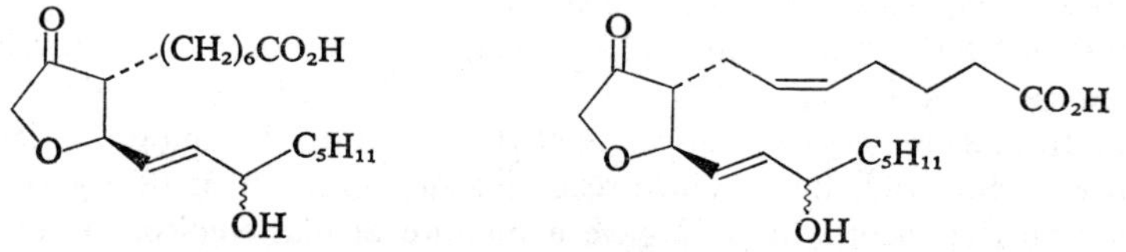

Reagents: i, OsO$_4$; ii, (HCHO)$_x$–HClO$_4$; iii, H$_2$–Pd/C; iv, ClCO$_2$Et–Et$_3$N then NaBH$_4$; v, DCC–DMSO; vi, C$_5$H$_{11}$COCHPO(OEt)$_2$; vii, ZnBH$_4$; viii, NaOH

Scheme 15

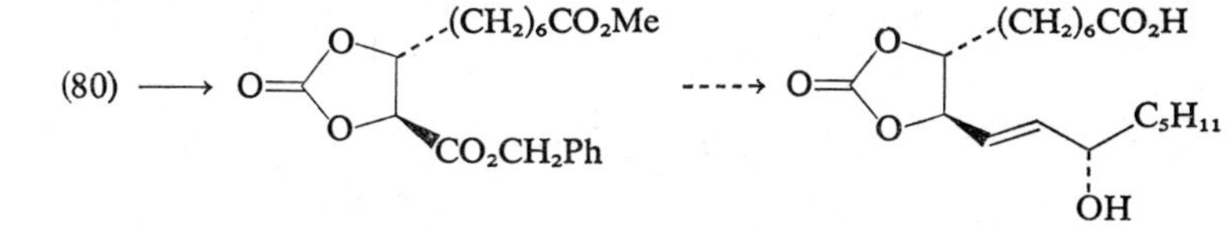

Chemistry of the Prostaglandins

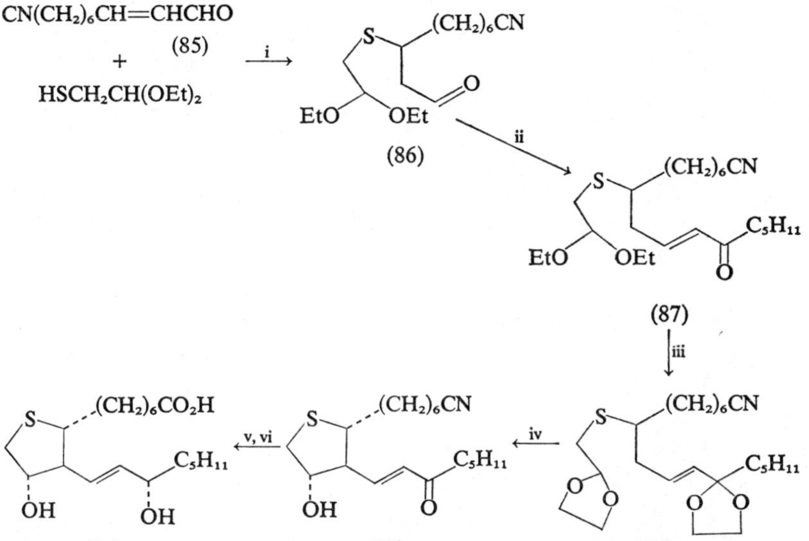

[Scheme showing structures (80) → (83) ----→ (84); (85) + HSCH$_2$CH(OEt)$_2$ →^{i} (86) →^{ii} (87) →^{iii} (88) →^{iv} (89) →^{v,vi} (90)]

Reagents: i, Et$_3$N; ii, C$_5$H$_{11}$COC̄HPO(OEt)$_2$; iii, (CH$_2$OH)$_2$–H$^+$; iv, p-TsOH; v, ZnBH$_4$; vi, KOH

Scheme 16

to (90) as shown. The same workers have also described the synthesis of the sulphoxide (91) of 9-thia-PG.[48]

7-Thiaprostaglandins. Fried and co-workers[49] have reported the synthesis of 7-thia-PGF$_1$ (98) and of *ent*-15-epi-7-thia-PGF$_1$ (99) (Scheme 17). Reaction of the epoxide (92) with the anion from 6-mercaptohexanoate followed by hydrolysis led to the hydroxy-acid (93). After conversion of the C-12 hydroxy-group into the corresponding bromide, reaction with the lithioacetylide (94) produced a mixture of diastereoisomers of (95). Esterification of (95), followed by cleavage of the t-butyl ether and reduction of the propargylic system, gave (96) as a mixture of diastereoisomers. Hydrogenolysis and oxidation of the resulting tetrol then led to (98). A similar series of transformations with (97) produced (99).

[48] I. Vlattas and L. DellaVecchia, *Tetrahedron Letters*, 1974, 4267.
[49] J. Fried, M. M. Mehra, and Y. Y. Chan, *J. Amer. Chem. Soc.*, 1974, **96**, 6759.

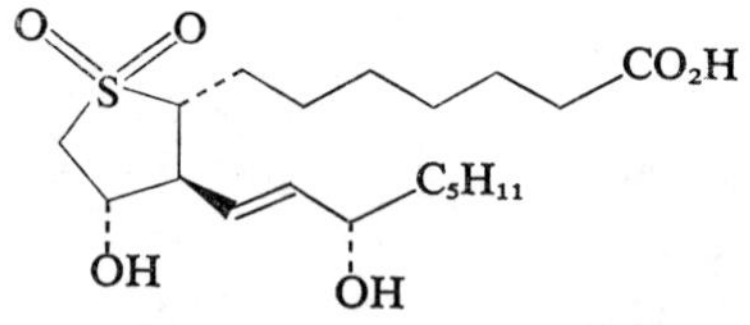

(91)

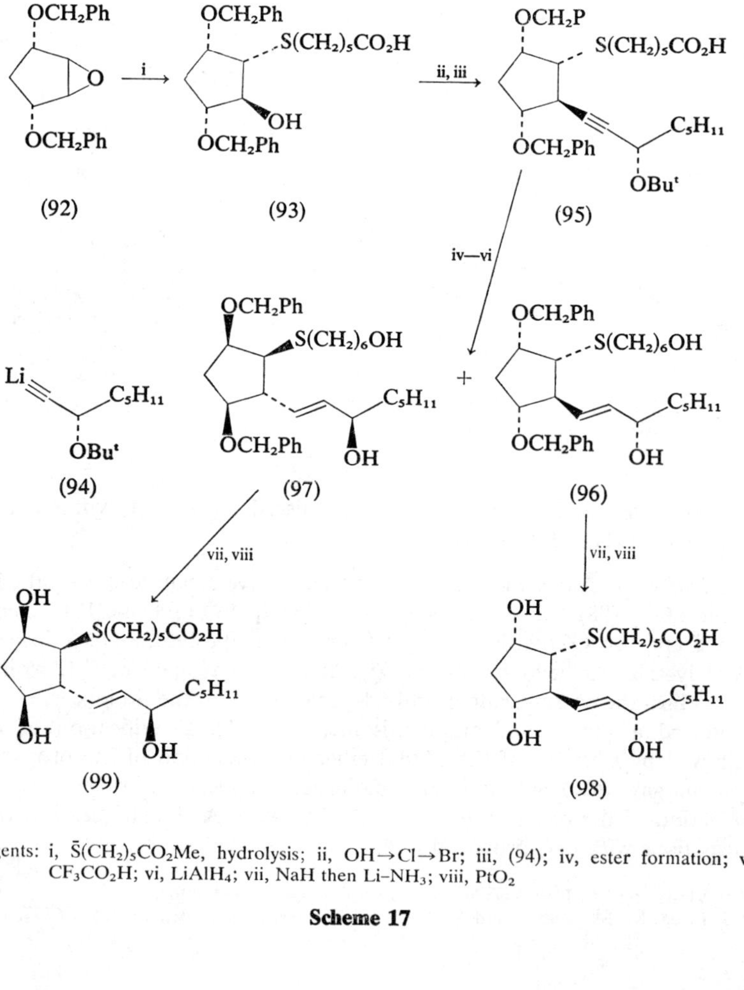

(92) (93) (95)

(94) (97) (96)

(99) (98)

Reagents: i, S̄(CH$_2$)$_5$CO$_2$Me, hydrolysis; ii, OH→Cl→Br; iii, (94); iv, ester formation; v, CF$_3$CO$_2$H; vi, LiAlH$_4$; vii, NaH then Li–NH$_3$; viii, PtO$_2$

Scheme 17

Chemistry of the Prostaglandins

Prostaglandin D_2.—Jenny and co-workers[50] have outlined a closely similar synthesis of PGD_2 to that described last year by Hayashi and Tanouchi (Vol. 3, p. 340).

Methylprostaglandins.—The Upjohn group[51] has described further fuller details of its syntheses of 15-methyl-PG's. The key intermediate employed in the syntheses was the lactone (101), available from (100). In each case the tertiary methyl group was introduced by Grignard reaction with methylmagnesium halide at $-78\,°C$ or by reaction with trimethylaluminium in benzene at ambient temperature. $(15S)$-15-Methyl-PG-$F_{2\alpha}$,-$F_{2\beta}$,-E_2,-A_2,-E_1,-$F_{1\alpha}$, and -A_1 and some 13,14-dihydro-analogues were prepared in this way.

(100)   (101).

Other Modified Prostaglandins.—Alternative synthetic routes to hydroxy-[52,53] and carboxy-PG's[54] have been published, and Sadahiko and co-workers[55] have outlined further aspects of their synthetic routes to 15-methyl- and 15,16-dimethyl-PG's.

7 Epiprostaglandins

An Upjohn group[56] has prepared all possible isomers of the enantiomeric $8\beta,12\alpha$-PG's A, E, and F in the parent series and in the 15-methyl-substituted series. The resolved lactone (102) was chosen as starting material for the syntheses, which proceeded *via* the key intermediates (103) and (104).

8 Prostaglandins in Coral

A review of PG's from marine sources has been published,[57] and the role of the symbiotic algae of *Plexaura homomalla* in PG biosynthesis has been discussed (see Section 10).

[50] E. F. Jenny, P. Schäublin, H. Fritz, and H. Fuhrer, *Tetrahedron Letters*, 1974, 2235.
[51] E. W. Yankee, U. Axen, and G. L. Bundy, *J. Amer. Chem. Soc.*, 1974, **96**, 5865.
[52] G. Doria and C. Gandolfi, *Farmaco, Ed. sci.* 1974, **29**, 327.
[53] P. L. Taylor and R. W. Kelly, *Nature*, 1974, **250**, 665.
[54] H. Miyake and M. Hayashi, *Prostaglandins*, 1973, **4**, 577.
[55] I. Sadahiko, F. Tanouchi, K. Kimura, and M. Hayashi, *Prostaglandins*, 1973, **4**, 535.
[56] E. L. Cooper and E. W. Yankee, *J. Amer. Chem. Soc.*, 1974, **96**, 5876.
[57] W. P. Schneider, L. E. Rhuland, R. D. Hamilton, G. L. Bundy, E. G. Daniels, F. H. Lincoln, and J. E. Pike, Proceedings of the 3rd Food–Drugs Society Conference, 1972.

9 Metabolism of Prostaglandins

The metabolism of PG's has been reviewed,[58] and Sun[59] has established that six urinary metabolities of PGF_2 in female rats have the constitutions (105)—(110). An enzyme '9-hydroxydehydrogenase', capable of oxidizing the 9-hydroxy-group of PGF_2 into 15-keto-13,14-dihydro-E_2, has been isolated.[60]

[58] K. Kunimoto, *Sanfujinka Chiryo*, 1974, **28**, 462.
[59] F. F. Sun, *Biochim. Biophys. Acta*, 1974, **348**, 249.
[60] C. P. Asciak and D. Miller, *Experientia*, 1974, **30**, 591.

10 Biosynthesis of Prostaglandins

Corey and Washburn[61] have reported the isolation and culture of the algae which coexist with the coral *Plexaura homomalla* in a symbiotic relationship. Attempts to incorporate eicosatrienoic acid into PGA in the algal culture were not successful, however, and it was therefore concluded that PG synthetase is not contained in the algae which coexist with coral.

Incubation of arachidonic acid with the microsomal fraction of a homogenate of the vesicular gland of sheep in the presence of *p*-mercuribenzoate has been shown to lead to the formation of the two *endo*-peroxides (111) and (112).[62]

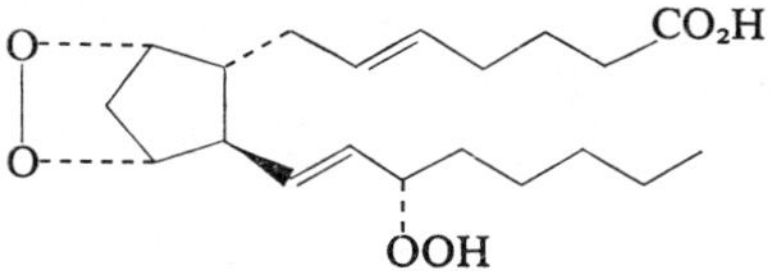

(111) 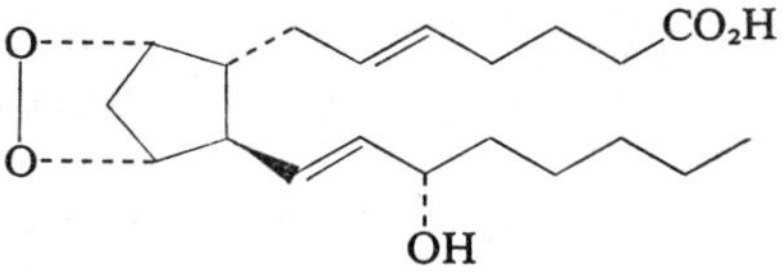 (112)

Sih[24] and Scott[63] and their respective collaborators have independently examined the role of epoxy polyunsaturated fatty acids in PG biosynthesis. Studies with radiolabelled epoxy-acids (46), (113), (114), and (115) established that it is highly unlikely that the four epoxy-acids are substrates or free intermediates in PG biosynthesis in sheep seminal vesicle.

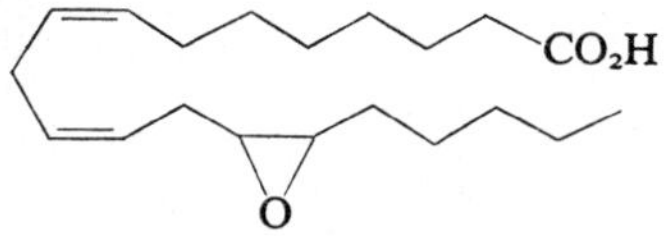

(113) 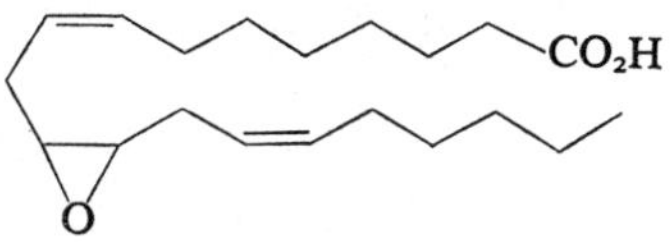   (114)

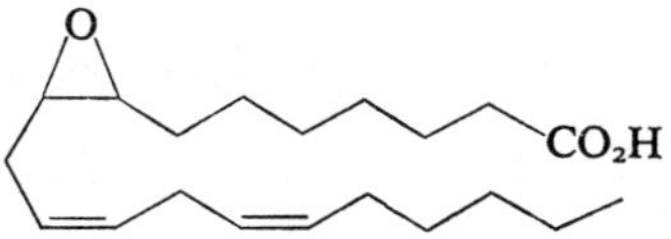

(115)

---

[61] E. J. Corey and W. N. Washburn, *J. Amer. Chem. Soc.*, 1974, **96**, 934.
[62] M. Hamberg, J. Svensson, I. Wakabayashi, and B. Samuelsson, *Proc. Nat. Acad. Sci. U.S.A.*, 1974, **71**, 345.
[63] S. K. Chung and A. I. Scott, *Tetrahedron Letters*, 1974, 3023.

Chemiluminescence has been observed during the incubation of polyunsaturated acids with the microsomal fraction from sheep vesicular glands but the origin of the light emission is not fully known.[64]

Two new reviews of the biosynthesis of PG's have been published,[65,66] and a paper dealing with the role of PG endoperoxides in the biological action of PG's has appeared.[67]

11 General

Crabbé has written a useful review of the application of physical methods to structural and stereochemical problems in the PG field.[68]

---

[64] L. J. Marnett, P. Wlodawer, and B. Samuelsson, *Biochem. Biophys. Res. Comm.*, 1974, **60**, 1286.
[65] E. J. Christ and D. H. Nugteren, *Chem. Weekblad*, 1974, **70**, 2.
[66] B. Samuelsson, *Prostaglandines, Semin. INSERM*, 1973.
[67] M. Hamberg, J. Svensson, and B. Samuelsson, *Proc. Nat. Acad. Sci. U.S.A.* 1974, **71**, 3824.
[68] P. Crabbé, *Tetrahedron*, 1974, **30**, 1979.

Author Index

Aason, A. J., 234
Abatjoglou, A. G., 199
Abdel- Maksoud, H. M., 107
Abe, K., 126, 227
Abell, P. I., 39
Aben, R. W., 42
Aberhart, J , 214
Abidi, S., 183
Abraham, N. A., 255
Abram, T. S., 18
Abramovitch, A., 14
Abramovitch, R. A., 61
Achenbach, H., 238, 240
Acheson, R. M., 12
Achini, R., 216
Acholonu, K. U., 139
Adam, W., 89
Adams, D. R., 234, 236
Agami, C., 86
Agawa, T., 41, 94
Ahmad, A., 111
Ahmed, M., 206
Aizawa, T., 184
Akasaka, K., 139, 227
Akimoto, H., 62
Akiyama, S., 156
Albonico, S. M., 15
Albright, J. D., 125
Alexander, M., 177
Alfaro, I., 48
Allen, L. E., 87
Allenmark, S., 193
Alley, W. D., 185
Allinger, N. L., 110
Allred, E. L., 42
Almqvist, S. O., 234
Al-Rawi, J. M. A., 223
Amar, F., 62
Amice, P., 77, 148
Amiel, Y., 18
Amosova, S. V., 8
Anand, N., 50
Andersen, N. H., 80
Anderson, J. E., 32
Anderson, N. H., 115
Anderson, R. J., 134
Anderson, W. A., 16
Ando, K., 225

Andrews, G. C., 74, 129, 149
Anetai, M., 242
Anghelide, N., 9
Angiolini, L., 157
Angres, I., 183
Anh, N. T., 155
Anselme, J.-P., 197
Antonini, P., 217, 231, 250
Aoki, K., 70
Aoyama, H., 118
Appleton, D. C., 196
Ara, A., 65
Arakawa, M., 82
Araki, M., 135
Aranda, V., 65
Arai, M., 226
Arase, A., 5, 113, 186
Araújo, H., 133
Arens, J. F., 10
Arhart, R. J., 178
Arison, B. H., 225
Arsenijevic, N. V., 97
Arzoumanian, H., 65
Asada, M., 167
Asami, Y., 140
Asciak, C. P., 264
Asensio, G., 65, 174
Ashby, E. C., 152, 155, 160, 161
Ashton, W., 48
Asveld, E. W. H., 43
Atal, C. K., 239
Atkins, T. J., 180
Atlani, P., 116, 170
Atlanti, P. M., 75
Atsumi, K., 102
Auerbach, J., 231
Aune, J. P., 65
Awang, D. V. C., 172
Axen, U., 263
Aya, T., 148

Babler, J. H., 175
Bach, R. D., 35
Bachand, C., 118
Back, R. A., 76
Baczynskyj, L., 214

Baese, H.-J., 178
Bagnell, L., 134
Bahurel, Y., 152, 229
Bailey, P. S., 72
Baird, M. S., 42
Baizer, M. M., 78
Bajorek, J. J. S., 174
Baker, D. J., 70
Baker, R., 68, 123, 203
Balakrishnan, M., 97
Balanson, R. D., 202
Balasubramanian, K., 155
Baldwin, J. E., 74, 142
Baldwin, S. W., 191
Balfour, W. J., 86
Ballarge, M., 149
Bampfield, H. A., 37
Ban, Y., 139, 142, 181, 227
Banerji, A., 239
Banger, J., 204
Banhidai, B., 162
Banno, K., 159
Bao, L. Q., 86
Barco, A., 197, 229
Barefield, E. K., 180
Barley, G. C., 206
Barluenga, J., 65, 174
Barnett, B. L., 199
Barnier, J. P., 77, 148
Barom Marszak, M., 33
Bartel, J., 119
Bartlett, A. J., 29
Bartmann, W., 257
Bartoletti, I., 67, 98
Barton, D., 191
Barton, D. H. R., 40, 113, 183
Barton, M. A., 112
Barton, T. J., 26
Basco, I., 32
Bates, G. S., 31, 155
Bates, R. B., 75
Battaglia, A., 157
Baudouy, R., 21, 33, 210
Bauer, L., 204
Baum, J., 12
Baum, K., 178
Bayet, P., 158
Beadling, L., 224

Beal, D. A., 187
Beames, D. J., 247
Beard, C. D., 178
Beauchamp, J. L., 189
Beavers, W. A., 75
Beck, A. K., 140, 158
Beck, B. R., 42
Beck, G., 257
Becker, G., 27
Becker, J. Y., 1, 170
Beckhaus, H., 162
Beckhaus, H.-D., 183
Beechey, R. B., 220
Behare, E. S., 96
Beier, J., 234
Bellassoued, M., 135
Bellucci, G., 62
Benetti, S., 197, 229
Bennett, M. J., 156
Benoiton, N. L., 191
Bentley, R. K., 206
Benz, J., 51
Bergbreiter, D. E., 30
Berger, K. R., 52
Berger, M., 116
Bergman, R. G., 21, 79
Bergson, G., 154
Bernardi, L., 181
Bernasconi, C. F., 62
Berscheid, H. G., 225
Berson, J. A., 46
Berthet, D., 234
Bertin, J., 97, 186
Bertrand, M., 34, 36, 38, 39, 101, 176
Bertsch, R. J., 65
Besl, H., 226
Bestmann, H. J., 83
Beumel. O. F., jun., 1
Beynon, P. J., 219
Bhat, S. V., 225
Bhatnagar, S. P., 234, 236
Bianchi, G., 51
Bickart, P., 46
Bickelhaupt, F., 199
Biehl, E. R., 52
Biellmann, J. F., 75
Biffin, M. E. C., 182
Bigami, M. K., 244
Billups, W. E., 211
Binamé, R., 141
Bindra, J. S., 243
Bindra, R., 243
Binger, P., 5, 191
Birkhofer, L., 15
Black, C. J., 86
Blackburn, G. M., 97
Bláha, K., 117
Blankenship, R. M., 166
Blanton, C. DeW., 180
Blaszczak, L. C., 152, 156
Block, A. McB., 24

Blomberg, C., 199
Bloomer, J. L., 95
Blount, J. F., 222
Blues, E. T., 198
Blum, D. M., 47, 187
Blum, J., 173
Blume, E., 81
Boche, G., 51
Bock, H., 27, 80
Bocksteiner, C., 182
Bodanszky, M., 111
Bodewitz, H. W. H. J., 199
Boeckman, R. K., 125, 143, 154, 155, 187, 215
Boeckman, R. K., jun., 47, 74, 77
Bogel, M. E., 177
Bogentoft, C., 21, 32, 176
Bohlmann, F., 205, 207, 208, 239
Bohn, B., 140
Boldrini, G. P., 171
Bolker, H. I., 191
Boll, P. M., 95
Bonet, G., 53
Bongini, A., 171
Bonjouklian, R., 48
Bora, J. M., 134
Borders, D. B., 211, 222
Borgen, G., 89
Borremans, F., 231
Bos, H. J. T., 10
Bosisio, G., 181
Boswell, G. A., 184
Botteghi, C., 21
Bourgeois, P., 3
Boutagy, J., 39, 163
Boutin, N. E., 184
Bowers, C. W., 184
Bowler, J., 244
Boxler, D., 151
Boyd, G. V., 15
Boyd, S. D., 196
Boyle, W. J., jun., 62
Bram, G., 103
Brändström, A., 112
Brandänge, S., 85
Brandsma, L., 10
Branfman, A. R., 217
Brattesani, D. N., 121
Braun, H., 28
Braverman, S., 29
Breslow, R., 186
Brettle, R., 174
Bridges, A. F., 183
Bridges, A. J., 39
Brieger, G., 130
Broaddus, C. D., 61
Brockmann, H., 202
Brocksom, T. J., 103
Brook, P. R., 37
Broughton, B. J., 115

Brown, C. A., 146, 191, 202
Brown, E. D., 244
Brown, G., 209
Brown, H. C., 1, 39, 40, 64, 156, 174
Brown, L. R., 113
Brown, R. F. C., 26
Bruggink, A., 105
Brugidou, J., 169
Brunelle, D. J., 147, 152, 255
Bruner, H. S., 113
Bruza, K. J., 74, 143
Bryce-Smith, D., 53, 198
Bryson, T. A., 149
Buchanan, G. L., 39
Buchi, G., 59, 113, 129, 194
Buckwalter, B., 74, 149
Buehler, C. A., 202
Büthe, I., 178
Bull, D. C., 196
Bundy, G. L., 263
Burdett, K. A., 166
Burger, K., 51
Burgi, H. B., 169
Burkholder, W. E., 85, 238
Burroughs, A. E., 177
Bushby, R. J., 10
Buter, J., 43
Butterick, J. R., 181
Byrd, L. R., 170
Byrne, K. J., 236

Cacchi, S., 158
Cadiot, P., 36
Cahiez, G., 187
Cain, E. N., 189
Calas, R., 3
Cama, L. D., 120
Cambie, R. C., 89
Cameron, A. F., 216
Campbell, C. B., 174
Cannon, J. B., 128
Cantrell, T. S., 53, 54
Cardillo, G., 43
Cardillo, R., 220
Cargill, R. L., 172
Carlson, B. A., 40
Carlson, R. M., 9, 95, 225
Carnahan, J. C., jun., 79
Carpenter, T. C., 183
Carroll, G. L., 171
Carruthers, W., 55
Carter, T. P., jun., 72
Cary, L. W., 223
Casals, P.-F., 139
Casanova, J., 42
Caserio, M. C., 169
Casey, C. P., 103
Casnati, G., 160
Cassady, J. M., 96

Author Index

Castro, B., 112
Caton, M. P. L., 255, 257
Catt, J. D., 183
Caubere, P., 1, 202
Cazes, B., 129
Cedar, F. J., 156
Čeković, Ž., 175
Ceraso, J. M., 199
Cerfontain, H., 42
Challis, B. C., 116
Chan, K. H., 73
Chan, T., 167
Chan, T. H., 43, 77
Chan, Y. Y., 261
Chandrasekhar, B. P., 134
Chang, E., 43
Chang, H.-H., 197
Chang, T. C., 167
Chao, L., 161
Chassin, C., 167
Chattopadhyaya, J. B., 112
Chen, M. F., 165
Chen, S. C., 203
Chen, S. L., 177
Cheng, Y. -M., 186
Chevalier, P., 45
Chevolot, L., 36
Chia, H.-A., 37
Chiang, Y., 116
Chidgey, R., 167
Chimiak, A., 111
Chip, G. K., 172
Chiusoli, G. P., 203
Chkir, M., 139
Chou, T. S., 196
Choudhary, G., 113
Chow, W. Y., 211
Chow, Y. L., 117, 126
Christen, H., 27
Christ, E. J., 266
Christol, H., 169
Christy, K. J., 236
Chua, C., 177
Chuit, C., 187
Chukovskaya, E. C., 187
Chuman, T., 234
Chung, S. K., 265
Ciereszko, L. S., 213
Cinquini, M., 193
Clady, J., 54
Claesson, A., 7, 21, 32, 176, 178
Clardy, J., 212, 215, 225
Clark, R. D., 44, 72, 126, 152, 166
Clark, R. T., 186
Clark, T., 55
Clarke, T. C., 79
Clarkson, R., 244
Cleary, J. J., 163
Clerici, A., 89
Clive, D. L. J., 60, 150

Closs, G. L., 72
Closson, W. D., 182
Clutter, D. R., 177
Coates, J., 231
Coates, R. M., 72, 88, 147
Coble, H. D., 113
Coffee, E. C. J., 255, 257
Cohen, J. F., 120
Cole, R. J., 225
Colón, M., 24
Colonna, S., 193
Colvin, E., 77, 143
Comninellis, C., 197
Conan, J. Y., 169, 201
Concepción, J. G., 24
Conia, J. M., 77, 148
Conlin, R. T., 198
Conover, W. W., 36
Contento, M., 43
Conway, W. P., 151
Cook, A. H., 68
Cook, F. L., 107, 184
Cook, J. C., 214
Cookson, R. C., 68, 123, 234, 236
Coombs, R. V., 32
Cooper, E. L., 263
Copenhafer, W. C., 117
Corcoran, R., 186
Corey, E. J., 1, 89, 95, 123, 140, 152, 163, 202, 247, 248, 254, 265
Cornforth, R. H., 173
Corkins, H. G., 194
Cossement, E., 141
Costa, D. J., 184
Costin, C. R., 80
Cotter, B. R., 56
Cottier, L., 152, 229
Coudert, G., 1
Coulson, D. R., 120
Courtois, G., 101
Coutrot, P., 196
Coyle, J. D., 39
Crabbé, P., 266
Cram, D. J., 110
Crandall, J. K., 36
Crawford, T. C., 175, 252
Cresp, T. M., 138
Crews, P., 213
Crichio, R., 219
Crist, D. R., 169
Crochet, R. A., 180
Cross, J. H., 211
Crossley, R., 196
Crumrine, A. L., 103
Csizmadia, I. G., 169, 170
Curran, A. C. W., 196
Cushman, M., 59, 113
Cutting, J. B., 73
Cuvigny, T., 121, 139, 146, 180, 229

Cyr, C. R., 254
Czapf, S. C., 118
Dagli, D. J., 102
Dahn, H., 139
Dailey, R. G., 217
Dale, J. A., 186
Dallwitz, E., 239
Dalton, D. R., 62
Daly, J. W., 209
D'Angelo, J., 159
D'Aniello, M. J., 180
Daniels, E. G., 263
Danishefsky, S., 50, 138
Dao, H. L., 139
Darby, N., 25
Dardioze, F., 113
Dauben, W. G., 47
Daum, H., 156
Davies, M., 66
Davis, R. E., 22
Day, J. C., 186
Dayer, F., 139
Dean, C. L., 60
Dean, F. M., 108, 222
Debal, A., 121
De Boer, C. D., 153
DeBrule, R. F., 64
DeBruyn, D. J., 158
De Clercq, P., 245, 249, 250
Decorzant, R., 138
de Graaf, C., 28
de Graaf, S. A. G., 146
Dehmlow, E. V., 200
de Jong, A. J., 10
Della Vecchia, L., 257, 259, 261
Delling, D. D., 135
De Mayo, P., 170, 214
De Micheli, C., 51
Demina, M. M., 10
Demmin, T. R., 119
Demole, C., 234
Demole, E., 234, 236
Den Elzen, R. V., 194
Dennis, J.-M., 27
Denny, R. W., 39
Deno, N. C., 31, 130
Denzel, Th., 83
Depezay, J.-C., 113, 143
Derocque, J.-L., 24
de Rossi, R. H., 99
Descotes, G., 45, 152, 229
Deshayes, H., 87
Deslongchamps, P., 116, 170
Des Marteau, D. D., 76
de Souza, N. J., 225
Dessau, R. M., 70, 89, 139
Dev, S., 127, 170, 242
Devlin, J. P., 115
Devon, T., 22

Dewar, M. J. S., 79
Dewhurst, B. B., 110
De Wolfe, R. H., 102
Dextraze, P., 257
Dhar, K. L., 239
Diamond, S. E., 112
Dietsche, T. J., 146, 151
Digenis, G. A., 89
Dillinger, H. J., 11
Dillon, J., 176
Dittrich, B., 12
Divakar, K. J., 95
Divekar, P. V., 225
Djerassi, C., 20, 178
Dodds, H. L. H., 97
Dodds, T. A., 22
Doleschall, G., 134
Doll, R. J., 191
Donnelly, S. J., 158
Doolittle, R. E., 237
Doria, G., 207, 263
Dormoy, J. R., 112
Douchkine, N., 125
Douglas., I B., 196
Dowd, P., 101
Doyle, K. M., 204
Doyle, M. P., 158
Drabowicz, J., 196, 197
Draghici, C., 9
Drake, D., 205
Drawert, F., 234
Drenth, W., 19
Dreux, M., 193
Driguez, H., 118
Drummond, P. C., 237
Dube, S., 75
DuBeshter, B., 117
Dubini, R., 59, 130
Dubois, J. E., 63
Duc, L., 139
Dürr, H., 15
Dufresne, R. F., 109
Duhamel, L., 141, 182
Duhamel, P., 141, 182
Duismann, W., 183
Dumont, W., 158
Duncan, D. J., 87
Durand, J., 155
Duréault, A., 165
Durst, H. D., 99, 103, 193
Durst, T., 194
Dušková, E., 117
Dye, J. L., 199
Dyott, T. M., 202

Eadon, G., 130
Eagleson, B. K., 13
Ebizuka, Y., 242
Eck, C. R., 127
Eckes, L., 79
Effenberger, F., 52
Ege, S. N., 13

Ehrig, V., 138
Eikenberry, J. N., 201
Eisch, J. G., 20
Eisenhardt, K. A., 130
Eisman, G., 62
Egan, R. S., 214
Eguchi, S., 28
Ekong, D. E. U., 238
Elbe, H.-L., 65
Elfert, K., 143
Elernter, H. G., 202
Eliel, E. L., 199
Ellestad, G. A., 222
Ellison, R., 91
Elvidge, J. A., 223, 231, 232
Emch, R., 177
Emoto, S., 82
Endo, T., 112
Engberts, J. F. N., 19
Enggist, P., 236
Ensslin, W., 27
Enzell, C. R., 234
Erickson, B. W., 163
Erickson, K. L., 212
Ermer, O., 79
Ernstbrunner, E. E., 169
Evans, D. A., 74, 129, 149, 171, 175, 252
Evans, N., 55
Evans, R. J. D., 32
Exner, O., 204

Fahey, D. R., 66, 126
Fahey, R. C., 17, 62
Fain, D., 63
Falk, J. C., 179
Farona, M. F., 22
Fatiadi, A. J., 204
Faulkner, D. J., 212, 213
Fayat, C., 55
Fayos, J., 215
Fedoryński, M., 201
Feeney, J., 110
Fehlhaber, H. W., 225
Fehr, T., 214
Feitler, D., 112, 197
Felix, A. M., 109
Fellows, R., 191
Felzenstein, A., 52
Fengler, G., 11
Fenical, W., 165, 212, 213
Fenton, D. M., 98
Ferard, J., 139
Ferguson, G., 66
Ferree, W., jun., 16
Ferreira, G. A. L., 133
Fetizon, M., 125
Fiaud, J.-C., 191
Ficini, J., 10, 14, 113, 165
Fienemann, H., 3

Fierz, G., 167
Filler, R., 125
Finkelhor, R. S., 16
Fischer, C. M., 72
Fisher, R. P., 5, 173
Fitzgerald, P. H., 116
Fizet, C., 119
Flanders, S. D., 163
Fleming, B., 191
Fleming, M. P., 157
Fletcher, V. R., 257, 258
Floyd, D. M., 3, 94
Floyd, J. C., 135
Floyd, M. B., 252
Fodor, G., 183
Folting, K., 71
Forbes, C. P., 76
Ford, M. E., 65, 127, 161, 190
Ford, P. W., 134
Ford, W. T., 184
Foucault, A., 55
Fougerousse, A., 257
Fowler, F. W., 12, 202
Franck, B., 139
Franz, J. A., 178
Fraser, R. R., 82
Frasnelli, H., 162
Fraunfelder, G. M., 128
Freasier, B. F., 119
Freer, A. A., 216
Fréhel, D., 116, 170
Freiberg, L. A., 214
Freidinger, R. M., 194
Freidlina, R. Kh., 187
Frejd, T., 27
Fried, J., 261
Fried, J. H., 255, 257
Friedman, A. R., 182
Friedman, H. Z., 13
Friedman, L., 187
Fries, R. W., 87
Friis, P., 119
Fritysche, U., 24
Fritz, H., 263
Fry, J. L., 122
Fuchs, P. L., 41, 95, 163
Fueno, T., 17, 35
Fuganti, C., 220
Fuhrer, H., 263
Fujii, N., 109
Fujii, S., 64, 181
Fujimori, T., 234
Fujimoto, T. T., 175, 252
Fujinami, T., 184
Fujita, T., 135, 179
Fujisawa, T., 40, 193, 200
Fujiwara, Y., 22, 39, 199
Fukuda, H., 112
Fukui, H., 85, 238
Fukuzumi, T., 234
Fullerton, T. J., 146

Author Index

Fuska, J., 215
Furtner, W., 242
Furukawa, J., 21, 176
Furukawa, K., 68
Furukawa, S., 100

Galantay, E., 32
Galle, J. E., 46
Galliaud, A., 103
Gallo, G. G., 217, 219
Gamba, A., 51
Gammill, R. B., 149
Gandhi, N. M., 225
Gandolfi, C., 257, 263
Gandolphi, R., 51
Ganem, B., 8, 125
Ganguly, A. K., 219
Garbarino, J. A., 113
Garcia, G. A., 159
Garnier, B., 148
Garratt, D. G., 60, 61
Garrett, P. J., 70
Garst, M. E., 163
Gaspar, P. P., 198
Gaston, L. D., 236
Gaudemar, M., 28, 113, 135
Gaudry, M., 160
Geiss, K., 74, 145
Gelin, R., 103
Gelin, S., 103
Genêt, J. P., 10, 113
Gennaro, G. P., 76
Gentry, C. R., 237
Georghiou, P. E., 31, 155
Gerlach, H., 89
Gerlach, O., 52
Germain, A., 136
German, A. L., 99
Ghera, E., 166
Ghiringhelli, D., 220
Ghosch, A., 238
Ghosch, P. G., 239
Ghose, B. N., 26
Ghosez, L., 48, 141
Giacomelli, G., 157
Giangrasso, D., 220
Giannoli, B., 158
Gieren, A., 51
Gifkens, K. B., 212
Giga, A., 195
Gilbert, A., 53
Giles, H. G., 170
Gilman., N., 3
Girard, C., 77, 148
Gladysz, J. A., 31, 200
Glass, R. S., 14
Gleason, J., 86
Goeke, G. L., 67
Goering, H. L., 201
Gogovac, M., 97

Gokel, G. W., 110
Gold, H., 139
Golden, H. J., 70
Golfier, M., 90, 125
Gomez Aranda, V., 174
Gompper, R., 138
Gonesnard, J. P., 50
Gonzalez, A., 3
Gonzalez, T., 99
Goosen, A., 76
Gordon, A. W., 183
Gore, J., 21, 33, 210
Gore, P. H., 204
Gorgues, A., 134
Gorski, R. A., 102
Goto, G., 238
Gotthardt, H., 54
Gottlieb, H., 118
Gotzler, H., 28
Gouin, L., 10
Gould, K. J., 6, 130
Graber, D. R., 182
Gracy, D. E. F., 214
Graf, W., 216
Grant, B., 20, 112, 178
Grant, J. A., 110
Grasselli, P., 220
Gravel, D., 169
Gravitz, N., 115
Gray, R. T., 236
Greaves, P. M., 113
Green, M. M., 175
Greene, A. E., 94
Greene, M. G., 75
Greenhill, J. V., 95
Greeves, D., 219
Gribble, G. W., 190
Grieco, P. A., 16, 44, 78, 93, 94, 95, 139, 151, 229
Griffin, A. C., 79
Griffith, R. C., 190
Griller, D., 80
Grimaldi, J., 36
Grimshaw, J., 172
Gröbel, B.-T., 141, 143
Gronowitz, S., 27
Grosjean, D., 80
Gross, P., 170
Grossert, J. S., 43
Gruber, L., 251
Gruetzmacher, R. R., 75
Grzonka, Z., 110
Guenzet, J., 60
Gudgeon, J. A., 223
Guibé, F., 103
Guida, W. C., 134
Guindon, Y., 257
Guinot, F., 169, 201
Guitard, J., 65
Gumulka, M., 139
Gutbrod, H.-D., 170

Guthikonda, R. N., 105
Guziec, F. S., jun., 40

Haber, A., 220
Hach, V., 64, 174
Haddad, Y. M. Y., 156
Hager, L. P., 212
Haidukewych, D., 86
Haink, H.-J., 27
Hajek, M., 70
Hall, A. J., 116
Hall, H. T., 199
Hall, P. L., 173
Hall, T.-W., 5
Halliday, D. E., 68
Halpern, B., 99
Hamada, M., 225
Hamberg, M., 265, 266
Hambrecht, J., 4
Hamilton, R. D., 263
Hamon, D. P. G., 202
Hampton, R. K. G., 85
Hammer, E., 175
Hanack, M., 79
Hanafusa, T., 40
Haneishi, T., 226
Hanessian, S., 257
Hanna, J., 91, 157
Hansel, R., 241
Hansen, P. E., 110
Hanson, J. R., 203
Hara, K., 54
Hara, M., 67
Harada, K., 108
Harada, N., 177
Harada, S., 215
Hargis, J. H., 185
Hargrove, R. J., 2
Harney, D. W., 178
Harpp, D. N., 86, 167
Harrington, K. J., 26
Harris, C. J., 163
Harris, C. M., 139
Harris, H. P., 99, 184
Harris, T. M., 139, 163
Harrison, C. R., 72, 89
Harrison, I. T., 257, 258
Hart, A. J., 202
Hart, D. W., 71, 185
Hart, H., 171
Hartmann, J., 142, 175
Hartzell, S. L., 100
Hartzler, H. D., 47
Haruki, E., 82
Harvie, I. J., 45
Hasegawa, T., 118
Hashimoto, K., 8
Hashimoto, T., 226
Haslett, R. J., 172
Hassner, A., 46, 61, 165
Hata, T., 215, 226
Hatanaka, S.-I., 110

Havel, J. J., 73
Haugh, M. J., 62
Hauptmann, H., 29
Hauri, R. J., 184
Hauser, C. F., 86
Hauser, F. M., 257
Haut, S. A., 191
Hawkins, E. G. E., 171
Hayashi, M., 157, 159, 263
Hayashi, S., 205
Hayashi, T., 140
Hayward, R. C., 89
Heasley, G. E., 64, 72
Heasley, V. L., 64, 72
Hearn, M. T. W., 206, 207
Heath, R. R., 237
Heathcock, C. H., 44, 72, 121, 126, 152, 166
Heather, J. B., 96
Hébert, J., 169
Heck, R. F., 67, 98, 136
Hedden, G., 117
Hedin, P. A., 107
Hegedus, L. S., 68, 136
Hehre, W. J., 187
Heiba, E. I., 70, 89, 139
Heikens, D., 99
Heinsohn, G., 152
Heissner, C. J., 220
Helboe, P., 109
Helgeson, R. C., 110
Helmchen, G., 177
Henbest, H. B., 156
Hendrickson, J. B., 39, 195
Henrick, C. A., 134
Henry, P. M., 66
Hensley, W., 31
Hepburn, D. R., 178
Hercouet, A., 139
Herriott, A. W., 201
Herrmann, J. L., 142
Hesp, B., 216
Hess, G. G., 64
Hesse, R. M., 183
Hester, J. B., jun., 13
Heuring, D. L., 36
Hibberty, P. C., 187
Hickmott, P. W., 204
Higa, T., 213
Higama, T., 255
Higashi, F., 97, 111
Higham, C. A., 206
Hill, K. A., 143
Hill, R. K., 57, 225
Hindley, K. B., 139
Hiraga, K., 238
Hirano, S., 255
Hirao, T., 203
Hiratani, K., 141
Hirowatari, N., 108
Hiyama, T., 34, 128, 147, 229

Hlubucek, J. R., 234
Ho, C.-T., 198
Ho, T.-L., 106, 115, 134, 138, 166, 172, 181, 227
Hockswender, T. R., jun., 63, 76
Hodge, P., 72,89
Hodge, V. F., 170
Hodges, R. V., 72
Höfle, G., 110,123
Höfle, G. A., 74, 142
Hoffman, D. H., 110
Hoffmann, H. M. R., 167
Hoffmann, R. W., 12
Hohorst, A., 76
Holcomb, W. D., 61
Holland, B., 3
Holland, H. L., 214
Holker, J. S. E., 223
Hollstein, U., 24
Holubka, J. W., 35
Hoobler, J. A., 117
Hooz, J., 152, 156
Hopf, H., 14
Hopkinson, A. C., 169
Hoppe, D., 122
Hoppe, I., 115
Horakova, K., 215
Hori, I., 140
Hornback, J. M., 56
Horng, A., 7, 130
Horton, E. W., 243
Hosokawa, T., 132, 225
Houdewind, P., 154
Houghton, E., 12
House, H. O., 24, 138, 156
Hovius, K., 19
Howe, W. J., 202
Howie, G. A., 96
Hseu, F.-H., 22
Hsieh, Z.-H., 169
Huang, B.-S., 107
Huang, F., 87
Hubert, P. R., 82
Huckin, S. N., 105
Hudec, J., 169
Hudlicky, M., 203
Hudrlik, P. F., 44, 91
Hudson, H. R., 178
Hünig, S., 24
Huet, J., 155
Huffman, R. G., 257
Hug, R. P., 191
Hughes, R. J., 173
Hullot, P., 180
Hummel, H. E., 236
Hunter, D. J., 127
Husband, S., 80
Husbands, J., 156
Hussain, S. A. M. T., 96
Hutch, H., 240

Hutchings, M. G., 5, 113
Hyatt, J. A., 54

Ididor, J. L., 225
Iguchi, T., 111
Ikan, R., 229
Ikeda, M., 64, 181
Imai, S., 40
Imamoto, T., 133
Imanaka, T., 22
Imoto, E., 82
Inaba, S., 179
Ingold, K. U., 80
Ingrosso, G., 62
Inoue, T., 25, 91
Inoue, I., 41
Ireland, C., 212
Irie, T., 212
Iriuchijima, S., 78, 127, 192
Irwin, J. G., 198
Isawa, K., 17
Ishibe, N., 8
Ishido, Y., 179
Ishikawa, H., 192
Ishikawa, K., 176
Ishimoto, S., 99
Isowa, Y., 179
Ito, I., 203
Ito, T., 82
Ito, Y., 102, 167, 203
Itoh, M., 5, 111
Ivanitskaya, L., 215
Iwai, K., 91
Iwaki, S., 203
Iwasaki, K., 215
Iwasaki, T., 108
Iwasawa, H., 151
Izawa, K., 35
Izawa, T., 99, 159, 170

Jackman, D. E., 167, 169
Jackowski, G., 32
Jackson, T. E., 172
Jacobs, T. L., 35
Jacobson, R. M., 132
Jacot-Guillarmod, A., 99
Jacquet, I., 157
Jaeger, C. W., 85
Jaenicke, L., 211
Jahngen, E. G. E., 83, 107
Jain, R. C., 50, 214
Jaiswal, D. K., 223
Jakubowski, A., 215
Jankowski, K., 50
Jaouen, G., 162
Jason, M. E., 67
Jasor, Y., 160
Javet, P., 197
Jeffrey, E. A., 134, 152, 171
Jencks, W. P., 115
Jenkins, J. K., 206

Author Index

Jenner, K., 47
Jenny, E. F., 263
Jensen, H. P., 72, 126
Jochims, J. C., 34
Joh, T., 156
Johansson, A., 117
John, J. P., 155
Johnson, B. F. G., 128
Johnson, C. R., 194
Johnson, F., 96
Johnson, R. D., 220
Jonas, V., 117
Jones, E. R. H., 206, 207
Jones, G., 48, 247
Jones, J. B., 177
Jones, J. H., 111
Jones, J. R., 223
Jones, M., 40
Jones, P. R., 141
Jones, R. L., 254
Jones, S. P., 116
Jones, T. H., 142
Jordan, G. J., 169
Josephson, S., 85
Julia, M., 149, 196
Julia, S., 38, 129
Jung, M. E., 50, 77, 143
Just, G., 110

Kaae, R. S., 236
Kagan, H. B., 97, 186
Kaiser, R., 231
Kakis, F. J., 125
Kakushima, M., 220
Kalicky, P., 186
Kalinowski, H.-O., 170
Kamata, K., 81
Kamio, K., 112
Kammerer, R. C., 35
Kaneda, K., 22
Kaneda, T., 25
Kaneko, H., 234
Kaneko, S., 110
Kane-Maguire, L. A. P., 130
Kang, K., 101
Kantlehner, W., 170
Kaplan, L., 110, 158
Kapoor, S. K., 190
Kappe, T., 204
Kappler, F. E., 95
Karaev, S. F., 25
Karich, G., 34
Kariyone, K., 97, 178
Karlsson, A., 219
Karunen, P., 242
Kashdan, D. S., 83
Kashima, C., 85
Kasina, S., 106
Kasuga, R., 234
Katagiri, K., 203

Kataoka, M., 165
Kato, T., 204
Kato, Y., 213
Katsuhara, Y., 54
Katsumata, H., 241
Katz, J.-J., 40
Katzenellenbogen, J. A., 41, 100, 103, 236
Katzer, E., 88
Kaufmann, D., 79
Kawai, M., 91, 183
Kawamoto, F., 22
Kawamoto, I., 141, 229
Kawana, M., 82
Kawatani, H., 109
Kawazu, K., 241
Kazantseva, V. M., 9
Kazaryan, S. A., 111
Kazinoti, P. I., 231
Kebarle, P., 86
Keeley, D. E., 163
Kees, F., 197
Keller, L. S., 58
Kellogg, R. M., 43
Kelly, R. C., 254
Kelly, R. W., 263
Kemp, T. J., 177
Kemp, T. R., 241
Kendall, P. M., 81, 157
Kerekes, I., 184
Kessler, H., 39, 170
Ketcham, R., 19
Kewley, R., 122
Khalil, M. H., 140
Khan, E. A., 3
Khashab, A.-I. Y., 72
Kho, E., 213
Kice, J. L., 196
Kieczykowski, G. R., 142, 193
Kiji, J., 21, 68
Kilgour, J. A., 26
Kim, C. U., 123
Kim, J. K., 169
Kim, M., 42, 183
Kimland, B., 234
Kimling, H., 5
Kimura, K., 263
Kimura, Y., 220
King, F. D., 180
King, R. B., 22
Kinloch, E.-F., 24, 156
Kinnick, M. D., 85
Kinoshita, F., 110
Kinoshita, H., 110
Kinoshita, M., 193
Kinumaki, A., 215
Kippling, J. W., 206
Kirchoff, R. A., 194
Kirk, B. E., 37
Kirksey, J. W., 225
Kirschner, S., 79

Kisfaludy, L., 112
Kishi, T., 215
Kiso, Y., 31
Kitahara, T., 50, 138
Kitamura, T., 156
Kitatani, K., 34, 147
Kitazawa, E., 112
Klaar, M., 242
Klausner, Y. S., 111
Klein, J., 1, 44, 75
Klein, K. P., 119
Kleinig, H., 238
Klindukhova, T. K., 9
Kloek, J. A., 38
Klopfenstein, C. E., 117
Kloster-Jensen, E., 27
Klumpp, G. W., 78
Knaus, G., 81
Knaus, G. A., 61
Knavel, D. E., 241
Knight, G. T., 56
Knights, E. F., 64, 174
Kobayashi, M., 99
Kobayashi, S., 45, 103, 112, 133
Kobayashi, T., 183
Kochi, J. K., 71, 108, 203
Kocienski, P. J., 2
Kocur, J., 208
Köbrich, G., 3, 65
König, D., 105, 139
Koenig, K. E., 120
Koenig, T., 117
Koermer, G. S., 201
Köster, R., 5
Koga, G., 197
Kogure, T., 101, 157
Kohen, S., 197
Kohimizu, K., 241
Kohl, H., 225
Kohl, W., 238
Kolb, M., 140, 141
Kollman, P., 117
Komin, J. B., 36
Komoda, Y., 217
Komura, H., 179
Kondo, H., 136
Kondo, K., 41, 91, 100, 118, 145
Kondo, S., 226
Kooistra, D. A., 158
Kop, J. M. M., 108
Koppel, G. A., 85
Koreeda, M., 177
Korenowski, T. F., 155
Kornblum, N., 85, 196
Korshimov, S. P., 9
Korzeniowski, S. H., 91
Kossmehl, G., 140
Koster, R., 191
Kosuge, Y., 225
Kosugi, H., 91

Author Index

Kotake, H., 110
Koul, A. K., 134
Kouwenhoven, C. G., 14
Kovacic, P., 183
Kovacs, G., 251
Kozikowski, A. P., 47
Kraemer, R., 91
Krapcho, A. P., 83, 107, 163
Krasniewski, J. M., jun., 31
Kraus, G. A., 159
Krebs, A., 5, 28
Kresge, A. J., 116
Krief, A., 158
Krippahl, G., 225
Kropp, P. J., 142
Krzyzanowski, S., 80
Kubota, K., 236
Kucherov, V. F., 19, 130
Kuhlmann, H., 144
Kuhr, I., 215
Kukhar, V. P., 119
Kumada, M., 31
Kumada, Y., 225
Kumagai, M., 121
Kunerth, D. C., 99
Kunieda, N., 193
Kunimoto, K., 264
Kunstmann, M. P., 222
Kunz, R. A., 45, 105, 118
Kupchan, S. M., 217
Kupper, R., 79
Kuramitsu, T., 22
Kurita, H., 179
Kuriyama, K., 96
Kurosawa, E., 212
Kurozumi, S., 99, 163, 255
Kursanov, D. N., 203
Kurz, W., 252
Kurzer, F., 204
Kuwajima, I., 151
Kyriakakou, G., 120

L'abbé, G., 51
Laemmle, J., 160, 161
Laird, T., 29
Lal, D., 80
Lam, J., 205
Lamaty, G., 169, 201
Lamm, B., 112
Lammek, B., 110
Lamparsky, D., 231
Lancaster, J. E., 222
Lancini, G., 219
Landini, D., 178, 184, 185, 196, 201
Landor, P. D., 113
Landor, S. R., 32, 113, 181
Lane, C. F., 174
Lange, G. L., 5
Långström, B., 154

Lansbury, P. T., 129
Lapierre Armand, J. C., 154
Larchevêque, M., 121, 139, 146, 229
Lardicci, L., 157
Large, R., 171
Larock, R. C., 64, 98
Larrahondo, J., 16
Larsen, P. O., 109
Lasne, M.-C., 35
Lattimer, C. J., 201
Laue, H. A. H., 76
Lauer, R. F., 150
LaVoie, E. J., 16
Lavrik, P. B., 171
Laws, D. R. J., 231, 232
Lawston, I. W., 198
Layton, R. B., 152
Leach, M., 222
Leandri, G., 34, 176
LeBelle, M. J., 194
Lebet, C. R., 216
Le Borgne, J.-F., 146
Lebreux, C., 116
Leclerc, G., 181
Le Corre, M., 139
Ledon, H., 197
Lee, A. O., 133, 257
Lee, D. G., 108, 169
Lee, D.-J., 17
Lee, Y.-J., 182
Leeney, T. J., 244
Lehn, J. M., 116, 169
Lelandais, D., 139
Le Merrer, Y., 143
Lemieux, R. U., 112
Lenich, F. T., 14
Lennon, M., 13
Lepschy, J., 110
Le Quesne, P. W., 14
Lerche, H., 105, 139
Lerch, U., 257
Lerner, L. M., 117
Leroux, Y., 193
Lessard, J., 118
Lethbridge, A., 65
Leung, T., 33
Leuschke, A., 241
Levene, R., 1, 44
Levenson, J. L., 158
Lever, O. W., jun., 74, 142
Levi, A., 169
Levitt, L. S., 31
Levy, A. B., 46
Lew, G., 100
Lewis, D. W., 201
Lewis, J., 128
Leworthy, D. P., 220
Leznoff, C. C., 204
Li, M. P., 167
Liang, G., 189
Liang, W. C., 138

Libman, J., 26
Liebeskind, L., 103
Liebfritz, D., 219
Lifson, S., 79
Lightner, D. A., 167, 169
Lilburn, J., 190
Lilenblum, W., 12
Liler, M., 117
Lilje, K. C., 34
Lin, G. H. Y., 212
Lin, H. C., 60, 136
Lin, Y. Y., 177
Lincoln, F. H., 263
Lindsey, J. J., 109
Lindstrom, M. J., 186
Linke, H. A. B., 214
Linstrumelle, G., 38, 74, 129
Liotta, C. L., 99, 107, 184
Lippke, W., 16
Lisy, J. M., 54
Litchman, W. M., 24
Little, R. D., 56
Liu, S., 186
Lizzani-Cuvelier, L., 191
Llenoyama, K., 209
Lloyd, R. V., 189
Lo, Y. S., 227
Loadman, M. J. R., 56
Lockhart, R. N., 126
Löw, M., 112
Lofgren, P. A., 22
Loim, N. M., 203
Lok, M. T., 199
Louw, R., 130
Lovey, A. J., 107
Luce, R., 50
Luche, J.-L. 97, 186
Luche, M.-J., 160, 194
Luck, E., 183
Ludwikow, M., 201
Luft, R., 191
Lundin, R. E., 241
Lwowski, W., 197
Lynch, T. R., 172
Lynen, F., 225
Lynd, R. A., 20

Ma, M. C., 85, 238
Mabuni, C., 110
MacAlpine, G. A., 96, 242
MacBrockway, N., 199
McCormick, J. P., 123
McCrearty, M. D., 201
McCurry, P. M., jun., 126, 160, 227, 228
McDaniel, L. E., 216
McDermott, M., 99
McDermott, J. X., 127
McDonald, F. J., 212
MacDonald, S. F., 165
Macdonald, T. L., 74, 149

Author Index

McEntire, E. E., 46
McGahren, W. J., 222
McGarrity, J. F., 197
McGinnety, J. A., 67
McGlotten, J., 219
McGuinness, J. D., 232
Machleder, W. H., 36
McInnes, A. G., 223
Maciorowski, C. A., 169
McIver, R. T., 177
McKelvey, J., 117
McKenna, J., 196
McKenna, J. M., 196
McKennis, J. S., 22
McKillop, A., 65, 105, 127, 190, 191
McKitrick, R. M., 76
McLean, A., 13
Maclean, D. B., 214
McLelland, R. A., 116, 117
McMahon, T. B., 86
McManus, S. P., 113, 189
McMillan, J. A., 212
McMullan, E. E., 221
McMullen, G. L., 26
McMurry, J. E., 132, 152, 156, 157
Macomber, R. C., 34
McPherson, C. A., 62, 162
McQuillin, F. J., 45, 66
McWatt, I., 13
Madhowan, S., 16
Maeda, K., 132, 226
Maehr, H., 222
Magnus, P. D., 113
Mahajan, J. R., 133
Mahany, P. G., 237
Major, E., 251
Makisumi, Y., 16
Makosza, M., 201
Malacria, M., 36
Malaval, A., 116, 170
Maldonado, L., 144
Malek, J., 70
Mallaby, R., 236
Mallion, K. B., 244
Malone, G. R., 81
Mancelle, N., 141
Manderville, W. H., 199
Maniwa, K., 78, 127, 192
Manne, R., 80
Mantzaris, J., 128
Marchand, A. P., 199
Marchand-Brynaert, J., 48
Marcuzzi, F., 17
Marino, A. F., 95
Marino, J. P., 3, 94, 160, 163
Marioni, F., 62
Markgraf, J. H., 31
Marnett, L. J., 266
Marquet, A., 160

Marschall, H., 95
Marshall, J. A., 91, 141
Marten, D. F., 103
Martin, A. R., 19
Martin, J., 96
Martin, J. C., 178
Martin, J. F., 216
Martin, M., 11
Martin, S. F., 138
Martin, T. I., 214
Martinelli, E., 217, 219
Marty, R. A., 170
Marumo, S., 203
Maruyama, H., 112
Masaki, Y., 44
Masamune, S., 31, 155
Masamune, T., 242
Masclet, P., 80
Massy-Westropp, R. A., 202
Mastrorilli, E., 62
Masuda, Y., 186
Masuoka, Y., 238
Masure, D., 87
Mataka, S., 197
Mathiaparanam, P., 110
Mathur, N. K., 134
Mathys, G., 51
Matier, W. L., 183
Matiskella, J. D., 181
Matser, H. J., 10
Matsueda, R., 112
Matsui, M., 236
Matsumura, F., 85, 238
Matsumura, N., 82
Matsumura, Y., 152
Matsuo, A., 205
Maurer, K. H., 214
Mechlinski, W., 214
Medvedeva, A. S., 10
Meese, C. O., 116
Mehra, M. M., 261
Mehrotra, A. K., 48, 146
Mehrotra, A. M., 244
Mehta, G., 59, 190
Meier, H., 22
Meijer, J., 28
Meinwald, J., 22
Meisters, A., 88, 134, 152, 171, 178
Melby, E. G., 189
Melloni, G., 17
Melton, J., 132
Melvin, L. S., 89
Melvy, L. R., 47
Menicagli, R., 157
Merault, G., 3
Merz, A., 201
Mesbergen, W. B., 163
Metzger, J., 65
Meyer, A., 162
Meyer, H., 95, 220
Meyer, R., 162

Meyers, A. I., 81, 86, 90, 157, 161, 217
Meyers, M., 16
Mhaskar, V. V., 242
Michaelson, R. C., 73
Michalak, R., 155
Michelot, D., 38, 129
Midland, M. M., 1
Midorikawa, H., 140
Miginiac, L., 101
Migliorese, K. G., 8
Mihelich, E. D., 81, 86, 90
Mikołajczyk, M., 196, 197
Miles, D. H., 107
Miller, D., 264
Miller, J., 182
Miller, J. A., 185
Miller, J. G., 252
Miller, J. S., 177
Miller, L. L., 100, 170
Miller, R. B., 56, 96, 121, 187
Miller, R. G., 70
Miller, S. I., 8
Minami, T., 41, 94
Minamikawa, J., 64, 181
Minisci, F., 89
Mioduski, J., 22
Mishima, T., 34, 147
Misiti, D., 158
Missakin, M. G., 19
Mistysyn, J., 211
Misumi, S., 25
Mitchell, E. R., 237
Mitchell, H. L., 199
Mitchell, R. H., 190
Mitchell, T. R. B., 156
Mitsudo, T., 105, 171
Mitsui, T., 241
Mitsuyasu, T., 67
Mittal, R. S. D., 96
Miura, I., 216
Miura, S., 99
Miyake, A., 136
Miyake, H., 263
Miyake, N., 31
Miyashita, M., 55, 78, 93, 95
Miyaura, N., 5
Mizuno, K., 54
Mo, Y. K., 189
Mocella, M. T., 180
Modena, G., 17, 169
Modi, M. N., 134
Modro, A., 61
Mody, N. V., 107
Mok, K.-L., 51
Mole, T., 88, 134, 152, 171, 178
Mondal, M. A. S., 99
Money, T., 127
Montanari, F., 178, 184, 201

Montgomery, F. C., 170
Monti, H., 34, 176
Montzka, T. A., 181
Monvier, G., 80
Moore, R. E., 211
Moreau, C., 170
Morelli, I., 62
Morgan, J. W., 57
Mori, F., 91
Mori, K., 236
Mori, Y., 135
Morin, J. G., 102
Moritani, I., 39, 132, 145, 180
Mornet, R., 10
Morrison, H., 16
Morrison, W. H., 122
Morton, G. O., 211, 222
Morton, J., 219
Moser, R., 78
Mosher, M. W., 31
Moss, R. A., 56, 204
Mossman, A. B., 89
Mosterd, A., 10
Motl, O., 214
Mourges, P., 125
Movsumzade, M. M., 25
Muchowski, J. M., 118
Müller, E., 4, 15
Müller, P., 106
Mukai, N., 85
Mukaiyama, T., 45, 91, 99, 102, 103, 112, 133, 135, 139, 141, 157, 159, 170, 192
Mukerjee, Y. N., 50
Mulheirn, L. J., 220
Muller, B., 216
Muller, D. G., 211
Muller, J.-C., 94
Mura, L. A., 195
Murahashi, S.-I., 132, 180
Murai, S., 148, 165
Muramatsu, S., 141, 229
Murphy, G., 225
Murphy, G. J., 161
Murray, S., 222
Murray, T. P., 139
Muthukrishnan, R., 175
Mychajlowskij, W., 77, 167
Mynderse, J. S., 213

Näf, F., 138, 202
Nagai, Y., 101, 121, 157, 183
Naganawa, H., 225, 226
Nagasawa, M., 250
Nagel, D. W., 222
Nakagawa, A., 215
Nakagawa, Y., 152
Nakai, T., 130, 140, 141
Nakamura, H., 97, 178

Nakanishi, K., 176, 216
Nakatani, Y., 236
Nakayama, K., 102
Nakayama, M., 205
Narang, C. K., 134
Narasaka, K., 139, 157, 159
Narducy, K. W., 179
Naruse, M., 1, 5, 138
Naso, F., 3
Natat, A., 169, 201
Nawata, Y., 225
Nazareth, J., 225
Neff, J. R., 75
Negishi, E., 6, 39, 41, 100, 155, 174
Nelson, N. A., 243
Nelson, R. B., 190
Nematollahi, J., 106
Nestrick, T. J., 130
Neuman, R. C., 117
Neumann, H., 129
Neumann, H. M., 160
Neunhoeffer, H., 9
Neunhoeffer, K., 3
Newcomb, M., 110
Newman, M. S., 128
Nickon, A., 39
Nicolaou, K. C., 89, 247
Nieberl, S., 54
Nieh, E., 72
Niimura, Y., 110
Niki, I., 41, 94
Ninomiya, K., 112
Nishida, S., 46, 55
Nishino, M., 136
Niwano, M., 111
Niznik, G. E., 122
Noe, E. A., 89
Noel, Y., 77, 143
Noguchi, M., 234
Nojima, H., 111
Nojima, M., 184
Nolen, R. L., 81, 90
Nordlander, J. E., 75
Norman, R. O. C., 65
Normant, H., 139, 146, 229
Normant, J. F., 87, 187
Norris, T., 15
Noszkó, L., 83
Notani, J., 111
Novak, L., 255
Nozaki, H., 1, 5, 34, 73, 100, 119, 128, 136, 138, 147, 174, 229, 255
Nugteren, D. H., 266
Nunn, M. J., 185
Nwaji, M. N., 129
Nye, M. J., 51
Nyéki, O., 112

Ochs, W., 8
Odaira, Y., 54

Odubela, A. A., 158
Odyek, O., 113
Ogasawara, K., 90
Ogawa, H., 109
Ogawa, T., 28
Ogura, K., 100, 108, 141
Ohashi, K., 35
Ohigashi, H., 241
Ohkata, K., 40
Ohloff, G., 202
Ohno, K., 67
Ohno, M., 165
Ohta, H., 40, 193, 200
Ohta, Y., 80
Oishi, T., 142
Ojima, I., 101, 121, 157
Okada, M., 140
Okamoto, T., 90
Okano, M., 186
Okawara, M., 112, 130, 140, 141
Okogun, J. I., 238
Okonogi, T., 41
Okuda, T., 215
Okuyama, T., 17, 35
Olah, G. A., 27, 60, 63, 76, 86, 136, 165, 178, 184, 187, 189
Oldenziel, O. H., 123
Ollinger, J., 147
Ollis, W. D., 29, 96, 115
Olsen, D. O., 175
Olsson, L. I., 7, 21, 32, 176, 178
Omura K., 176
Omura, S., 215, 225
Ono, N., 196
Onoe, A., 186
Onyiriuka, O. S., 129
Oosterhoff, P. E. R., 146
Orger, B., 53
Oritani, T., 236
Oshima, K., 100, 174
Osselton, M. D., 220
Otake, N., 215
Otsuji, Y., 82
Ouellette, R. J., 65
Ourisson, G., 94
Overberger, C. G., 204
Overman, L. E., 174, 177
Oyama, K., 61
Oyler, A. R., 9, 95

Pac, C., 54
Padegimas, S. J., 178
Padgett, H., 132
Paglietti, G., 12
Palmer, D. N., 177
Palmertz, I., 112
Pande, P. S., 11
Pandit, U. K., 154
Pankow, L. M., 135

Author Index

Panunzio, M., 171
Paraskewas, S., 112
Parish, E. J., 107
Park, B. K., 108
Parker, D. G., 66
Parker, T., 257
Parker, V. D., 73
Parnes, Z. N., 203
Partyka, R. A., 181
Pasternak, V. I., 119
Patell, J. R., 225
Paton, J. M., 118
Patra, A., 238
Pattenden, G., 229, 231
Patterson, J. W., 255
Patronik, V. A., 102
Patsiga, R. A., 87
Patwardhan, S. A., 170
Pau, J. K., 169
Paukstelis, J. V., 76, 183
Paul, D. B., 182
Paul, I. C., 212
Pauson, D. R., 72
Pavlin, M., 61
Pavlov, S., 97
Payne, M. T., 17
Pearson, D. E., 178, 202
Pearson, H., 32
Pechet, M. M., 183
Peel, R., 247
Peet, N. P., 172
Pelegrina, D. R., 72, 149
Pelter, A., 5, 6, 113, 130, 173
Pendarvis, R. O., 85
Pensak, D. A., 202
Perfetti, R. B., 173
Periasamy, M. P., 123
Périé, J.-J., 79
Perret, F., 3
Perrin, C. L., 117
Perriot, P., 28
Person, H., 55
Peterson, D., 44
Petragnan, N., 103
Petrillo, E. W., 46
Petroff, O. A. C., 95
Pettit, R., 22
Pettus, J. A., jun., 211
Pews, R. G., 100
Phillips, S., 66
Picker, D., 201
Pigott, H. D., 147
Pieter, R., 74, 145
Pike, J. E., 243, 263
Pillai, C. N., 173
Pillay, K. S., 126
Pines, H., 39
Pinke, P. A., 70
Pipithakul, T., 202
Pirisi, F. M., 201
Pisareva, V. S., 9

Pitts, J. N., jun., 62
Pizzorno, M. T., 15
Plamondon, J. E., 7, 130
Plattner, E., 197
Pochat, F., 165
Pochini, A., 160
Pogonowski, C. S., 94, 139, 151, 229
Pohl, D. G., 31, 130
Pollak, A., 17
Pollini, G. P., 197, 229
Polonski, T., 111
Pond, D. M., 172
Porta, O., 89
Posner, G. H., 147, 152, 255
Posvic, H., 42
Potdar, M. A., 239
Potter, B. E., 13
Potts, K. T., 12
Poulton, G. A., 221
Prager, R. H., 166
Prangé, T., 90, 125
Pratt, G., 174
Proctor, G. R., 13
Proverb, R. J., 16
Pryde, A., 16

Quast, H., 197
Quillinan, A. J., 3

Raban, M., 89
Raber, D. J., 134
Rabine, J.-P., 191
Rabone, K. L., 48
Radlick, P., 113
Radom, L., 202
Ragonnet, B., 39, 176
Railenau, D., 9
Rakhit, S., 215
Ramachandran, V., 100
Ramegowda, N. S., 134
Rance, M. J., 11
Ranganathan, D., 48, 146, 244
Ranganathan, S., 48, 146, 244
Rao, A. S., 95
Rao, A. V. R., 112
Rao, G. V., 97
Rao, Y. S., 125
Raphael, R. A., 96, 242, 244, 247
Rapp, U., 214
Rasmussen, J. K., 165
Rathke, M. W., 100
Rautenstrauch, V., 161
Ravid, U., 229
Rawls, H. R., 243
Reap, J. J., 94
Rebek, J., 89, 112, 197

Reboul, O., 87
Redjal, A., 120
Reed, J. O., 197
Rees, C. W., 11
Reetz, M. T., 59
Reeves, P. C., 52
Regan, J. P., 32
Regan, M. T., 87
Regen, S. L., 173
Regitz, M., 11
Reich, H. J., 18, 150, 152
Reich, I. L., 150
Reichardt, C., 140
Reichenbach, H., 238
Reichenbach, T., 187
Reiffers, S., 83
Reilly, J. J., 87
Reinhoudt, D. N., 14
Reinshagen, H., 118
Rej, R. N., 239
Relles, H. M., 202
Rempel, G. L., 130
Renga, J. M., 150, 152
Renge, T., 148
Restivo, R., 66
Reuss, R. H., 61, 165
Reynolds, W. F., 117
Rhee, S.-G., 20
Rhodes, J. E., 129
Rhuland, L. E., 263
Rich, D. H., 110
Richardson, W. H., 170
Richman, J. E., 180
Ried, W., 8, 91
Riehl, J.-P., 191
Riess, J. G., 184
Riley, R. G., 140, 238
Rimmelin, J., 47
Rinehart, K. L., jun., 214, 220
Ripka, W. C., 184
Ripoll, J.-L., 48
Risch, N., 202
Robbins, M. D., 61
Robert, J. L., 216
Roberts, G. C. K., 110
Roberts, J. D., 190
Roberts, J. L., 89
Roberts, S. M., 247
Robinson, D. A., 85
Robinson, G. E., 244
Rodé-Gowal, H., 139
Rodewald, P. G., 70, 89
Rodrigues, R., 103
Rogers, D., 42
Rogers, H. R., 42
Rogers, M. T., 189
Rogers, N. A. J., 48
Rogers, R. J., 199
Rogers, T. E., 196
Rogić, M. M., 119
Roitburd, G. V., 19, 130

Rolla, F., 178, 184, 185, 196
Romanet, R. F., 141, 142, 231
Ronlán, A., 73
Ronzini, L., 3
Rose, C. B., 174
Rosebery, G., 110
Rosen, M. H., 53
Rosenblum, L. D., 134
Rosenblum, M., 203
Rossi, R. A., 99
Rothenberg, S., 117
Rouot, B., 181
Roush, W. R., 197
Roy, D. N., 12
Ruasse, M.-F., 63
Rubottom, G. M., 72, 149
Ruden, R. A., 26, 48
Rudnick, L. N., 91
Rüchardt, C., 183
Ruge, B., 15
Rusch, G. M., 136
Russell, B. R., 117
Ruppert, J. F., 99
Rutledge, P. S., 89
Ryback, G., 236
Ryu, I., 148

Sadahiko, I., 263
Sadar, M. H., 108, 169
Sadowski, J. S., 190
Saegusa, T., 102, 167, 203
Sagredos, A. N., 78
Saigo, K., 159, 170
Saito, T., 203
Saito, Y., 97
Sakai, N., 119
Sakai, S., 184
Sakakibara, T., 78, 192
Sakamoto, N., 156
Sakan, T., 135
Sakata, S., 135
Sakurai, H., 54, 100, 118
Salaun, J., 148
Salbaum, H., 83
Salerno, G., 160
Sales, R., 169
Salomon, C., 21
Salomon, R. G., 71, 76
Sam, D. J., 184
Samuelsson, B., 265, 266
Sandefur, L. O., 147
Sanders, J. A., 19, 32
Sandri, S., 43
Sane, P. P., 95
Sankawa, U., 242
Sannes, K. N., 204
Sano, H., 60
Santelli, M., 38, 39, 176
Santi, R., 157
Santopietro-Amisano, A., 220

Santosusso, T. M., 194
Saran, M. S., 22
Sarel, S., 52
Sargent, M. V., 138
Sarre, O. Z., 219
Sarthou, P., 103
Sartori, G., 219
Sasaki, H., 225
Sasaki, T., 28
Sasson, Y., 13, 130, 173
Satchell, D. P. N., 116, 204
Sato, T., 91, 157
Sato, Y., 181, 193
Sauer, J., 80
Saussine, L., 196
Sauvêtre, R., 87
Savignac, P., 193, 196
Saville, B., 56
Savoia, D., 171
Savoie, J. Y., 112
Schäublin, P., 263
Schaffner, C. P., 214
Scharf, H.-D., 204
Scharf, V., 139
Schauble, J. H., 102
Schaumann, E., 117
Scheeren, J. W., 42
Schegolev, A. A., 19, 130
Scheinmann, F., 3
Schener, P. J., 213
Scherer, K. V., 198
Schill, G., 141
Schilling, P., 60
Schinco, F. P., 169
Schlessinger, R. H., 141, 142, 231
Schletter, I., 254
Schlosser, M., 142, 175
Schlott, R. J., 179
Schluenz, R. W., 202
Schmid, G. H., 60, 61
Schmid, H., 16
Schmidt, E. A., 167
Schmidt, H., 28
Schmidt, W., 15
Schmitz, F. J., 212, 213
Schmitz, R. F., 78
Schneider, P., 263
Schöllkopf, U., 81, 115, 162
Schön, I., 112
Schoenberg, A., 67, 98, 136
Schouteeten, A., 149
Schrameyer, M., 139
Schreckenberg, M., 143
Schreurs, H., 28
Schröder, R., 81
Schultz, A. G., 153
Schultz, G., 118
Schulz, J., 241
Schumann, D., 11
Schwartz, A. L., 117
Schwartz, J., 71, 128, 185

Schwartz, R. D., 155
Schweig, A., 28
Scorrano, G., 169
Scott, A., 85
Scott, A. I., 224, 265
Scouten, C. G., 64, 174
Scribner, R. M., 184
Seebach, D., 74, 80, 95, 129, 138, 156, 158, 220
Seger, D., 29
Seitz, D. E., 141
Seki, Y., 165
Selikson, S. J., 122
Semmelhack, M. F., 199
Sensi, P., 219
Sepp, D. T., 198
Sethi, S. C., 127
Seto, H., 222, 223
Setton, R., 97, 186
Seuleiman, A., 33
Seuring, B., 74, 145
Severin, T., 105, 139
Seyden-Penne, J., 120
Seyferth, D., 161, 186
Shahak, I., 13, 40
Shambu, M. B., 89
Shani, A., 187
Shanker, R., 173
Shannon, P. V. R., 232
Sharma, A. K., 203
Sharma, C. S., 127
Sharpless, K. B., 72, 73, 126, 150
Sharts, C. M., 184
Shaw, A., 96, 242
Shaw, C. C., 217
Shaw, J. E., 99
Shaw, J. R., 80
Sheehan, J. C., 227
Sheppard, W. A., 47, 184
Shetty, R. V., 57
Shiekh, M. Y., 130
Shigematsu, Y., 236
Shigemitsu, Y., 54
Shim, S. C., 171
Shimamura, T., 180
Shimizu, N., 46, 55
Shimoji, K., 100
Shiono, H., 130
Shiono, M., 141
Shiori, T., 102, 112
Shirai, H., 181
Shoer, L. I., 31
Shono, T., 152
Shorey, H. H., 236
Short, F. W., 107
Shoua, S., 166
Shue, H.-J., 202
Siegfried, B., 106
Siehl, H.-U., 79
Sieveking, M. F., 117
Sih, C. J., 96, 250

Author Index

Silhavy, P., 70
Silverstein, R. M., 140, 236, 238
Silvon, M. P., 163
Simalty, M., 33
Simmons, H. E., 184
Simpson, T. J., 223
Sims, J. J., 212
Sinclair, J. A., 1
Singh, A. N., 242
Singh, B. P., 59
Singh, G., 11
Singh, J., 77, 154, 239
Singh, K., 215
Singh, R. K., 160, 228
Singh, U. P., 12
Sinou, D., 45
Sinoway, L., 152
Sisti, A. J., 136
Sivaram, S., 191
Skatova, N. N., 8
Skell, P. S., 186
Skidgel, R. A., 64
Slates, H. L., 249
Sloane, R. B., 72
Smit, W. A., 19, 130
Smith, A. B., 198
Smith, B. F., 121
Smith, C., 96
Smith, D. G., 223
Smith, D. L., 14
Smith, H. E., 183
Smith, K., 39, 99, 173
Smith, P. J., 195
Smith, R. A., 62, 86
Smith, S. G., 184
Smith, T. N., 68
Smith, W. N., 1
Smolikiewicz, A., 136
Smudin, D. J., 56
Snider, B. B., 58, 186, 248
Sniegowski, P. J., 98
Soai, K., 139
Solly, R. K., 189
Solomon, P. H., 216
Soma, G., 110
Sondheimer, F., 25
Sonnet, P. E., 236
Sonoda, N., 100, 118, 148, 165
Sonola, O. O., 181
Sood, R., 250
Soulie, J., 36
Souma, Y., 60
Sousa, L. R., 110
Southwick, P. L., 109
Sower, L. L., 237
Spagnolo, P., 11
Spear, R. J., 27
Spencer, R. L., 13
Spencer, T. A., 163
Spinelli, H. J., 130

Sprangers, W. J. J. M., 130
Springer, J. P., 225
Sprung, J. L., 62
Sreekumar, C., 173
Spridharan, N. S., 202
Staab, H. A., 3
Stähle, M., 142
Stafforst, D., 81
Staley, R. H., 189
Stallard, M. O., 212
Stang, P. J., 2, 39
Stanley, G., 234
Staral, J., 189
Stauffer, R. D., 70
Steglich, W., 110, 226, 242
Steinwand, P. J., 98
Stempel, A., 222
Stephen, A., 118
Sterling, J. J., 147
Stetter, H., 143, 144
Stevenson, G. R., 24
Steyn, P. S., 222
Stiggall, D. L., 170
Stilke, R., 15
Still, W. C., 74, 149
Stille, J. K., 87
Stillwell, M. A., 220, 231
Stirling, J. J., 255
Stiverson, R. K., 68, 136
Stoddart, J. F., 96
Stoltz, L. P., 241
Stoochnoff, B. A., 191
Storer, R., 229, 231
Stork, G., 77, 105, 120, 143, 144, 154, 159, 252
Storr, R. C., 11
Stotter, P. L., 143
Stoute, V. A., 170
Stowell, J. C., 178
Strating, J., 83
Straub, H., 4
Strawson, C. J., 216
Strege, P. E., 67, 151
Streib, W. E., 71
Streith, J., 119
Strickland, D., 64
Stridsberg, B., 193
Strobl, G., 28
Strunz, G. M., 220, 231
Studt, W. L., 95
Su, H. C. F., 237
Su, Y. Y., 76
Suga, K., 135, 179
Sugawara, Y., 215
Sugimoto, K., 40, 193, 200
Sullivan, D. F., 100
Sullivan, F. R., 183
Sun, F. F., 264
Sunami, M., 8
Sunderman, F., 117
Sundermann, F.-B., 24
Sunthankar, S. V., 134

Sutherland, J. K., 174, 247
Sutton, J. R., 174
Suwita, A., 207
Suzuki, A., 5, 186, 225
Suzuki, H., 40
Suzuki, K., 215
Suzuki, M., 141, 212, 215
Suzuki, Y., 40
Svanholm, U., 73
Svendsen, A., 95
Svensson, J., 265, 266
Swaminathan, S., 155
Swenton, J. S., 54, 166
Swern, D., 118, 194, 203
Synerholm, M. E., 57
Szabolics, A., 83
Szammer, J., 83
Szantay, C., 255
Szendrey, L. M., 89
Szilagyi, I., 214
Szirtes, T., 112
Szmulewicz, S., 219

Tabacchi, R., 99
Taddia, R., 197
Tagaki, W., 41
Taguchi, H., 100, 136
Tahara, R., 236
Taillefer, R., 116
Takada, S., 16
Takahagi, H., 112
Takahashi, Y., 135
Takami, Y., 82
Takamori, I., 215
Takano, S., 90
Takasugi, M., 242
Takayanagi, H., 200
Takechi, H., 142
Takeda, T., 102
Takegami, Y., 105, 171
Takei, H., 45, 103, 112, 135, 170
Takeuchi, T., 225
Takita, T., 226
Talaty, E. R., 118, 135
Tam, J., 110
Tam, J. N. S., 117
Tamao, K., 31
Tamborini, G., 219
Tamm, C., 214, 216
Tamura, C., 226
Tamura, S., 220, 225
Tamura, Y., 64, 181
Tanabe, M., 222, 223
Tanaka, H., 90
Tanaka, K., 97, 178
Tanaka, M., 171
Tanaka, S., 73, 174
Tanaka, T., 99
Tancrede, J., 22
Tang, Y.-N., 76

Author Index

Tanny, S. R., 12
Tanouchi, F., 263
Tarasova, O. A., 8
Tasker, P. A., 12
Tatchell, A. R., 181
Taub, D., 249
Taube, H., 112
Tavernier, D., 250
Taylor, A. W., 96, 242
Taylor, D. R., 37
Taylor, E. C., 65
Taylor, P. L., 263
Taylor, S. K., 174
Taylor, W., 222
Tehan, F. J., 199
Telang, S. G., 134
Temple, D. L., 81, 86
Temple, P., 40
Tendil, J., 17
Terahara, A., 226
Teranishi, S., 22
Tezuka, T., 126
Thaller, V., 206, 207
Thalmann, A., 89
Theiling, L. F., 86
Thenn, W., 51
Theobald, N., 240
Thijs, L., 140
Thil, L., 136
Thomas, A. F., 59, 130
Thomas, C. B., 65
Thomas, E. J., 40
Thomas, H. G., 88
Thomas, R., 39, 163, 223
Thomas, R. C., 175, 252
Thompson, D. J., 128
Thorpe, J. E., 115
Thuillier, A., 35
Thummel, R. P., 48
Tichý, M., 117
Tidwell, T. T., 60, 61
Tietze, L. F., 50
Timko, J. M., 110
Tin, K.-C., 194
Tippett, J. M., 166
Tishler, M., 215
Tömösközi, I., 251
Toh, H. T., 183
Tokuyama, T., 209
Tom, G. M., 112
Tomassini, T. 95,
Tominaga, M. 236
Tomita, H., 68
Tomita, T., 5, 138
Tonnard, F., 55
Torii, S., 90
Toru, T., 99
Touchard, D., 118
Toullec, J., 63
Toumi, A., 60
Toumi, M., 60
Touzin, A. M., 14

Tramontini, M., 157
Trisler, J. C., 119
Trocha-Grimshaw, J., 156
Trofimova, A. G., 8
Trofimov, B. A., 8
Tronchet, J. M. J., 3
Tropitzsch, R., 43
Trost, B. M., 45, 67, 105, 118, 140, 146, 151, 163, 255
Truce, W. E., 36, 194, 195
Truesdale, L. K., 171
Trueman, R. E., 32
Truxillo, V., 182
Tsuchihashi, G., 78, 100, 108, 127, 141, 192
Tsuda, Y., 85
Tsui, E. Y., 13
Tsuji, J., 67, 200
Tsuji, R., 54
Tsukanaka, M., 128, 229
Tuddenham, R. M., 236
Tuggle, R. M., 156
Tuller, F. N., 91
Tullock, C. W., 184
Tumlinson, J. H., 237
Tunemoto, D., 41
Turner, J. L., 207
Tyrrell, H., 53
Tyssee, D. A., 78

Ubik, K., 203
Uchida, I., 96
Uchio, Y., 205
Uda, H., 91
Uemura, S., 186
Ullman, R., 72
Umani-Ronchi, A., 171
Umezawa, H., 225, 226
Ungaro, R., 160
Unrau, A. M., 181
Untch, K. G., 252
Upadhye, A. B., 242
Uramoto, M., 215
Utawanit, T., 41, 100
Utekhina, N. V., 9
Utermoehlen, C. M., 118, 135
Utimoto, K., 1, 5, 119, 138

Vallén, S., 85
van den Bossche, R., 231
Vanderah, D. J., 213
van der Gen, A., 146
van der Meer, R., 99
van der Wielen, F. W. M., 42
Vander Zwan, M. C., 128
Vandewalle, M., 244, 245, 249, 250
Vanhaelen-Fastré, R., 207
Van Haver, P., 250

van Hooland, J., 245
van Leusen, A. M., 123
van Noort, P. C. M., 42
Van Peppen, J. F., 119
van Sweeten, A. P., 130
Van Tamelen, E. E., 31, 200
Van Verth, J. E., 127
van Wageningen, A., 42
Varma, V., 156
Vazquez, M. A., 72, 149
Vedejs, E., 44, 149
Vederas, J. C., 216
Vemey, M., 17
Venkatasubramanian, N., 97
Verhelst, W. F., 19
Vermeer, P., 28
Verzele, M., 231
Vessiere, R., 17
Vicens, J. J., 75
Vick, K. W., 237
Vidyarthi, S. K., 76
Vigneron, J. P., 157
Villieras, J., 87, 187
Vilsmaier, E., 43
Vining, L. C., 223
Vinson, J. R., 68, 123
Viola, A., 16
Visaisouk, S., 86
Vlattas, I., 133, 257, 259, 261
Vleggaar, R., 222
Vogel, D., 208, 225
Vogel, E., 95
Vogel, P., 138
Volhardt, K. P. C., 21
Voorhees, K. J., 42
Vopilina, L. A., 9
Voronkov, M. G., 10
Vostrowsky, O., 43
Vowinkel, E., 119, 178, 196
Vrkoč, J., 203
Vuitel, L., 99

Wada, M., 91
Wagner, F., 22, 180
Wagner, G., 80
Wagnon, J. C., 22
Wakabayashi, I., 265
Wakamatsu, T., 139, 227
Wakatsuki, Y., 22, 70
Walborsky, H. M., 87, 108, 122, 123
Walling, C., 186
Walter, G. J., 102
Walter, J. A., 223
Walter, W., 116, 117
Walton, D. R. M., 26, 180
Wang, A. H.-J., 180
Wang, G. L., 110
Waraszkiewicz, S. M., 212
Ward, A. D., 115

Author Index

Ward, J. W., 72
Ware, D. W., 189
Wareing, J., 195
Warheit, A. C., 196
Warren, P. J., 115
Washburn, W. H., 214, 265
Wasserman, H. H., 58, 127
Wat, C. K., 223
Watanabe, S., 135, 179
Watanabe, Y., 105, 171
Watkins, D. A. M., 55
Watkins, G. L., 255, 257
Watt, D. S., 121, 122
Watts, P. C., 96
Watts, W. E., 18
Weber, H., 51
Weber, W. P., 120, 198
Wedegaertner, D. K., 139
Weeks, P. D., 138
Weiler, L., 105
Weininger, S. J., 197
Weinreb, S. M., 231
Weinstein, B., 197
Weiss, B., 15
Weissberger, E., 128
Welbourn, M. J., 177
Welch, J., 86, 165, 178
Wemple, J., 102
Wendler, N. L., 249
Wenkert, E., 162
Wepplo, P. J., 142
Wermuth, C. G., 181
Werner, G., 9
Wernick, D. L., 201
West, C. T., 158
Westerman, P. W., 27, 60, 189
Wetzel, E. R., 211
Weyerstahl, P., 95
Whalley, A. R., 196
White, D. M., 14
White, D. R., 81
White, J. D., 42, 99
White, R. C., 130
White, R. H., 212
White, R. J., 217, 219
Whitesides, G. M., 30, 127, 190, 199, 201
Whitfield, F. B., 234
Whitfield, G. F., 118
Whitham, G. H., 39, 40
Whitkop, P. G., 16
Whitlock, B. J., 85
Whitlock, H. W., 85
Whitman, P. J., 140
Whitmarsh, D., 55
Whitten, C. E., 147, 255
Wiberg, K. B., 67
Wicha, J., 125
Wiessler, M., 182
Wieting, R. D., 189
Wild, H.-J., 96
Wild, H. K., 242
Wildgoose, J., 56
Wilding, H. F., 31
Williams, J. W., 72, 88
Williams, K., 99
Williams, M. E., 221
Williams, R. M., 6
Williams, T. H., 222
Williams, V. P., 236
Williamson, K. L., 177
Willis, C., 76
Wilschowitz, L., 110
Wilson, R. M., 56
Wing, R. M., 212
Winnik, M. A., 32, 170
Winter, W., 15
Winterfeld, E., 11
Wipff, G., 116, 169
Wipke, W. T., 67, 202
Witkop, B., 209
Wittel, K., 80
Wlodawer, P., 266
Wolf, G. C., 36, 47
Wolf, S. F., 89
Wolff, C., 178, 196
Wollenberg, R. H., 1, 152
Wong, C. M., 106, 134, 138, 166, 172, 181
Wood, D. E., 189
Wood, R. J., 115
Woodruff, R. A., 161
Wooldridge, K. R. H., 115
Woolsey, N. F., 140
Woon, P. S., 22
Wright, D. E., 115
Wright, J. L. C., 223
Wright, S., 247
Wu, D. K., 81
Wu, E. S. C., 44
Wu, S.-M., 119
Würthwein, E.-U., 140
Wuest, H., 59, 113, 129
Wunderly, S. W., 56
Wunz, T. P., 87
Wyatt, M., 70
Wynberg, H., 83

Yajima, H., 109
Yamada, M., 203
Yamada, S., 102, 112
Yamamoto, H., 73, 100, 136, 174
Yamomoto, K., 25, 31, 68
Yamamoto, O., 176
Yamamoto, S., 141
Yamamoto, Y., 145
Yamashita, K., 236, 242
Yamashita, M., 105, 141, 171
Yamazaki, H., 22, 70, 97, 111
Yamazaki, T., 135
Yamdagni, R., 86
Yandovski, V. N., 9
Yang, N. C., 26
Yankee, E. W., 263
Yano, Y., 41
Yasuda, A., 174
Yates, B. L., 16
Yates, J. A., 206
Yates, K., 61, 79
Yazawa, H., 97, 178
Yelvington, M. B., 170
Yen, S.-C., 202
Yip, R. W., 117
Yoder, C. H., 117
Yokayama, T., 102
Yonce, C. E., 237
Yonezawa, K., 102, 167
Yoshida, T., 6, 100, 155
Yoshikawa, S., 21
Yoshikoshi, A., 55
Yoshinari, T., 5
Yoshino, T., 179
Young, F., 195
Young, R. N., 113
Youssef, A.-H., 107
Yovell, J., 52
Yu, S. H., 187
Yura, Y., 141, 229
Yus, M., 174

Zarecki, A., 125
Zdero, C., 205, 207
Zeck, O. F., 76
Zeisberg, R., 205
Zelawski, Z. S., 249
Zieger, H. E., 183
Ziegler, E., 204
Ziegler, F. E., 38, 95
Zimmerby, V. A., 217
Zimmerman, H. E., 56
Zipfel, K., 226
Zonnebelt, S. M., 158
Zsindly, J., 16
Zubrick, J. W., 193
Zudan, M., 17
Zuech, E. A., 66, 126
Zumwald, J.-B., 3
Zune, A. E., 24
Zwanenburg, B., 140
Zweifel, G., 5, 7, 20, 33, 130, 173

QD
300
A58
v.4
1974

MAR 24 1977